Recovering Australian Threatened Species

A BOOK OF HOPE

Editors: Stephen Garnett, Peter Latch, David Lindenmayer and John Woinarski

© Stephen Garnett, John Woinarski, David Lindenmayer and Peter Latch 2018

All rights reserved. Except under the conditions described in the *Australian Copyright Act 1968* and subsequent amendments, no part of this publication may be reproduced, stored in a retrieval system or transmitted in any form or by any means, electronic, mechanical, photocopying, recording, duplicating or otherwise, without the prior permission of the copyright owner. Contact CSIRO Publishing for all permission requests.

The moral rights of the author(s) have been asserted.

National Library of Australia Cataloguing-in-Publication entry

> Recovering Australian threatened species : a book of hope /
> Stephen Garnett, Peter Latch, David Lindenmayer and John Woinarski, editors.
>
> 9781486307418 (paperback)
> 9781486307425 (epdf)
> 9781486307432 (epub)
>
> Includes bibliographical references and index.
>
> Wildlife recovery – Australia.
> Wildlife management – Australia
> Endangered species – Australia.
> Endangered plants – Australia.
> Wildlife conservation – Australia.
> Plant conservation – Australia
>
> Garnett, Stephen, 1955– editor.
> Latch, Peter, editor.
> Lindenmayer, David, editor.
> Woinarski, J. C. Z. (John Casimir Zichy), 1955–

Published by

CSIRO Publishing
36 Gardiner Road, Clayton VIC 3168
Private Bag 10, Clayton South VIC 3169
Australia

Telephone: [+613] 9545 8555
Email: csiropublishing@csiro.au
Website: www.publishing.csiro.au

Front cover: (top, left to right) trout cod (photo: Gunther Schmida), Lord Howe Island phasmid (photo: Dean Hiscox), forty-spotted pardalote (photo: Andrew Browne); (bottom, left to right) mountain pygmy possum (photo: Linda Broome), sandhill spider orchid (photo: Matt Cameron), juvenile western swamp tortoise (photo: Gerald Kuchling).
Back cover: (left to right) volunteers surveying an area for pygmy bluetongue lizard burrows (photo: Mark Hutchinson), measuring a fledgling Norfolk Island green parrot (photo: Mark Delaney), volunteers helping to rehabilitate glossy black-cockatoo habitat (photo: Colin Wilson).

Set in 10.5/12 Minion Pro and Stone Sans
Edited by Peter Storer
Cover design by Andrew Weatherill
Typeset by Desktop Concepts Pty Ltd, Melbourne
Index by Max McMaster
Printed by Ingram Lightning Source

CSIRO Publishing publishes and distributes scientific, technical and health science books and journals from Australia to a worldwide audience and conducts these activities autonomously from the research activities of the Commonwealth Scientific and Industrial Research Organisation (CSIRO). The views expressed in this publication are those of the author(s) and do not necessarily represent those of, and should not be attributed to, the publisher or CSIRO. The copyright owner shall not be liable for technical or other errors or omissions contained herein. The reader/user accepts all risks and responsibility for losses, damages, costs and other consequences resulting directly or indirectly from using this information.

Foreword

Since my appointment as Australia's first Threatened Species Commissioner in 2014, I have met thousands of environmentally conscious citizens and worked on conservation projects with people from all walks of life. Volunteers, mums, dads, aunties, uncles, research scientists, farmers, politicians, Indigenous peoples, school kids – the list goes on. Despite our diversity, interests and opinions something that unites us is hope. Hope in the face of extinction. Hope in spite of unprecedented change and uncertainty. And hope that what we are doing will make a difference to create a brighter future for the people and the species we care about.

Hope is powerful, but it needs regular renewal. This is not always easy. There is plenty of bad news out there. When we open a paper or click on a link we can easily see dire warnings or heartbreaking images of loss. It's not difficult to become worn down and disengaged. This is why celebrating success and sharing good news stories is so important. As an advocate for Australia's threatened species, it's a big part of what I do. These stories remind us of what's important. They inspire us. And they instil a sense of optimism that if we get involved and keep working hard good things will come. The uplifting stories in *Recovering Australian Threatened Species: A Book of Hope* are an antidote to conservation fatigue. I loved reading them and could not help feeling reinvigorated by the incredible work and achievements they describe. If you are looking for some motivation or a positive role model to follow, look no further.

We need books like this to help engage and inspire the next generation of Australian conservation leaders. We are going to need their help. The future of our ecological systems and wildlife species are now, more than any time in our history, bound to the actions of people. No matter how powerful our technology or political and economic systems may be, they alone will not be able to achieve environmental sustainability. It is up to us. We will need engaged and motivated citizens if we are going to flight extinction to ensure the future for our precious native plants and animals. The stories of perseverance and success contained in this book are an inspiration and timely reminder that it takes action to create positive change. I hope they serve as rallying call for all Australians.

Gregory Andrews

Contents

	Foreword	iii
	Acknowledgements	ix
	List of contributors	xi
1	**Turning threatened species around: celebrating what we have done well** Stephen T. Garnett, Peter Latch, David B. Lindenmayer and John C.Z. Woinarski	1
2	**Recovery of Australian subpopulations of humpback whale** Peter L. Harrison and John C.Z. Woinarski	5
3	**Eradication of invasive species on Macquarie Island to restore the natural ecosystem** Keith Springer	13
4	**Management of seabird bycatch leads to sustainable fisheries and seabird populations** G. Barry Baker and Graham Robertson	23
5	**Mary's Famous Five: a story of connection, commitment and community in the recovery of threatened aquatic species in the Mary River catchment, Queensland** Tanzi Smith and Marilyn Connell	33
6	**Spiny rice-flower: small, unassuming but with many friends** Vanessa Craigie, Debbie Reynolds, Neville Walsh, Steve Mueck, Liz James and Pauline Rudolph	43
7	**Saving the pygmy bluetongue lizard** C. Michael Bull and Mark N. Hutchinson	55
8	**Malleefowl: answering the big questions that guide all malleefowl management** Sharon Gillam, Tim Burnard and Joe Benshemesh	65
9	**From the brink of extinction: successful recovery of the glossy black-cockatoo on Kangaroo Island** Karleah Berris, Michael Barth, Trish Mooney, David Paton, Martine Kinloch, Peter Copley, Anthony Maguire, Gabriel Crowley and Stephen T. Garnett	75

10 Science, community and commitment underpin the road to recovery for the red-tailed black-cockatoo 85
Vicki-Jo Russell, Richard Hill, Tim Burnard, Bronwyn Perryman, Peter Copley, Kerry Gilkes, Martine Maron, David Baker-Gabb, Rachel Pritchard and Paul Koch

11 Collaborative commitment to a shared vision: recovery efforts for noisy scrub-birds and western ground parrots 95
Allan Burbidge, Sarah Comer and Alan Danks

12 Back from the brink – again: the decline and recovery of the Norfolk Island green parrot 105
Luis Ortiz-Catedral, Raymond Nias, James Fitzsimons, Samantha Vine and Margaret Christian

13 Progress in the conservation of populations of the eastern bristlebird from central coastal New South Wales and Jervis Bay Territory 115
David B. Lindenmayer, Chris MacGregor and Nick Dexter

14 Tasmania's forty-spotted pardalote: a woodland survivor 125
Sally L. Bryant

15 Broad-scale feral predator and herbivore control for yellow-footed rock-wallabies: improved resilience for plants and animals = Bounceback 135
Robert Brandle, Trish Mooney and Nicki de Preu

16 Recovering the mountain pygmy-possum at Mt Blue Cow and Mt Buller 147
Linda Broome, Dean Heinze and Mellesa Schroder

17 Wild orchids: saving three Endangered orchid species in southern New South Wales 159
Helen P. Waudby, Matt Cameron, Geoff Robertson, Rhiannon Caynes and Noushka Reiter

18 Population enhancement plantings help save the Tumut grevillea 169
John Briggs and Dave Hunter

19 The spiny daisy: the disappearance and re-emergence of a unique Australian shrub 179
Doug Bickerton, Erica Rees, Tim Field, Amelia Hurren and Christophe Tourenq

20 The path to recovery for the 'extinct' Lord Howe Island phasmid 189
Hank Bower, Nicholas Carlile, Rohan Cleave, Chris Haselden, Dean Hiscox and Lisa O'Neill

21 Against the flow: the remarkable recovery of the trout cod in the Murray–Darling Basin 199
Jarod P. Lyon, Mark Lintermans and John D. Koehn

| 22 | Underbelly: the tale of the threatened white-bellied frog | 207 |

Manda Page, Kay Bradfield and Kim Williams

| 23 | Western swamp tortoise: slow and steady wins the race | 217 |

Gerald Kuchling, Andrew Burbidge, Manda Page and Craig Olejnik

| 24 | Twenty-five years of helmeted honeyeater conservation: a government–community partnership poised for recovery success | 227 |

Dan Harley, Peter Menkhorst, Bruce Quin, Robert Anderson, Sue Tardif, Karina Cartwright, Neil Murray and Merryn Kelly

| 25 | Bringing back warru: return of the black-footed rock-wallaby to the APY Lands | 237 |

John Read, Peter Copley, Matt Ward, Ethan Dagg, Liberty Olds, David Taggart and Rebecca West

| 26 | Recovery of the mainland subspecies of eastern barred bandicoot | 249 |

Richard Hill, Amy Coetsee (nee Winnard) and Duncan Sutherland

| 27 | Arid Recovery: a successful conservation partnership | 259 |

Katherine Moseby, Peter Copley, David C. Paton and John L. Read

| 28 | Effective conservation of critical weight range mammals: reintroduction projects of the Australian Wildlife Conservancy | 269 |

John Kanowski, David Roshier, Michael Smith and Atticus Fleming

| 29 | The contribution of captive breeding in zoos to the conservation of Australia's threatened fauna | 281 |

Dan Harley, Peter R. Mawson, Liberty Olds, Michael McFadden and Carolyn Hogg

| 30 | Mobilising resources for the recovery of threatened species | 295 |

Samantha Vine, Linda Bell and Allan Williams

| 31 | Reporting on success in threatened species conservation: the national policy context | 305 |

Peter Latch

| 32 | More than hope alone: factors influencing the successful recovery of threatened species in Australia | 315 |

Stephen T. Garnett, Peter Latch, David B. Lindenmayer, David J. Pannell and John C.Z. Woinarski

Index   325

Acknowledgements

This book arose from a workshop convened by the Threatened Species Recovery Hub of the National Environmental Science Programme and held in March 2016. The workshop sought to gauge the extent to which threatened species recovery was working in Australia and to identify the underlying drivers of success in threatened species recovery. It drew together people from around the country with experience in threatened species conservation, including representatives of all the states and territories as well as the federal agency. We recognise, of course, that those present at this workshop (and those contributing to this book) represent only the latest generation of people committed to the conservation of our biodiversity, and only a subset of those currently saving threatened species. The feeling at the workshop was that Australia's achievements in threatened species recovery receive far too little acknowledgement. This book emerged as a way to give back to the many people, groups and agencies that have supported the conservation of threatened species and communities over many decades – to show that the investment has been worthwhile. So, first, as representatives of the many authors who have contributed to the book, we would like to acknowledge the substantial and ongoing support provided by individuals, private organisations and the Australian taxpayers through successive governments to retain our biological inheritance in its entirety.

Second, as editors, we would like to thank our fellow authors, not just for contributing to this volume but for what has often been a lifetime of work nurturing threatened species so that they can continue to live successfully in this country and be enjoyed by future generations. We write in the concluding chapter about the importance of champions. We suspect that their role is understated in the chapters because many of our authors are those very champions. For such people, threatened species conservation is a vocation, and one for which they give far more than they are paid to perform, if indeed they are paid at all. We feel privileged to have been working in such company. We thank also the authors for their forbearance when dealing with editors who sought to constrict their lifetime's work to only a few pages of text.

Third, books are partly inspiration, partly hard graft. For the workshop, we thank David Pannell for his skilful facilitation and Jane Campbell for her organising skills. Once we got under way with the book, Claire Johnston was instrumental in coordination and finalisation of the texts and imagery, Ian Leiper used his mastery of GIS to create standardised maps and, in the final days, Hayley Geyle's meticulous attention to detail brought the book to production standard. We are most grateful to all three. At CSIRO Publishing, John Manger patiently negotiated a contract that would give us many more words than we initially anticipated, Tracey Kudis ran the production process and Peter Storer edited both thoroughly and sensitively.

The project is funded under the Threatened Species Recovery Hub of the National Environmental Science Programme within the Commonwealth Department of the Environment and Energy. We appreciate the vision of then Minister for the Environment, Greg Hunt, in establishing such a program within the Department and ensuring that threatened species were given special attention. We are also grateful for Australia's first Threatened Species Commissioner, Gregory Andrews, for his Foreword and his ongoing commitment and enthusiasm for threatened species conservation generally. Additional funding has been provided by Charles Darwin University and The Australian National University. Numerous other organisations and individuals have donated time, data and images and to all we are grateful.

Stephen Garnett, Peter Latch, David Lindenmayer and John Woinarski
May 2017

List of contributors

Robert Anderson
Friends of the Helmeted Honeyeater, PO Box 131, Woori Yallock, Victoria 3139, Australia

G. Barry Baker
Latitude 42, 114 Watsons Road, Kettering, Tasmania 7155, Australia

David Baker-Gabb
Elanus Pty Ltd, PO Box 131, St Andrews, Victoria 3761, Australia

Michael Barth
Natural Resources Kangaroo Island, Department of Environment, Water and Natural Resources, Kingscote, South Australia 5223, Australia

Linda Bell
Office of Environment and Heritage, PO Box A290 Sydney South, New South Wales 1232, Australia

Joe Benshemesh
Malleefowl Recovery Team and Department of Zoology, La Trobe University, Bundoora, Victoria 3068, Australia

Karleah Berris
Natural Resources Kangaroo Island, Department of Environment, Water and Natural Resources, Kingscote, South Australia 5223, Australia

Doug Bickerton
Conservation, NRM and Protected Areas Policy Branch, Department of Environment, Water and Natural Resources, 81–95 Waymouth Street, Adelaide, South Australia 5001, Australia

Hank Bower
Lord Howe Island Board, Lord Howe Island, New South Wales 2898, Australia

Kay Bradfield
Perth Zoo, 20 Labouchere Road, South Perth, Western Australia 6151, Australia

Robert Brandle
Natural Resources, South Australian Arid Lands, Department of Environment, Water and Natural Resources, 9 Mackay Street, Port Augusta, South Australia 5700, Australia

John Briggs
Ecosystems and Threatened Species Team, Regional Operations – South East Region, NSW Office of Environment and Heritage, PO Box 733, Queanbeyan, New South Wales 2620, Australia

Linda Broome
Office of Environment and Heritage, PO Box 733, Queanbeyan, New South Wales 2620, Australia

Sally L. Bryant
Tasmanian Land Conservancy, PO Box 2112, Lower Sandy Bay, Tasmania 7005, Australia

C. Michael Bull (deceased)
School of Biological Sciences, Flinders University, Adelaide, South Australia 5000, Australia

Allan Burbidge
Department of Biodiversity, Conservation and Attractions, Locked Bag 104, Bentley Delivery Centre, Western Australia 6983, Australia

Andrew Burbidge
87 Rosedale Street, Floreat, Western Australia 6014, Australia

Tim Burnard
National Malleefowl Recovery Team, Casterton, Victoria 3311, Australia

Matt Cameron
Ecosystems and Threatened Species, South West Region, Regional Operations Division, Office of Environment and Heritage, Albury, New South Wales 2640, Australia

Nicholas Carlile
Office of Environment and Heritage, PO Box 1967, Hurstville, New South Wales 1481, Australia

Karina Cartwright
Healesville Sanctuary, Zoos Victoria, Healesville, Victoria 3777, Australia

Rhiannon Caynes
Murray Local Land Services, 931 Garland Avenue, North Albury, New South Wales 2640, Australia

Margaret Christian
WildMob, PO Box 1724, Milton, Queensland 4064, Australia; and Norfolk Island Flora & Fauna Society, PO Box 702, Norfolk Island, New South Wales 2899, Australia

Rohan Cleave
Melbourne Zoo, Elliott Avenue, Parkville, Victoria 3052, Australia

Amy Coetsee (nee Winnard)
Zoos Victoria, PO Box 74, Parkville, Victoria 3052, Australia

Sarah Comer
Department of Biodiversity, Conservation and Attractions, 120 Albany Highway, Albany, Western Australia 6330, Australia

Marilyn Connell
Tiaro and District Landcare, PO Box 6, Tiaro, Queensland, 4650 Australia; and Research Institute for Environment and Livelihoods, School of Environment, Charles Darwin University, Darwin, Northern Territory 0909, Australia

Peter Copley
Conservation, NRM and Protected Area Policy, Department of Environment, Water and Natural Resources, 81–95 Waymouth Street, Adelaide, South Australia 5000, Australia

Vanessa Craigie
Department of Environment, Land, Water and Planning, Level 2, 8 Nicholson Street, East Melbourne, Victoria 3002, Australia

Gabriel Crowley
The Cairns Institute, James Cook University, PO Box 6811, Cairns, Queensland 4870, Australia

Ethan Dagg
Anangu Pitjantjatjara Yankunytjatjara Land Management UMUWA, via Alice Springs, Northern Territory 0872, Australia

Alan Danks
Department of Biodiversity, Conservation and Attractions, 120 Albany Highway, Albany, Western Australia 6330, Australia

Nicki de Preu
Ardeotis Biological Consultants Pty Ltd, Skillogallee Creek Road, Watervale, South Australia 5452, Australia

Nick Dexter
Parks Australia, Village Road, Jervis Bay Territory 2540, Australia

Tim Field
Banrock Station, Holmes Road, Kingston-on-Murray, South Australia 5331, Australia

James Fitzsimons
The Nature Conservancy, Suite 2-01, 60 Leicester Street, Carlton, Victoria 3053, Australia; and School of Life and Environmental Sciences, Deakin University, 221 Burwood Highway, Burwood, Victoria 3125, Australia

Atticus Fleming
Australian Wildlife Conservancy, 5/280 Hay Street, Subiaco, Western Australia 6008, Australia

Stephen T. Garnett
NESP Threatened Species Recovery Hub, Research Institute for the Environment and Livelihoods, Charles Darwin University, Casuarina, Northern Territory 0909, Australia

Kerry Gilkes
BirdLife Australia, c/o PO Box 37, Avenue Range, South Australia 5273, Australia

Sharon Gillam
National Malleefowl Recovery Team, Department of Environment, Water and Natural Resources, 81–95 Waymouth Street, Adelaide, South Australia 5000, Australia

Dan Harley
Wildlife Conservation and Science, Zoos Victoria, Healesville, Victoria 3777, Australia

Peter L. Harrison
Institute for Development, Environment and Sustainability, Marine Ecology Research Centre, School of Environment, Science and Engineering, Southern Cross University, PO Box 157, Lismore, New South Wales 2480, Australia

Chris Haselden
Lord Howe Island Board, Lord Howe Island, New South Wales 2898, Australia

Dean Heinze
Ecology Links, 9 Mill Road, Collinsvale, Tasmania 7012, Australia

Richard Hill
Department of Environment, Land, Water and Planning, 147 Bahgallah Road, Casterton, Victoria 3311, Australia

Dean Hiscox
PO Box 38, Lord Howe Island, New South Wales 2898, Australia

Carolyn Hogg
School of Life and Environmental Sciences, Faculty of Veterinary Science, The University of Sydney, New South Wales 2006, Australia; and Zoo and Aquarium Association Australasia, Mosman, New South Wales 2088, Australia

Dave Hunter
Ecosystems and Threatened Species Team, Regional Operations – South East Region, NSW Office of Environment and Heritage, PO Box 733, Queanbeyan, New South Wales 2620, Australia

Amelia Hurren
Spiny Daisy Recovery Team, Trees For Life, 5 May Terrace, Brooklyn Park, South Australia 5032, Australia

Mark N. Hutchinson
South Australian Museum, North Terrace, Adelaide, South Australia 5000, Australia

Liz James
Royal Botanical Gardens Victoria, Private Bag 2000, Birdwood Avenue, South Yarra, Victoria 3141, Australia

John Kanowski
Australian Wildlife Conservancy, 5/280 Hay Street, Subiaco, Western Australia 6008, Australia

Merryn Kelly
Department of Environment, Land, Water and Planning, Heidelberg, Victoria 3084, Australia

Martine Kinloch
Natural Resources Kangaroo Island, Department of Environment, Water and Natural Resources, Kingscote, South Australia 5223, Australia

Paul Koch
Greening Australia, 5 Fitzgerald Road, Pasadena, South Australia 5042, Australia

John D. Koehn
Applied Aquatic Ecology, Arthur Rylah Institute for Environmental Research, Department of Environment, Land, Water and Planning, 123 Brown Street, Heidelberg, Victoria 3084, Australia

Gerald Kuchling
Western Australian Department of Biodiversity, Conservation and Attractions, Parks and Wildlife Service, Swan Coastal District, 5 Dundebar Road, Wanneroo, Western Australia 6065, Australia

Peter Latch
Department of the Environment and Energy, John Gorton Building, King Edward Terrace, Parkes, ACT 2600, Australia

David B. Lindenmayer
NESP Threatened Species Recovery Hub, Fenner School of Environment and Society, The Australian National University, Canberra, ACT 2601, Australia

Mark Lintermans
Institute for Applied Ecology, University of Canberra, Canberra, ACT 2601, Australia

Jarod P. Lyon
Applied Aquatic Ecology, Arthur Rylah Institute for Environmental Research, Department of Environment, Land, Water and Planning, 123 Brown Street, Heidelberg, Victoria 3084, Australia

Chris MacGregor
Fenner School of Environment and Society, The Australian National University, Canberra, ACT 2601, Australia

Anthony Maguire
Natural Resources Kangaroo Island, Department of Environment, Water and Natural Resources, Kingscote, South Australia 5223, Australia

Martine Maron
School of Earth and Environmental Science, The University of Queensland, Brisbane, Queensland 4072, Australia

Peter R. Mawson
Perth Zoo, South Perth, Western Australia 6151, Australia

Michael McFadden
Taronga Conservation Society Australia, Taronga Zoo, Mosman, New South Wales 2088, Australia

Peter Menkhorst
Arthur Rylah Institute for Environmental Research, Department of Environment, Land, Water and Planning, 123 Brown Street, Heidelberg, Victoria 3084, Australia

Trish Mooney
Natural Resources, South Australian Arid Lands, Department of Environment, Water and Natural Resources, 9 Mackay Street, Port Augusta, South Australia 5700, Australia

Katherine Moseby
Arid Recovery, PO Box 147, Roxby Downs, South Australia 5725, Australia; and the University of New South Wales, Kensington, Sydney, New South Wales 2052, Australia

Steve Mueck
Biosis Pty Ltd, 38 Bertie Street, Port Melbourne, Victoria 3207, Australia

Neil Murray
Department of Ecology, Environment and Evolution, La Trobe University, Victoria 3086, Australia

Raymond Nias
Island Conservation, PO Box 1260, Sutherland, New South Wales 1499, Australia

Liberty Olds
Zoos SA, Frome Road, Adelaide, South Australia 5000, Australia

Craig Olejnik
Western Australian Department of Biodiversity, Conservation and Attractions, Parks and Wildlife Service, Swan Coastal District, 5 Dundebar Road, Wanneroo, Western Australia 6065, Australia

Lisa O'Neill
Office of Environment and Heritage, PO Box 1967, Hurstville, New South Wales 1481, Australia

Luis Ortiz-Catedral
Institute of Natural and Mathematical Sciences, Massey University, Private Bag 102-904 North Shore Mail Centre, Auckland, New Zealand

Manda Page
Western Australian Department of Biodiversity, Conservation and Attractions, Parks and Wildlife Service, Science and Conservation Division, 17 Dick Perry Avenue, Kensington, Western Australia 6151, Australia

David J. Pannell
School of Agriculture and Environment, The University of Western Australia, Crawley, Western Australia 6009, Australia

David Paton
School of Biological Sciences, The University of Adelaide, North Terrace, South Australia 5005, Australia

Lynn Pedler
School of Biological Sciences, The University of Adelaide, North Terrace, South Australia 5005, Australia

Bronwyn Perryman
BirdLife Australia, c/o PO Box 320, Port MacDonnell, South Australia 5291, Australia

Rachel Pritchard
Department of Environment, Land, Water and Planning, 12 Murray Street, Heywood, Victoria 3304, Australia

Bruce Quin
Department of Environment, Land, Water and Planning, Woori Yallock, Victoria 3139, Australia

John Read
Ecological Horizons, PO Box 207, Kimba, South Australia 5641, Australia; and The University of Adelaide, North Terrace, Adelaide, South Australia 5005, Australia

Erica Rees
Northern & Yorke, Trees For Life, 5 May Terrace, Brooklyn Park, South Australia 5032, Australia

Noushka Reiter
Royal Botanic Gardens Victoria, Cnr of Ballarto Road and Botanic Drive, Cranbourne, Victoria 3977, Australia; and Evolution, Ecology and Genetics, Research School of Biology, The Australian National University, Canberra, ACT 2601, Australia

Debbie Reynolds
Trust For Nature, Level 5/379 Collins Street, Melbourne, Victoria 3002, Australia

Geoff Robertson
Ecosystems and Threatened Species, South West Region, Regional Operations Division, Office of Environment and Heritage, Queanbeyan, New South Wales 2620, Australia

Graham Robertson
Australian Antarctic Division, 203 Channel Highway, Kingston, Tasmania 7050, Australia

David Roshier
Australian Wildlife Conservancy, 5/280 Hay Street, Subiaco, Western Australia 6008, Australia

Pauline Rudolph
Department of Environment, Land, Water and Planning, Level 2, 8 Nicholson Street, East Melbourne, Victoria 3002, Australia

Vicki-Jo Russell
Trees for Life, 5 May Terrace, Brooklyn Park, South Australia 5032, Australia

Mellessa Schroder
National Parks and Wildlife Service, PO Box 2228, Jindabyne, New South Wales 2627, Australia

Michael Smith
Australian Wildlife Conservancy, 5/280 Hay Street, Subiaco, Western Australia 6008, Australia

Tanzi Smith
Mary River Catchment Coordinating Committee, PO Box 1027, Gympie, Queensland 4570, Australia

Keith Springer
Pines Beach, North Canterbury, New Zealand

Duncan Sutherland
Phillip Island Nature Parks, 1019 Ventnor Road, Summerlands, Victoria 3922, Australia

David Taggart
University of Adelaide, Adelaide, South Australia 5005, Australia

Sue Tardif
Friends of the Helmeted Honeyeater, PO Box 131, Woori Yallock, Victoria 3139, Australia

Christophe Tourenq
Banrock Station, Holmes Road, Kingston-on-Murray, South Australia 5331, Australia

Samantha Vine
BirdLife Australia, Suite 2-05, 60 Leicester Street, Carlton, Victoria 3053, Australia

Neville Walsh
Royal Botanical Gardens Victoria, Private Bag 2000, Birdwood Avenue, South Yarra, Victoria 3141, Australia

Matt Ward
Department of Environment, Water and Natural Resources, 81–95 Waymouth Street, Adelaide, South Australia 5000, Australia

Helen P. Waudby
Ecosystems and Threatened Species, South West Region, Regional Operations Division, Office of Environment and Heritage, Albury, New South Wales 2640, Australia

Rebecca West
University of New South Wales, PO Box 600, Broken Hill, New South Wales 2880, Australia

Allan Williams
Department of Environment and Heritage Protection, Level 5, 400 George Street, Brisbane, Queensland 4000, Australia

Kim Williams
Western Australian Department of Biodiversity, Conservation and Attractions, Regional and Fire Management Division, South West Highway, Bunbury, Western Australia 6230, Australia

John C.Z. Woinarski
NESP Threatened Species Recovery Hub, Research Institute for the Environment and Livelihoods, Charles Darwin University, Casuarina, Northern Territory 0909, Australia

1

Turning threatened species around: celebrating what we have done well

Stephen T. Garnett, Peter Latch, David B. Lindenmayer and John C.Z. Woinarski

The rate of change in nature is escalating as human impacts become more pervasive and intensive. Many of those who care for nature and recognise the value of a healthy and diverse natural environment lament the losses and may be demoralised by the erosion of biodiversity. With conservation failures often highlighted, policy makers may see conservation of highly imperilled species as a lost cause – and be reluctant to invest resources if little return is likely – and so may look elsewhere for policy wins. But the story of conservation has another, less well-reported, side – one that gives hope. This book tells that other side for Australia's threatened species. We choose to focus here on threatened species because to a large degree they are at the forefront of the conservation challenge and because our regard, or disregard, for them may have dramatic and irretrievable consequences. Although the focus here is typically on individual threatened species, in many cases these individual species also represent swathes of other biodiversity values and issues: actions taken, or not taken, for their care will help or hinder many other species.

For millennia, Australian biodiversity has been managed by Indigenous landowners. The effects of the many purposes and outcomes of such management cannot be disentangled but the empirical result is that, for many thousands of years, people and the species now considered threatened existed alongside each other across much of the Australian continent and many of its offshore islands. Most Australian plant and animal species persist still, despite more than 200 years of extensive modification to the environment by Europeans and other colonists. The first element of hope for the conservation future of Australia is that many Australian species are thriving.

However, many other species have suffered marked declines since European settlement. As at April 2017, 1717 Australian species are formally listed as threatened and another 91 are formally recognised as having become extinct since 1788. These are disconcertingly large tallies – for example, the number of Australian plant and mammal extinctions over the last 200 years surpasses that of any other country – and it is likely that both tallies are also severe under-estimates of the actual numbers of threatened and extinct Australian species.

Against a tide of environmental degradation, there is also a long history of attempts to protect species by many individual Australians, and some government authorities. The providence petrel (*Pterodroma solandri*) of Norfolk Island was first subject to a conservation order in 1802 when hunting was banned there (Bonyhardy 2000). That failed, but the species did persist on Lord Howe Island and, nearly two centuries after that order was imposed, the species returned to an island in the Norfolk group that had been set aside for conservation. That is the second reason to have hope: our conservation reserves are acting as havens within which species can adapt on their own.

However, such stories of self-recovery of threatened species or ecological communities are rare. Most require substantial input of sweat and funding, knowledge and dedication, and, above all, time. Too much time had passed for some threatened species by the time the Australian Government began to accept an obligation for the protection of threatened species, in the late 1980s. By then it became apparent that the previously limited and *ad hoc* approach to threatened species conservation needed more strategic planning and investment. Action plans were written and a major conference in 1996 was able to report substantial adoption of recovery planning as an organising principle for saving species (Stephens and Maxwell 1996).

Governance for threatened species and ecological communities has advanced. Australia ratified the Convention on Biological Diversity in 1992. By the end of the 20th century, legislation designed to help the country meet its international obligations – the *Environment Protection and Biodiversity Conservation Act 1999* – had come into force. This was the first time threatened species were recognised as Matters of National Environmental Significance, meaning that applications for development had to prove that they would have no significant impact on listed species.

Legislative achievements would not have been possible had there not been a widespread public aversion to extinctions. But such support would not have happened had all the stories been of loss. Probably equally important were impressive and inspiring accounts of conservation successes. Campaigns to prevent the extinction of animals such as the giant panda or to discourage the hunting of big cats for fashion inspired a generation of conservation activists. Wildlife programs on television brought battles to save species and the environment into suburban living rooms. It was this period that spawned many of the threatened species recovery projects described in this book.

So what is 'success' in threatened species recovery? Ideally, success is when a species is fully restored to the habitats and numbers it had before it was affected by the threats that led to its imperilment (Redford *et al.* 2011). Alternatively, a more modest 'success' may simply be that almost certain extinction has been prevented, and there is the potential for a reduction in extinction risk in the future. In some cases, success may also be viewed as a social process whereby the human community is sufficiently organised and effective for a species' persistence to be probable. All are valid, all can be seen as stages in threatened species recovery, and all are exemplified in this book. In selecting chapters, we allowed a definition of success to emerge in the stories so that we can draw the elements of success together at the end.

To find appropriate examples, we first asked a range of government agencies and non-government organisations to volunteer examples from their area of influence that they considered successful. The examples were far from exhaustive – we sought novelty and diversity. Some success stories have already been well told – such as the brilliant work undertaken to recover Gould's Petrel in New South Wales (Priddel and Carlile 2009). We aimed to have a good mixture of plants, vertebrates and invertebrates, and government, non-government and private endeavours. Although 'hope' is a central theme of this book, and a unifying feature across the case studies, these examples demonstrate that many qualities and ingredients are needed by those people who recover threatened species. Some examples hinge on successful application of policy and legislation and some on advances in *ex situ* management. Most, however, relate to on-ground actions – of people on the ground caring for their country and its species, uncovering the factors most threatening to a species, determining what to do about those threats and then carefully and assiduously applying informed management. Some of these case studies have been going for 50 years, others for less than a decade. None of the people involved would consider the work complete, but all would assert that their work has substantially reduced the risk of extinction for their target species. What they demonstrate above all is that extinction can be prevented.

At the end of the book, we draw together the main messages to emerge from these stories, on the basis that such lessons can help with many more species. In doing so, we aim to throw light on the following seven questions:

1. What characteristics of individual threatened species help or hinder recovery efforts? Some species may be relatively easy to recover, but others not. Some species may attract attention and empathy, but others not (or not immediately). Our case studies include a wide range of life forms and life histories but are there elements that the selected species have in common?
2. Has the community contributed to the recovery? What sorts of people were involved? What were their motivations? How did they organise to achieve beneficial outcomes? Much is made of the role of citizen science in conservation, but to what extent and under what circumstances have members of the general public contributed to success in threatened species recovery and what roles have they played?
3. How important were committed individuals to the trajectories of species? Effective leadership of recovery programs is known to be an important element of successful conservation (Black *et al.* 2011), but are there other aspects of governance and decision making that characterise successful recovery programs?
4. What was the function of good policy and governance? Did recovery require a strong legal framework and a commitment in policy to be effective? Are there improvements to the laws and the policing of compliance that could make recovery faster and more secure?
5. How much does it cost to recover a species successfully? Although there are many variables that affect the cost of recovery, from the biology of the species to the nature of the threats they face and the location where the threats are occurring, we aim to provide some broad documentation of what has been spent – an empirical measure of the scale of investment needed to make a difference.
6. Have the actions taken to date in these case studies entrenched recovery? Or is the fight for these species against extinction a never-ending commitment?
7. Much effort is needed for the recovery of threatened species. To what extent can we learn lessons from these cases that will reduce the decline of many additional species before they too become threatened and hence need intensive management response?

This book is not just about science and management. It is also a celebration of extraordinary achievements, often by extraordinary people and organisations. Initially we asked the authors to write more about themselves, but most were too self-effacing to do justice to what they had accomplished not just for their target species but for the wider Australian society. Our authors are among the heroes of threatened species conservation in Australia and deserve far wider recognition: they are altruistic, committed, caring, resourceful, admirable and expert. The people in these case studies want to make a difference: to leave this world at least as healthy, diverse and beautiful as that they were born into. For very few is the work on threatened species just a job. Rather it is a vocation for which some have dedicated most of their lives.

In the century since Gandhi is attributed as saying that 'a nation's greatness is measured by how it treats its weakest members', we have learnt that our concept of greatness must encompass not only a nation's people but also its environment. And the weakest members of the environment are those species we have brought close to the abyss of extinction. Often we have failed such species, just as we have failed the poor. But we need do so no longer. Chapter after chapter in this book demonstrates what can be achieved with vision and dedication, persistence, money, insight and innovation. This is a 'book of hope', not just for our most vulnerable species but for the wider Australian environment. Ultimately it is also a book of hope for our society.

References

Black SA, Groombridge JJ, Jones CG (2011) Leadership and conservation effectiveness: finding a better way to lead. *Conservation Letters* **4**, 329–339. doi:10.1111/j.1755-263X.2011.00184.x

Bonyhardy T (2000) *The Colonial Earth*. Melbourne University Press, Melbourne.

Priddel D, Carlile N (2009) Key elements in achieving a successful recovery programme: a discussion illustrated by the Gould's Petrel case study. *Ecological Management & Restoration* **10**, S97–S102. doi:10.1111/j.1442-8903.2009.00460.x

Redford KH, Amato G, Baillie J, Beldomenico P, Bennett EL, Clum N, *et al.* (2011) What does it mean to successfully conserve a (vertebrate) species? *Bioscience* **61**, 39–48. doi:10.1525/bio.2011.61.1.9

Stephens S, Maxwell S (Eds) (1996) *Back from the Brink: Refining the Threatened Species Recovery Process*. Surrey Beatty & Sons, Chipping Norton, NSW.

2

Recovery of Australian subpopulations of humpback whale

Peter L. Harrison and John C.Z. Woinarski

Summary

The problem
1. Populations of most large cetacean species, including humpback whales (*Megaptera novaeangliae*), were severely depleted to critically low levels by extensive and unsustainable whaling programs.
2. Although humpback whales in Australia were subject to some coastal shore-based whaling mortality, the near catastrophic decline in their abundance was caused by massive illegal Soviet whaling in waters south of Australia and New Zealand during the 1959–1961 summer seasons. This resulted in the near extinction of East coast humpback whales and collapse of the Australian and New Zealand coastal whaling operations.

Actions taken to manage the problem
1. Compilation of records of reported commercial whaling catches and recognition of declines in whale abundance by the International Whaling Commission, indicating critical status, and its causality.
2. Protection of humpback whales by international agreement globally and in Australia with strong ongoing protection through explicit provisions in law and policy at national and state/territory level.
3. Ongoing and robust monitoring of population size.

Markers of success
1. The previous population decline has been reversed with substantial population increases. The two Australian breeding populations of humpback whales are now recovering strongly, with abundance estimated to be approaching natural population sizes before whaling exploitation last century.
2. The primary cause of decline was identified and is being effectively managed.

Reasons for success
1. Commercial whaling was the major threat, so cessation of whaling was an effective remedy.
2. Changing community attitudes from hunting whales to valuing living humpback whales as an iconic species. This led to strong community support and international collaboration for protecting whales.
3. Governments took strong and concerted action in international forums to protect whales.
4. Funding for management, research and monitoring was sustained over many decades.

Introduction

This chapter focuses on the two breeding subpopulations of humpback whale occurring in Australian waters – the East coast and West coast Australian subpopulations (designated as E1 and D, respectively, by the International Whaling Commission (IWC); Fig. 2.1). However, a broader regional Southern Hemisphere context is also relevant given the extensive migrations undertaken by these, and most other, subpopulations of humpback whale – up to ~18 000 km (Robbins *et al.* 2011) from their winter breeding and calving grounds in warm subtropical and tropical waters to summer feeding grounds in high latitude cold waters (Chittleborough 1965; Bannister 2008).

The primary threat driving the decline of the humpback whale (and many other cetaceans) has long been well established: unsustainable rates of whaling (Clapham *et al.* 2008; Clapham and Baker 2009). Compared with many other threatened species, such a simple threat context should allow for ready remedial response. However, actions to ameliorate this threat have not been straightforward, because much of the threat operated on the high seas beyond the territorial boundaries of Australia, or any other single nation, and much of the management of commercial whaling and take was poorly regulated, leading to unsus-

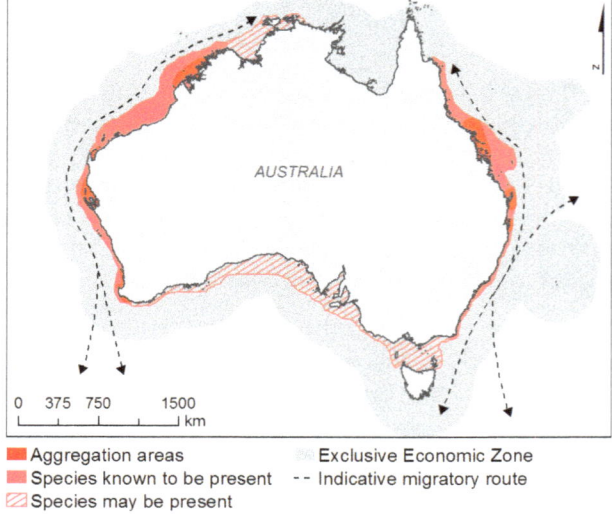

Fig. 2.1. Distribution of East and West coast humpback whale subpopulations around coastal Australia.

Fig. 2.2. The Australian population of the humpback whale (*Megaptera novaeangliae*) was reduced to less than 5% of its pre-whaling abundance before hunting ceased (photo: P. Harrison).

tainable scales of industrial whaling and illegal whaling. Global catch data for industrial whaling operations show that nearly 2.9 million large whales were killed last century, including more than 2 million whales killed in the Southern Hemisphere (Rocha *et al.* 2015). Soviet illegal whaling over three decades from 1947 to 1973 contributed substantially to declines of whale populations, with more than 178 000 of the total USSR global catch of 534 204 whales not reported to the IWC (Rocha *et al.* 2015).

Extensive commercial pelagic and coastal whaling in the 19th and 20th centuries caused a 95% global reduction in humpback whale abundance and extirpation of some subpopulations (Clapham *et al.* 2008). More than 200 000 humpback whales were killed in the Southern Hemisphere from 1904 to 1983 (Clapham and Baker 2009). Before the 1950s, there was relatively little exploitation of humpback whales within Australian waters, but shore-based coastal whaling stations were established at Point Cloates in 1949, Carnarvon in 1950 and Albany in 1952 (south-western Australia), and in eastern Australia at Tangalooma (Moreton Island) in 1952 and Byron Bay in 1954, with these stations each taking between ~100 to 1000 humpback whales per year during their migrations through western and eastern Australian coastal waters (Clapham *et al.* 2009). Chronic impacts of many decades of whaling and acute impacts from illegal and massively under-reported Soviet whaling led to the collapse of the two humpback whale populations in Australian waters and more broadly in the Southern Ocean by the early 1960s. It is now known that Soviet fleets illegally killed more than 25 000 humpback whales in waters south of Australia and New Zealand during two austral summer whaling seasons in 1959–1960 and 1960–1961 (Rocha *et al.* 2015). Consequently, by the mid-1960s, the East coast Australian subpopulation was estimated to have been reduced to a few hundred individuals or less (Bannister and Hedley 2001; Jackson *et al.* 2013), and the total Australian population reduced to ~3.5–5% of its pre-whaling abundance (DEH 2005a; Fig. 2.2).

Conservation management

People, agencies, governance and accountability

Although some regulation of commercial whaling was initiated with the establishment of the IWC in 1946, the scale of catches increased over subsequent decades and sequential depletion of great whales continued until whale abundance and whaling catches were seriously diminished through over-exploitation (Clapham and Baker 2009; Rocha *et al.* 2015). The collapse of the coastal whaling industry in Australia and in many other regions resulted in increasing recognition of the need to conserve remaining whale stocks. This coincided with changing public and political attitudes to whales and dolphins and increasing awareness of their extraordinary biology and ecology and their intelligence and complex cognitive abilities and social interactions (e.g. McIntyre 1974). People began to value these mammals more highly alive than dead, leading to increasing concern over whaling and by-catch of dolphins revealed in the Eastern Pacific Ocean tuna fishery in the 1960s. This in turn led to increased advocacy for conserving whales and dolphins and direct action anti-whaling campaigns, such as that by Greenpeace in Albany in 1977, which galvanised the conservation movement in Australia and elsewhere (Day 1987). Particular priority was given to the conservation of some of the most imperilled and iconic species, such as the humpback whale.

Planning and policy

The IWC banned killing of the humpback whale in the Southern Hemisphere in 1963, but substantial illegal Soviet whaling catches continued for several years thereafter. In 1972, the United Nations Conference on the Human Environment in Stockholm recommended a 10-year ban on commercial whaling, and in 1986 the IWC introduced an international moratorium on commercial whaling (commercial catch limits were set to zero) that continues today (Rocha *et al.* 2015).

In addition to the protective measures developed through the IWC, other international measures that provide global protection of humpback whales included their listing on Appendix I of the Convention on International Trade in Endangered Species (CITES) and on Appendix II of the Convention on Migratory Species, and general and explicit conservation protective measures under the Convention of Antarctic Marine Living Resources, the Antarctic Treaty Consultative Meetings and the South Pacific Regional Environment Programme.

Reflecting and responding to this community opinion, Australian governments were active at national and international levels in seeking tighter regulation or cessation of the whaling industry. Largely due to the population crash of the two Australian humpback whale subpopulations (and hence loss of economic viability), the Tangalooma and Byron Bay whaling stations closed in 1962 and the Carnarvon and Albany stations in Western Australia stopped hunting humpback whales in the same year (ultimately closing entirely in 1978) (Clapham *et al.* 2009). Consistent with the IWC, the Australian Government banned hunting of humpback whales in Australian waters in 1963. A notable landmark was an independent inquiry (Frost 1978) established by the Australian Government, which resulted in the *Whale Protection Act 1980*.

Under the *Environment Protection and Biodiversity Conservation Act 1999*, humpback whales are given protection as a listed threatened species (as Vulnerable) and as a migratory species. The Act also provides some explicit protection for cetaceans, and provides for the establishment and management of the Australian Whale Sanctuary, which protects all cetaceans found in Australian waters, including all Commonwealth (federal) waters from the 3 nautical mile state waters limit out to the boundary of Australia's Exclusive Economic Zone. In addition, management measures to support the recovery and protect important habitats for these whales have included the development of guidelines to manage acoustic disturbance from offshore seismic operations (DEWHA 2008), the development of national standards for managing entanglements, guidelines for interactions with whales during whale watching (DEH 2005b) and identifying habitat requirements for humpback whales in the planning, establishment and management of marine protected areas (e.g. DSEWPC (2012)). To some extent, these diverse aspects of the conservation of humpback whales have been coordinated through a national recovery plan (DEH 2005a), although this plan is no longer in force.

Research, biology, identification of key threats

Long-lasting national and international research programs have provided robust evidence on the humpback whale's life history and population structure, dispersal (including the identification of breeding grounds in tropical Australian waters), ecology and threats. Although demographic data clearly identified the main historic threat (whaling) and helped to ensure that this threat was effectively managed, other threats may affect the species currently or in the future. The diet of humpback whales in the Southern Hemisphere is almost exclusively Antarctic krill (*Euphausia superba*) (Chittleborough 1965); hence over-harvesting may become a potential future threat if the krill fishery is not managed effectively (Nicol *et al.* 2012). Global climate change may further affect this key food resource (Nicol *et al.* 2008; Kawaguchi *et al.* 2011), alter trophic interactions and availability and distribution of krill and other potential prey resources, and may have other effects on migratory and breeding habitats. Research has provided some information on other threats and indicated mechanisms for their management. These threats include anthropogenic noise (particularly from seismic surveys military active sonar) (Nowacek *et al.* 2007), entanglement in fishing gear and shark netting (Cassoff *et al.* 2011), and vessel strike (Laist *et al.* 2001; Redfern *et al.* 2013).

Monitoring

Results from long-standing monitoring programs for the East coast and West coast Australian subpopulations of humpback whales have provided classic examples of recovery of a species following control of a principal threat. The East coast Australian subpopulation has been monitored at Point Lookout, North Stradbroke Island, since 1978 (Noad *et al.* 2011, 2016), with largely consistent increases over this period of ~11% per year – close to the maximum plausible 11.8% annual rate of growth of humpback whale populations (Best 1993; Brandao *et al.* 2000; Bannister and Hedley 2001) (Fig. 2.3). By 2015, the estimate for the eastern Australian subpopulation was ~24 500 individuals, a marked increase from its mid-1960s low of a perhaps a few hundred individuals, and somewhere between 58–98% of pre-whaling population size (Noad *et al.* 2016). Trajectories have been similar for long-term monitoring data for the West coast Australian subpopulation (Bannister and Hedley 2001), with 2008 population estimates of ~28 000 individuals (Hedley *et al.* 2011) and

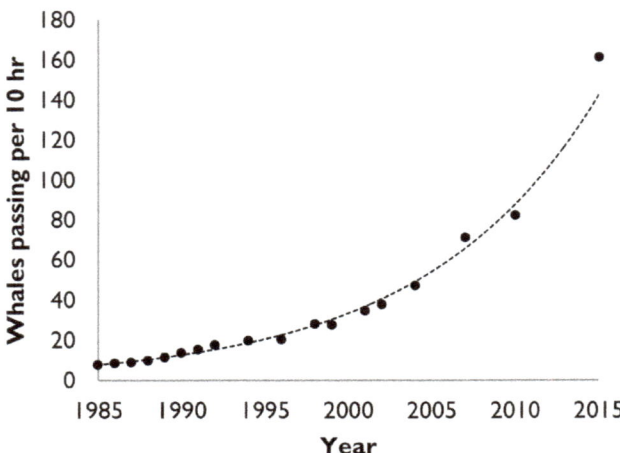

Fig. 2.3. Monitoring results from Point Lookout, North Stradbroke Island, showing increase in relative abundance of humpback whales from 1984 to 2015 (adapted from Noad *et al.* 2016).

~26 100 (Salgado Kent *et al.* 2012), indicating a recovery to ~90% of pre-exploitation levels (Müller and Butterworth 2012).

The future

The recovery of Australian humpback whales has been achieved largely because of international cooperation in banning whaling, and subsequent ongoing management. Such cooperation may not last, and this recovery cannot be assumed to be secured. Indeed, and perversely, the extraordinary success of the recovery may mean that there may be some future pressure to resume commercial whaling or kill humpback whales under the IWC Article VIII Special Permit Whaling. Furthermore, other current or potential threats are likely to intensify. Most notably, the humpback whale's main food resource could become depleted by expanding krill fishing pressure and the impacts of global climate change. Furthermore, increases in whale populations and shipping traffic are likely to lead to increased incidence of collisions, whose economic and human costs may potentially dampen some public sentiment for whale conservation or require some more interventionist management (Ritter 2012).

Conclusion

As for many other cetaceans, human action (whaling) directly caused the imperilment of the humpback whale. With changing community attitudes, catalysed by the realisation that extinction was otherwise a likely outcome for some of the largest and most iconic animals ever to have lived on Earth, human actions have led to the extraordinary recovery of Australian breeding populations of humpback whales.

The recovery of Australian populations of the humpback whale represents a remarkable conservation success. Whereas many Australian threatened species are highly localised and their conservation is governed through carefully framed recovery plans and dependent upon local champions and multi-stakeholder recovery teams, the humpback whale represents a notable contrast. Its extensive migratory dispersal spanning and beyond national borders has demanded that its conservation has been dependent upon policies and agreements established at international level. The reversal of fortune for the humpback whale involved a recognition of the impacts of unsustainable whaling and population collapse, sustained advocacy campaigns and recognition of whales as intelligent and valuable marine mammals, complex and long-lasting international negotiations, explicit legislative responses, and a very robust evidence base developed through collaborative research and monitoring.

References

Bannister JL (2008) *Great Whales*. CSIRO Publishing, Melbourne.

Bannister JL, Hedley SL (2001) Southern Hemisphere Group IV humpback whales: their status from recent aerial survey. *Memoirs of the Queensland Museum* **47**, 587–598.

Best PB (1993) Increase rates in severely depleted stocks of baleen whales. *ICES Journal of Marine Science* **50**, 169–186. doi:10.1006/jmsc.1993.1018

Brandao A, Butterworth DS, Brown MR (2000) Maximum possible humpback whale increase rates as a function of biological parameter values. *The Journal of Cetacean Research and Management* (Supplement 2), 192–193.

Cassoff RM, Moore KM, McLellan WA, Barco SG, Rotstein DS, Moore MJ (2011) Lethal entanglement in baleen whales. *Diseases of Aquatic Organisms* **96**, 175–185. doi:10.3354/dao02385

Chittleborough RG (1965) Dynamics of two populations of the humpback whale, *Megaptera novaeangliae* (Borowski). *Australian Journal of Marine and Freshwater Research* **16**, 33–128. doi:10.1071/MF9650033

Clapham PJ, Baker CS (2009) Whaling, modern. In *Encyclopedia of Marine Mammals*. (Eds WF Perrin, B Würsig and JGM Thewissen) pp. 1239–1243. Academic Press, Amsterdam, Netherlands.

Clapham PJ, Aguilar A, Hatch LT (2008) Determining spatial and temporal scales for management: lessons from whaling. *Marine Mammal Science* **24**, 183–201. doi:10.1111/j.1748-7692.2007.00175.x

Clapham PJ, Mikhalev YA, Franklin W, Paton D, Baker CS, Ivashchenko YV, et al. (2009) Catches of humpback whales, *Megaptera novaeangliae*, by the Soviet Union and other nations in the Southern Ocean, 1947–1973. *Marine Fisheries Review* **71**, 39–43.

Day D (1987) *The Whale War*. Sierra Club Books, San Francisco CA, USA.

DSEWPC (2012) 'Species group report card – cetaceans. Supporting the marine bioregional plan for the North-west Marine region.' Department of Sustainability Environment Water Population and Communities, Canberra.

DEH (2005a) 'Humpback Whale Recovery Plan 2005–2010.' Australian Government Department of Environment and Heritage, Canberra.

DEH (2005b) 'Australian National Guidelines for Whale and Dolphin Watching.' Department of the Environment and Heritage, Canberra.

DEWHA (2008) 'EPBC Act Policy Statement 2.1 – Interaction between offshore seismic exploration and whales.' Department of the Environment Water Heritage and the Arts, Canberra.

Frost S (1978) *Inquiry into Whales and Whaling*. Australian Government Publishing Service, Canberra.

Hedley SL, Dunlop RA, Bannister JL (2011) Evaluation of WA Humpback surveys 1999, 2005, 2008: where to from here? Australian Marine Mammal Centre, Hobart.

Jackson JA, Zerbini A, Clapham P, Constantine R, Garrigue C, Hauser N, et al. (2013) 'Population modelling of humpback whales in East Australia (BSE1) and Oceania (BSE2,

BSE3, BSF2)'. Paper SC/65a/SH07 presented to the International Whaling Commission Scientific Committee, Cambridge, UK.

Kawaguchi S, Kurihara H, King R, Hale L, Berli T, Robinson JP, *et al.* (2011) Will krill fare well under Southern Ocean acidification? *Biology Letters* **7**, 288–291. doi:10.1098/rsbl.2010.0777

Laist DW, Knowlton AR, Mead JG, Collett AS, Podesta M (2001) Collisions between ships and whales. *Marine Mammal Science* **17**, 35–75. doi:10.1111/j.1748-7692.2001.tb00980.x

McIntyre J (1974) *Mind in the Waters*. Charles Scribner's Sons, New York, USA.

Müller A, Butterworth DS (2012) 'Initial population model fits to the humpback breeding stocks D, E1 and Oceania'. Paper SC/64/SH29 presented to the International Whaling Commission Scientific Committee, Cambridge, UK.

Nicol S, Worby A, Leaper R (2008) Changes in the Antarctic sea ice ecosystem: potential effects on krill and baleen whales. *Marine and Freshwater Research* **59**, 361–382. doi:10.1071/MF07161

Nicol S, Foster J, Kawaguchi S (2012) The fishery for Antarctic krill – recent developments. *Fish and Fisheries* **13**, 30–40. doi:10.1111/j.1467-2979.2011.00406.x

Noad MJ, Dunlop RA, Paton D, Cato DH (2011) Absolute and relative abundance estimates of Australian east coast humpback whales (*Megaptera novaeangliae*). *The Journal of Cetacean Research and Management* **Special Issue 3**, 243–252.

Noad MJ, Dunlop RA, Bennett L, Kniest H (2016) 'Abundance estimates of the east Australian humpback whale population (BSE1): 2015 survey and update'. Paper SC/66b/SH/21 submitted to the International Whaling Commission, Cambridge, UK.

Nowacek DP, Thorne LH, Johnston DW, Tyack PL (2007) Responses of cetaceans to anthropogenic noise. *Mammal Review* **37**, 81–115. doi:10.1111/j.1365-2907.2007.00104.x

Redfern JV, McKenna MF, Moore TJ, Calambokidis J, DeAngelis ML, Becker EA, *et al.* (2013) Assessing the risk of ships striking large whales in marine spatial planning. *Conservation Biology* **27**, 292–302. doi:10.1111/cobi.12029

Ritter F (2012) Collisions of sailing vessels with cetaceans worldwide: first insights into a seemingly growing problem. *The Journal of Cetacean Research and Management* **12**, 119–127.

Robbins J, Dalla Rosa L, Allen JM, Mattila DK, Scchi ER, Friedlaender AS, *et al.* (2011) Return movement of a humpback whale between the Antarctic Peninsula and American Samoa: a seasonal migration record. *Endangered Species Research* **13**, 117–121. doi:10.3354/esr00328

Rocha CR, Clapham PJ, Ivashchenko YV (2015) Emptying the oceans: a summary of industrial whaling catches in the 20th century. *Marine Fisheries Review* **76**, 37–48. doi:10.7755/MFR.76.4.3

Salgado Kent C, Jenner C, Jenner M, Bouchet P, Rexstad E (2012) Southern Hemisphere Breeding Stock D humpback whale population estimates from North West Cape, Western Australia. *The Journal of Cetacean Research and Management* **12**, 29–38.

3

Eradication of invasive species on Macquarie Island to restore the natural ecosystem

Keith Springer

Summary

The problem

1. From the 1950s, it was increasingly recognised that exotic animals were having major negative impacts on the vegetation and fauna of Macquarie Island.
2. Rabbits removed native vegetation cover and accelerated erosion, while removing vegetation cover from seabird burrows and nests.
3. Cats and rodents predated on seabirds and invertebrates.
4. Rodents consumed fruits and seeds, impacting native plants.

Actions taken to manage the problem

1. Rabbits, rats and mice were targeted in a combined eradication operation that commenced in 2007 and was completed successfully in 2014.
2. Aerial baiting was designed to eradicate rats and mice and to remove most rabbits.
3. Detection dogs, field huts and hunting equipment were provided to support hunting teams for a 3-year period to remove surviving rabbits.

Markers of success

1. The previous declining trend of many threatened species was first stabilised and is now reversing.
2. The conservation status of several species has been down-listed.
3. Causes of decline were identified correctly and are now being managed effectively.

Reasons for success

1. The adoption of a carefully planned and peer-reviewed strategy for operational planning.
2. Commitment of funding from the outset for a multi-year project to employ dedicated staff in both planning and field roles and deliver on the project objectives.

3. Tenacity, persistence and creativity to resolve problems that arose during the project.

Introduction

Macquarie Island is located in the Southern Ocean, 1500 km south-east of Hobart – about half way between New Zealand and Antarctica. The island, located at 54°30′ south, comprises 12 865 ha, with many small offshore rock stacks. It is a Tasmanian Nature Reserve and a World Heritage Site. It provides a rare speck of land in the Southern Ocean where sea birds and marine mammals breed.

Introduced mammals became established on Macquarie Island through human activity soon after its discovery in 1810. Cats (*Felis catus*) became feral by 1820, house mice (*Mus musculus*) may have established by 1830, rabbits (*Oryctolagus cuniculus*) and weka (*Gallirallus australis*) were introduced as a food source around 1879, and ship rats (*Rattus rattus*) were recorded in the early 20th century. Dogs (*Canis familiaris*) established a feral population early in the 19th century, but subsequently died out. It is likely that they depredated most of the larger tubenose seabird species, such as wandering albatross (*Diomedea exulans*), which are now restricted to around 20 breeding pairs.

Conservation management

The combined impacts of invasive species were probably significant early in their establishment, but, as is often the case, it is not until people with an interest in natural history record what they found that these impacts caused concern. An example of this is the visit by A. Hamilton in 1894, and his intent to collect specimens of the Macquarie Island parakeet (*Cyanoramphus erythrotis*), only for the resident penguin-oiling crew to say that, although they used to be numerous, none had been seen in the 2 years they had been there. This was the first recognition that the bird was in fact extinct, probably by 1890, along with the only other endemic land bird – the Macquarie Island rail (*Rallus philippensis macquariensis*) (Taylor 1979).

Scientists regularly visited the island to study flora and fauna after the establishment of the Australian Antarctic Division station in 1948, by when it is likely that rats and cats had already caused the extirpation of some species of burrow-nesting and surface-nesting petrels from the main island, such as the grey petrel (*Procellaria cinerea*), storm petrels (*Oceanites* spp.), the soft-plumaged petrel (*Pterodroma mollis*), the blue petrel (*Halobaena caerulea*) and the cape petrel (*Daption capense*). Blue petrels were able to maintain a small local population on rat-free offshore rock stacks, while colonies of Antarctic prion and white-headed petrels maintained colonies on the plateau, well above the tussock areas that formed the predominant rat habitat (Brothers and Bone 2008).

Botanists raised concerns about significant vegetation damage caused by rabbit overgrazing soon after establishment of the Australian Antarctic Division station, with several papers and reports produced through the 1950s and 1960s (e.g. Costin and Moore 1960; Taylor 1955; 2005). Management of rabbits began in the early 1960s with poisoning trials and initial release of the myxoma virus (Johnston 1966). Both trials were unsuccessful in achieving large-scale rabbit control. Cat control commenced in the mid-1970s after researchers estimated that feral cats were killing ~60 000 seabirds annually (Jones 1977).

Throughout the 1970s, annual releases of European rabbit fleas (*Spilopsyllus cuniculi*) were made until flea distribution was island-wide, whereupon the myxoma virus (Lausanne strain) was released in December 1978. Rabbit numbers decreased dramatically within the next few years, and stayed at relatively low levels for ~15 years (Copson and Whinam 2001). Weka (*Gallirallus australis*) were eradicated by 1988 by shooting, after their numbers dropped following reduction of the rabbit population (which may have led to increased cat predation on weka).

After some 20 years of cat control, the management goal shifted to eradication. With additional funding, a concerted program to remove them commenced in 1998 and was completed by 2001. Grey petrels showed an immediate response, with a small number of chicks fledged in 2000 being the first confirmed breeding of this species in about a century. White-headed petrels and Antarctic prions were also considered to benefit from the removal of cats (Brothers and Bone 2008).

Although some researchers claimed that cat predation was the limiting factor on rabbit population and thus removal of cats was the primary reason for the increase in rabbits in the early 2000s (Bergstrom *et al.* 2009), this simplistic theory is implausible when it is considered that rabbits established and expanded from the late 1870s in the face of a cat population that had been there for 60 years, and that rabbit overgrazing was already considered extreme by the 1950s, when cats were abundant. In addition, rabbit count areas established in the 1970s showed a higher rabbit population in the early 1970s than in the mid-2000s (Terauds *pers. comm.*), yet at that time there was no cat control, so the rabbit population in the 1950s to 1970s period was obviously not limited by cat predation. Rabbit numbers had increased significantly since 1999, probably because the myxoma virus was attenuating and partly because, after low rabbit numbers for 20 years, there was a healthy vegetation cover to support an expanding population as food availability increased (Dowding *et al.* 2009). Taylor (1955), Costin and Moore (1960) and Jenkin (1975) emphasised the massive damage done to vegetation by rabbits – when cats were not controlled at all. After removal of cats, attention then turned to the eradication of the remaining introduced species – rabbits, ship rats and mice, primarily driven by the very visual impacts of rabbit damage to vegetation.

A decade or so earlier, eradication of pest animals on this scale was considered impossible (Copson and Whinam 2001). However, advances in methodology and technology had developed to the point where a project on the scale of Macquarie Island was considered challenging but feasible. The eradication of Norway rats on New Zealand's Campbell Island in 2001 provided a relevant precedent (McClelland 2011), as did eradication of rabbits from Saint-Paul Island (French Southern and Antarctic Territories) (Micol And Jouventin 2002)and Enderby Island (NZ sub-Antarctic) (Torr 2002).

The Macquarie Island Pest Eradication Project was established by the Tasmania Parks and Wildlife Service and was clear in its goals and objectives. Restoration of the Macquarie Island ecosystem was the long-term goal, to be achieved by the eradication of ship rats, mice and rabbits. A Tasmanian NGO (Tasmanian National Parks Association) was a strong supporter, and partnered with WWF-Australia and Macquarie Island researcher Dr Jenny Scott to lobby for the eradication plan to be implemented. Funding of A$24.6 million was jointly committed by the Tasmanian and Australian Governments in June 2007, following a joint A$100 000 donation by WWF-Australia and Peregrine Adventures.

Following some scoping work in 2004–2005, further detailed planning commenced in October 2006. A two-phase methodology was proposed – aerial baiting to eradicate both rodent species and remove most rabbits, followed by a hunting phase using hunters and rabbit detection dogs to track down and remove the small number of rabbits expected to survive aerial baiting. To support the hunting phase, 11 dogs were trained to detect rabbits (Fig. 3.1). Five field huts and all ancillary equipment were procured and installed at strategic locations around the island. The aerial baiting was postponed a year after shipping delays and bad weather stopped a 2010 attempt but was completed in 2011, with hunting

Fig. 3.1. The dogs were trained not only to search for rabbits, rats and mice, but also not to molest ever inquisitive birds such as (A) king penguin chicks (*Aptenodytes patagonicus*) and (B) brown skuas (*Catharacta lonnbergi*) (photos: Karen Andrew).

and monitoring completed by March 2014. The eradication of all three species was declared successful by the Parks and Wildlife Service in April 2014.

There were many responses of the Macquarie Island environment to introduced species eradication. Some responses were relatively rapid: for instance, blue petrels had established breeding burrows on the main island in the first summer after aerial baiting removed the rodents, and the proportion of Antarctic terns breeding on beaches, instead of on rock stacks, increased to 45% from 5% on a previous survey. Grey petrel breeding success also increased following rodent eradication, and cape petrels are slowly re-establishing.

The vegetation response was immediately evident as it began to regenerate in the absence of grazing pressure (Fig. 3.2). Heavily grazed species such as Macquarie Island cabbage (*Stilbocarpa polaris*), *Pleurophyllum hookeri* and *Poa foliosa* began establishing across the island, as did the introduced annual meadowgrass (*Poa annua*). Threatened species such as the orchids *Nematoceras sulcatum* and *N. dienemum* and the lycopodium *Huperzia australiana* began to increase in both abundance and distribution.

Fig. 3.2. Images (A) 2005, (B) 2017 taken from Razorback Spur at the base of The Isthmus, Macquarie Island showing recovery from rabbit grazing. Rabbits were largely removed in 2011 and their eradication declared successful in April 2014 (photos: A, Keith Springer; B, Chris Howard).

Rather than aiming to protect a single threatened species, the project to eradicate introduced pests from Macquarie Island recognised that each pest species had different impacts on Macquarie Island ecosystems: rabbits impacted vegetation by overgrazing and destabilising steep slopes; ship rats through predation of burrowing seabird eggs and chicks and impeding seedling recruitment by eating seeds and fruits; and mice by impacting invertebrate populations and also eating seeds and fruits. Several threatened plant and animal (mostly bird) species were unable to maintain healthy (or, in some cases, any) populations in the face of these impacts. A relatively short-term detrimental impact was also sustained by another threatened species, northern giant petrels (*Macronectes halli*) that – along with the non-threatened skua (*Catharacta lonnbergi*), kelp gull (*Larus dominicanus*) and black duck (*Anas superciliosa superciliosa*) – were impacted by primary or secondary poisoning following the aerial baiting phase of the project.

The eradication project on Macquarie Island was logistically complex and required a high degree of detailed planning. Some challenges were regulatory. For example, there was no provision to allow the predicted non-target mortality of a threatened species (northern giant petrel) under a conservation action in the *Environment Protection and Biodiversity Conservation Act 1999*. The project also needed exemptions under legislation for some of the hunting methods used. Other challenges were logistical. For example, the ship chosen to transport the expedition to Macquarie Island was not available when required, creating a month's delay in 2010. Other challenges were operational, such as insects in the bait arriving from New Zealand requiring the whole shipment to be quarantined, and the extended period of bad weather that caused the abandonment of the 2010 baiting program. Extensive liaison was required with the Australian Antarctic Division, which operates the station on Macquarie Island and with whose operational procedures the project needed to comply. Substantial time each year was required to recruit and train staff, including the hunter and dog handler positions for year-long deployments on the island. In total, 60 people were employed directly in the implementation of the eradication project – the core planning team of four plus people employed for the aerial baiting phase and the annual hunting/monitoring teams.

Monitoring

Following the aerial baiting phase, hunting teams searched the island for surviving rabbits, working from field huts for month-long rosters. Eleven rabbits were removed in this phase, including four juveniles – the only evidence of post-baiting breeding. After 2 years, the rabbit-hunting team (typically 13 staff plus 12 rabbit detection dogs) was joined by a rodent team of two staff and three rodent detection dogs, whose job was to search the island to determine whether rodent eradication had been successful. Between the rabbit and rodent detection teams over the 3 years of fieldwork, hunters covered ~94 000 km of search effort – on an island 32 km long by ~4 km wide.

Annual counts since 2011 have shown that the breeding northern giant petrel population is nearly back to pre-baiting levels (Parks and Wildlife Service *pers. comm.*). Other bird species were monitored, including skua populations that have declined as expected with the removal of a preferred prey item (rabbits). Recovery of plant species and their distributions have been monitored and invertebrate surveys undertaken.

Factors in success

Success of a complex and long-term project such as the pest eradication project on Macquarie Island does not happen by chance. There are only two possible project outcomes. To achieve success, every single individual of the three pest species must be killed. Anything less is a project failure. Numerous factors contributed to project success. Some of these are typical of successful eradication projects, while others reflect project management principles generally.

Funding commitment

Securing the funding commitment for a project with an 8-year timeframe was hugely important. Not only did it mean that project planning could proceed seamlessly and multi-million dollar contracts could be signed knowing the funding would be in place, but it also saved considerable time each year through not having to prepare submissions for annual competitive funding bids. The latter type of funding is more typical of government process, but highly risky for eradication projects because work achieved over several years may – if the funding is not approved for the closing stages of the project – result in project failure and the waste of resources invested in the project to that date. Such long-term (whole of project) commitment of financial resources is rare for conservation projects.

Funding level

The project received the level of funds required to implement the project as planned, following scrutiny of the proposed budget by Department of Environment and Heritage finance staff in early 2007. This meant project staff did not have to 'cut corners' because of cost constraints – an important consideration when eradication is the goal. A contingency amount of 20% of the budget agreed with Department of Environment and Heritage (now Department of the Environment and Energy) proved invaluable when a baiting attempt in 2010 failed due to bad weather and had to be repeated in 2011, incurring additional costs for shipping, helicopters, bait and bait pods.

Management structure

The project had a Steering Committee comprised of four people from stakeholder agencies with a genuine desire to see the project succeed. The project was managed by the Tasmania Parks and Wildlife Service and had a small team (four full-time staff with occasional part-time assistance) all of whom were focused solely on delivering a successful project outcome. A rare feature was that these four staff were 'quarantined' from other duties, notably firefighting, because it was recognised that extended firefighting duties could hamper project momentum. The project manager was responsible to the Parks and Wildlife Service General Manager, but was empowered to manage design and implementation of the project.

Clear objective

The project had a clearly defined objective and this allowed project staff to fully focus on delivering that objective.

Detailed planning

Comprehensive planning is critical to the success of a large eradication project, especially with the logistical challenges of this isolated island location. Planning included developing the training requirements for the 12 dogs used on the project, 2 years in advance of the required delivery date, plus a system of evaluation to ensure dogs were trained to the contracted standards.

Adherence to international best practice

There were numerous risks to the project, and one way to minimise these was to apply proven techniques from previous eradication projects, notably in New Zealand. Lessons from previous eradications for each of the three target species were examined to see what had worked and what had not, and best practice methods were adapted to the Macquarie Island situation.

Peer review

Tasmania Parks and Wildlife Service obtained strategic and operational support for the eradication program from the Island Eradication Advisory Group within the New Zealand Department of Conservation. This group of six experienced eradication staff had been involved in numerous eradication projects globally and formed an invaluable peer review group with unparalleled experience.

Economic and employment agreements with New Zealand

The project benefitted significantly from close economic ties established between Australia and New Zealand through being able to source appropriate goods and services, as well as experienced staff, from New Zealand. For example, two of three dog-trainers and the dog training coordinator were from New Zealand. Bait was sourced from New Zealand, as were helicopter pilots experienced in aerial baiting techniques and specialised baiting equipment. Indeed, half of the 60 project staff were recruited from New Zealand.

Adapt to changing circumstances

On several occasions, planned operational activities faced challenges as circumstances changed. Each time a solution needed to be found, whether by changing aspects of the original plan, or introducing new components to the plan. A good example of this was the decision to introduce rabbit haemorrhagic disease virus before baiting in 2011, as a measure to reduce non-target mortality among seabirds during the baiting phase to follow – a suggestion that came from the Island Eradication Advisory Group in New Zealand. The flexibility to adapt the project methodology was an important part of the overall project success.

Dedication of staff – belief in project outcomes

A key feature of the project was the dedication of staff involved in the project, and their willingness to 'go the extra mile' to meet project goals. The small project team delivered an outstanding result to meet all regulatory, procurement, logistical and operational requirements on time for the ship to depart for the island as scheduled. The aerial baiting phase in 2011 went well, largely due to the contribution of the bait-loading teams, helicopter pilots and engineers. The rabbit and rodent hunters also needed to be highly motivated to search for the target species in Macquarie Island's cold, wet and windy conditions, especially when numbers dropped to zero after several months.

Measures of success

The Macquarie Island Pest Eradication Project can be gauged as a success from several different viewpoints. First, the objective of eradicating rabbits, ship rats and house mice from the island was met, and this remains the largest island worldwide where this has been achieved for each of these three species.

A second objective was to increase the capacity of the Parks and Wildlife Service to undertake island pest eradications. This has also been achieved, with cat and rodent eradications achieved or underway on Tasman Island and Big Green Island, closer to Tasmania.

The goal of restoring natural ecosystems on Macquarie Island is well underway. Although it can never be 'pristine', because some human impacts remain – including those associated with changing climate, introduced plants and off-island fisheries impacts on seabirds – the island is free from the pressures of overgrazing of vegetation and predation of seabird and invertebrate fauna (Springer 2016). Vegetation is regenerating rapidly, both in abundance and distribution of many species and in density of groundcover. Invertebrate populations are more abundant in the absence of rodent predation, and populations of many seabird species are either increasing (e.g. grey petrels, cape petrels), adapting their breeding behaviour (Antarctic terns (*Sterna vittata*), probably a precursor to improved breeding success) or newly (re)establishing on the island (blue petrels, soft-plumaged petrels, diving petrels). In November 2016, just 5 years after the last pest individual was detected, BirdLife Australia downlisted eight seabird taxa on the Australian IUCN Red List because the threats justifying a higher threat category were no longer operating (BirdLife Australia 2016).

Tasmania Parks and Wildlife Service produced a Monitoring Report on the project, which confirmed the project had been successful and assessed the eradication as effective in delivery of project objectives and outcomes. The report is available at <http://www.parks.tas.gov.au/file.aspx?id = 31160>.

The Macquarie Island project was completed a year ahead of schedule. Although the aerial baiting took a year longer than planned due to abandoning the initial attempt in 2010, the hunting and monitoring phase was completed 2 years ahead of schedule. This helped contribute to the project coming in under budget – with a final expenditure of about A$20 million, or nearly A$5 million less than the approved budget of A$24.7 million.

Finally, the success of the project has been formally recognised. It won the 'Project of the Year' in the Sustainable Projects category in both the Tasmanian and Australian Project Management Achievement Awards in 2015 (Australian Institute of Project Management), and the Banksia Natural Capital Award (Banksia Foundation) in 2015.

The future

With introduced pests removed from Macquarie Island, ecosystems are recovering, and there should be no need for further management interventions. The key concern now is to ensure that pests do not re-establish. Although that is unlikely to occur accidentally for rabbits, rodents are adept stowaways, and stringent biosecurity measures have been established and need to be maintained.

The two extinct endemic land bird species both have close relatives elsewhere, and consideration could be given to introducing these extant subspecies to the island, to fill the ecological niches once occupied by the Macquarie Island parakeet and the Macquarie Island rail.

Acknowledgements

The removal of invasive pests from Macquarie Island was the outcome of efforts from numerous people, including those who lobbied to get government support for the project and members of the steering and technical advisory committees. Peter Mooney provided a steady hand supporting the project in Tasmania, while Veronica Blazely carried the flag in Canberra. Keith Broome and the Island Eradication Advisory Group in New Zealand provided excellent technical support. The project planning team included Keith Springer, Yeutha May, Geoff Woodhouse, Graeme Beech, Noel Carmichael, Luke Gadd and Andrea Turbett. Contractors contributed significantly to the project including: dog trainers Steve Austin, Guus Knopers and John Mead; dog training coordinator John Cheyne; Helicopter Resources Ltd; and P&O Shipping. The late John Oakes played a significant role as senior pilot for the aerial baiting work in 2010 and 2011. The annual field teams on Macquarie Island, led by Peter Preston, Peter Kirkland and Stephen Horn, did a fantastic job over 3 years to remove surviving rabbits and then search the island for any sign of rabbits and rodents. Hunters worked in often adverse weather conditions and the success of the project was largely due to their commitment. Robb Clifton and Don Hudspeth from the Australian Antarctic Division were unstinting in coordinating AAD logistics support for the project.

References

Bergstrom DM, Lucieer A, Kiefer K, Wasley J, Belbin L, Pedersen TK, *et al.* (2009) Indirect effects of invasive species removal devastate World Heritage Island. *Journal of Applied Ecology* **46**, 73–81. doi:10.1111/j.1365-2664.2008.01601.x

BirdLife Australia (2016) *Macquarie Island Celebrates 5 years of Being Rabbit-free*. BirdLife Australia, Melbourne, <http://www.birdlife.org.au/media/macquarie-island-celebrates-5-years-of-being-rabbit-free/>

Brothers N, Bone C (2008) The response of burrow-nesting petrels and other vulnerable bird species to vertebrate pest management and climate change on Subantarctic Macquarie Island. *Papers & Proceedings of the Royal Society of Tasmania* **142**, 123–148.

Copson G, Whinam J (2001) Review of ecological restoration programme on subantarctic Macquarie Island: pest management progress and future directions. *Ecological Management & Restoration* **2**, 129–138. doi:10.1046/j.1442-8903.2001.00076.x

Costin AB, Moore DM (1960) The effects of rabbit grazing on the grasslands of Macquarie Island. *Journal of Ecology* **48**, 729–732. doi:10.2307/2257346

Dowding JE, Murphy EC, Springer K, Peacock AJ, Krebs CJ (2009) Cats, rabbits, myxoma virus, and vegetation on Macquarie Island: a comment on Bergstrom et al. (2009). *Journal of Applied Ecology* **46**, 1129–1132. doi:10.1111/j.1365-2664.2009.01690.x

Jenkin JF (1975) Macquarie Island, Subantarctic. In *Structure and Function of Tundra Ecosystems*. (Eds T Rosswall and OW Heal) pp. 375–397. Ecological Bulletin, Stockholm.

Johnston GC (1966) Macquarie Island and its rabbits. *Tasmanian Journal of Agriculture* **37**, 277–280.

Jones E (1977) Ecology of the Feral Cat, *Felis catus* on Macquarie Island. *Australian Journal of Wildlife Research* **4**, 249–262. doi:10.1071/WR9770249

McClelland PJ (2011) Campbell Island – pushing the boundaries of rat eradications. In *Island Invasives: Eradication and Management*. (Eds CR Veitch, MN Clout MN and DR Towns) pp. 204–207, IUCN, Gland, Switzerland and Auckland, New Zealand.

Micol T, Jouventin P (2002) Eradication of rats and rabbits from Saint-Paul Island, French Southern Territories. In *Turning the Tide: the Eradication of Invasive Species*. (Eds CR Veitch and MN Clout) pp. 199–205. IUCN, Gland, Switzerland.

Springer K (2016) Methodology and challenges of a complex multi-species eradication in the sub-Antarctic and preliminary effects of invasive species removal. *New Zealand Journal of Ecology* **40**, 273–278. doi:10.20417/nzjecol.40.30

Taylor BW (1955) 'The flora, vegetation and soils of Macquarie Island'. ANARE Reports Series B(2) No. 19. Antarctic Division, Commonwealth of Australia, Department of External Affairs, Melbourne.

Taylor BW (2005) The critical problem of rabbits on Macquarie Island. *Aurora* **25** No. 2 December 2005.

Taylor RH (1979) How the Macquarie Island Parakeet became extinct. *New Zealand Journal of Ecology* **2**, 42–45.

Torr N (2002) Eradication of rabbits and mice from subantarctic Enderby and Rose Islands. In *Turning the Tide: the Eradication of Invasive Species*. (Eds CR Veitch and MN Clout) pp. 319–328. IUCN, Gland, Switzerland.

4

Management of seabird bycatch leads to sustainable fisheries and seabird populations

G. Barry Baker and Graham Robertson

Summary

The problem
1. Bycatch of seabirds from longline fisheries became a major conservation concern in the late 1980s following evidence of high bycatch rates of seabirds, particularly albatrosses and larger petrels, in Australian waters and elsewhere
2. This threat has been particularly well documented for longline fisheries, but mortality associated with trawl, gillnet and purse-seine fisheries has been increasingly recognised over the last decade

Actions taken to manage the problem
1. As a consequence of the initial research highlighting the problem, seabird bycatch in longline fishing was listed a key threatening process under the EPBC Act and a Longline Fishing Threat Abatement Plan (TAP) developed
2. Ongoing research helped develop cost-effective mitigation measures that can be deployed by fishers in ways that both reduce bycatch and maintain or increase fishing productivity
3. These mitigation measures were deployed by fishers in Australian waters, and in many other parts of the world. Although development of effective mitigation for all gear types that catch seabirds remains a challenge, particularly for gillnets and artisanal fisheries, some are known to be highly effective and have greatly reduced bycatch
4. Monitoring was undertaken on a scale that increased understanding of the extent and nature of the problem and ensured effectiveness of and compliance with the measures recommended under the TAP

Markers of success
1. Bycatch levels in longline fisheries have been reduced to fewer than 50 seabirds per year – a significant reduction from earlier estimates of thousands.

2. An effective TAP has been delivered and maintained through three iterations and continues to drive conservation improvements.

Reasons for success

1. Development of the TAP, and the ongoing support for an annual meeting and review of progress by a TAP Team whose members have shown a long-term commitment to improvement. This process affords all stakeholders the opportunity to contribute to management decisions and ensures rapid response to implementation issues as they arise.
2. A substantial commitment to monitoring of both seabird bycatch and compliance with regulations so that there has been a capacity for adaptive management of mitigation measures and an assurance that they were working.
3. Fortuitously, imposition of regulations to protect southern bluefin tuna (*Thunnus maccoyii*) resulted in a shift in fishing location for part of the fleet, which benefited flesh-footed shearwaters (*Puffinus carneipes*) and albatrosses.
4. Adequate research was undertaken on seabird movements to understand the timing and nature of overlaps with high risk fishing grounds so that regulations can match biological reality.

Introduction

Seabirds are killed in fisheries throughout the world, with fisheries-related mortality being responsible for population decreases in many species (Baker *et al.* 2002; Phillips *et al.* 2016). This threat to seabirds has been well documented for longline fisheries, where birds drown after being accidentally caught while scavenging on baited hooks set to target pelagic and demersal fish (Baker *et al.* 2002), but it has become apparent in the last decade that trawl, gillnet and purse-seine fisheries also kill many birds (Baker *et al.* 2007; Phillips *et al.* 2016). In trawl fisheries, seabirds can be struck by the warp lines and drown, are killed or injured from collisions with other vessel cables or become entangled in the mesh of nets at the sea surface. In gillnet and purse-seine fisheries, seabirds can become entangled in the mesh of nets and drown, either accidentally, or while actively diving and feeding around the nets (Žydelis *et al.* 2013).

Bycatch of seabirds in fisheries became a major conservation concern in the late 1980s (Weimerskirch and Jouventin 1987; Brothers 1991). Initial evidence came from recoveries of wandering albatrosses (*Diomedea exulans*) ringed at South Georgia (Croxall and Prince 1990), and estimates of very high bycatch from the Japanese tuna fishery off Australia (Brothers 1991). High rates of seabird bycatch were subsequently confirmed in a wide range of longline fisheries in Australian waters (Gales *et al.* 1998) and elsewhere (Phillips *et al.* 2016). Although attention focused initially on industrial fishing, bycatch in artisanal fleets has also been identified as a major source of mortality for many seabirds (Phillips *et al.* 2016). Most affected are albatrosses and larger petrels (Families Diomedeidae and Procellariidae), species with long life spans, low rates of natural mortality, high recruitment rates and low productivity (Baker *et al.* 2002). These bird species start breeding from 7 to 12 years of age and produce only one chick per breeding cycle, which in many species is only every 2 years at most (Baker *et al.* 2002; Phillips *et al.* 2016). These demographic characteristics mean that even slight reductions in survival of adults caused by incidental mortality in fisheries can lead to severe population declines over a period of 20–30 years (Phillips *et al.* 2016). The migratory nature of albatrosses and larger petrels further complicates conservation management. Satellite-tracking studies have shown that many species breeding in Australian and New Zealand waters regularly migrate thousands of kilometres to other ocean basins where they come into contact with fishing fleets managed by other jurisdictions (e.g. Baker *et al*. 2007; Phillips *et al*. 2016). Management of such species needs to occur in many countries if the many threats they face are to be reduced across their range.

Heightened awareness of the plight of albatrosses and petrels resulted in the Australian Government adopting a three-pronged approach to seabird conservation (Baker *et al.* 2002):

1. Assessment of the conservation status of seabirds, with the subsequent listing of 17 species of albatrosses and two species giant petrels as threatened species under Commonwealth (federal) legislation (Baker *et al*. 2002), and development of a national recovery plan (DSEWPC 2011).
2. Identification of threats to seabirds and the subsequent (1995) national listing of the 'incidental catch (or by-catch) of seabirds during oceanic longline fishing operations' as a key threatening process, leading to development of a national Longline Fishing Threat Abatement Plan (TAP) (Commonwealth of Australia 2014).
3. Facilitation and implementation of cooperative international action to complement domestic conservation actions.

The first and third of these approaches dealt with more than just fishing-related threats to seabirds, such as the need to remove invasive alien species from key breeding sites (see Chapter 3). To achieve the last of these objectives, Australia played a lead role in nominating all albatrosses to the appendices of the Convention on the Conservation of Migratory Species of Wild Animals (CMS) with subsequent development, in 2004, of the Agreement for the Conservation of Albatrosses and Petrels (ACAP). This globally significant action recognised that only concerted and cooperative action by all the jurisdictions in which a species feeds and breeds can ensure effective conservation for these birds (Cooper *et al*. 2006).

This chapter focuses on the second approach described above: management of bycatch in Australian longline fisheries. Most Commonwealth managed longline fisheries now minimise bycatch within their operations, which has significantly reduced the estimated incidental mortality (DSEWPC 2011). The chapter reviews what has been achieved so far, and highlights what remains to be done if the threat of extinction to migratory seabirds from fishing operations is to be eliminated.

Conservation management

Conservation management has required: gathering data on the extent of bycatch in fisheries; modelling of impacts on populations to place the perceived threat in perspective; trials of technical solutions; establishment of legislative and regulatory frameworks to require implementation of mitigation; and outreach and support for fishers to encourage voluntary adoption of measures.

Planning and policy

The Longline Fishing TAP has been critical to reducing longline fishery-related mortality. The Threat Abatement Plan for nationally coordinated action to alleviate the impact of longline fishing activities on seabirds in Australian waters was first released in 1998 and has since been reviewed and updated twice (Commonwealth of Australia 2014).

The objective of the TAP is to reduce seabird bycatch in all fishing areas, seasons or fisheries to fewer than 0.05 birds per 1000 hooks for pelagic fisheries, and to lower rates for demersal fisheries. The TAP aims to reduce bycatch of seabirds in the Australian Fishing Zone by:

- prescribing appropriate modifications to fishing practices or equipment (mitigation measures)
- providing for development of new mitigation measures
- educating fishers and the public
- collecting information necessary to improve knowledge of seabird-longline fishery interactions.

The TAP was initially prepared after a series of focus group workshops with key stakeholders including the fishing industry, non-government groups, research organisations and state and Commonwealth government agencies that aimed to clarify the threats and possible mechanisms for managing them.

Initially, the Commonwealth Environment Minister established a ministerial advisory body (the 'TAP Team') with a broad membership to provide advice on the development and implementation of the TAP. The TAP Team has met annually since its inception in 1997 to review progress in mitigating seabird bycatch in Australian fisheries, and has been instrumental in efforts to achieve continual progress in reducing bycatch. Many of the original TAP Team members are still actively involved, providing a level of 'corporate memory' not available in many conservation initiatives.

Data gathering

A precursor to tackling bycatch in any area or fishery is an assessment of the level of mortality that is occurring. This is essential if fishers associated with lethal fishing techniques are to be convinced that their activities are affecting seabirds at a population level, especially because birds are rarely observed actually being caught. However, even low observed capture rates, when multiplied by the magnitude of fishing effort, can result in significant mortalities. For example, pelagic longlining may involve setting a single longline up to 130 km in length holding between 600 and 3000 hooks; in demersal or bottom longlining up to 20 000 hooks are set per day; autoline demersal vessels can set 1000 hooks in 10 minutes.

The impact of fishery-related mortality is assessed by multiplying the number of birds killed by fishing effort to derive the total number of birds caught each year. Such assessments rely on the use of fisheries effort and observer data, combined with knowledge of the geographic foraging distributions of a species, to evaluate the spatial and temporal overlap with a fishery. Early studies reported that 75% of the birds killed and retained by Japanese longliners operating in the Australian Fishing Zone were albatrosses (Gales et al. 1998). Subsequent data from domestic pelagic vessels indicated that in fact albatrosses formed fewer than 10% of the species killed in Australian longline fisheries, with flesh-footed shearwaters (*Puffinus carneipes*) and other petrels dominating the catch (Baker and Wise 2005; Trebilco et al. 2010). These studies confirmed that longline bycatch was essentially a problem in waters below latitude 25° south, and was lowest in winter and highest in spring, which was important in developing temporal and spatial conservation measures.

Catch data are complemented by modelling the relative contribution that vital demographic rates (survival, fecundity, immigration and emigration) make to population growth rate. Such models highlighted the impact of the longline bycatch problem in Australia's Eastern Tuna and Billfish Fishery (ETBF), particularly for shearwaters (Baker and

Wise 2005), potentially threatening the survival of the Lord Howe Island population of flesh-footed shearwater (Baker and Wise 2005).

Development of mitigation solutions

The early work of Agnew *et al.* (2000), Melvin and Parrish (2001), Robertson *et al.* (2006), and the Commission for the Conservation of Antarctic Marine Living Resources (CCAMLR) (Waugh *et al.* 2008) showed that changes to fishing practices and gear could substantially reduce bycatch. This early work led to the prescriptions in the TAP requiring the mandatory use of bird-scaring lines, night-setting, use of line-weighting and management of offal. However, voluntary adoption in the absence of strict compliance measures, such as high levels of observer coverage, remained problematic.

Significant reduction in bycatch in the ETBF was achieved inadvertently through the implementation of spatial and temporal closures to avoid the catch of southern bluefin tuna (*Thunnus maccoyii*) by vessels not holding quota for that species (Larcombe and McLoughlin, 2007). This effectively shifted much fishing effort to the north of the fishery, thus reducing the overlap between effort and the pelagic distribution of many at-risk seabird species at critical times of the year. It provided valuable breathing space for the development of other technical mitigation solutions at a time when proven and accepted seabird avoidance measures required substantial improvement.

Within the ETBF, night-setting remains unpopular for operational reasons, and the current line weighting regimes used in the fishery, although largely effective, appear to be unable to reduce bycatch below the Longline Fishing TAP-prescribed level at times of high seabird abundance (B. Baker unpublished). This has provided an incentive for gear technologists and researchers to develop and trial new approaches to mitigation. These have included technologies such as 'safe leads' that improve safety for fishers while improving line-sink rates (Sullivan *et al.* 2012), hook-protection systems such as the Smart Tuna Hook (http://www.oceansmart.com.au/1142/smart-tuna-hook.aspx) and the Hook Pod (http://www.hookpod.com), and an underwater bait setter (Robertson *et al.* 2015), all of which provide viable mitigation solutions in high-value tuna fisheries. However, more work is needed to develop affordable solutions for artisanal longline fisheries in developing countries, where value of the fish catch is low.

Monitoring

Fishing is dynamic, with changes in effort, gear and practices driven by market forces. Bycatch is inevitable and understanding the causes of these factors, and finding suitable approaches to manage them, will require good data collection and ongoing collaboration among fishers, fishery managers, seabird biologists and gear technologists. This can be expensive. However, recent moves by the Australian Fisheries Management Authority to replace costly observer programs with electronic monitoring systems on all fishing vessels is cheaper, improving compliance and collecting data more effectively.

A mandatory requirement of the Longline Fishing TAP is that fishers retain all seabirds killed in fishing operations and return carcasses for analysis to determine species, age, sex, breeding status and, where possible, which breeding population they derive from. The resulting information has been essential in assessing risk to species and improving knowledge of fishery impacts (Trebilco *et al.* 2010).

Infrastructure

Almost all species of albatross and large petrel have been tracked at some stage while breeding, and many during the non-breeding season (Phillips *et al.* 2016). These studies, although not directly prescribed by the Longline Fishing TAP, have helped understand differences in foraging distributions at each stage of the annual (breeding and non-breeding) cycle and the relative overlap with high-risk fisheries (Phillips *et al.* 2016).

Time, temperature and depth recorders have also shown that albatrosses are poorer divers than *Procellaria* petrels (Phillips *et al.* 2016). This has implications for the degree of overlap, and hence risk of bycatch in different fisheries (Phillips *et al.* 2016). Although deploying telemetry equipment is expensive, it is important for ongoing management of bycatch, particularly assessment of risk in new and developing fisheries and as the climate changes.

Money

The success achieved so far has largely been achieved with funds from the Australian Government, particularly the employment of observers by the Australian Fisheries Management Authority and ongoing support for ACAP and the Longline Fishing TAP, including modelling and monitoring, from the Department of Energy and the Environment and its predecessors. Research has also been substantially supported by fishing industries, Australian Government agencies, and philanthropic organisations in Australia and overseas. Over the last 5 years, government support has probably been between A$1 and $2 million with industry supplying at least A$0.5 million for R&D each year. This support will need to continue if success is to be sustained and extended to other industries.

The future

Bycatch levels in Australian longline fisheries are likely to remain low now that significant reductions have been achieved through the implementation of mitigation measures and a change in the spatial component of fishing effort. In 2015, fewer than 50 birds were observed being caught in Australian longline fisheries (AFMA *pers. comm.*), a reduction of over 97% from the estimated number of 1794–4486 flesh-footed shearwaters killed each year between 1998 and 2002 (Baker and Wise, 2005).

However, total elimination of bycatch in longline fisheries remains unlikely. Mitigation still presents challenges to gear technologists and fishery managers, and currently no single measure can reliably prevent incidental mortality of seabirds in most pelagic longline fisheries. Although hook-protection systems would appear capable of doing this, they have not been universally adopted at this stage, most likely because of concerns over costs of materials and an unwillingness to change existing practices.

Although this chapter has focused on longline fisheries, seabird bycatch is also problematic in trawl and other gear types in Australia (Baker *et al.* 2007; Fig. 4.1). Some steps have been taken to introduce mitigation approaches for trawl fisheries, but the methods proposed have not been extensively tested and therefore their efficacy is unknown. In addition, the most effective measure for trawl fisheries – retention of offal and discards – is rarely practiced as existing vessels do not have enough space on board to hold waste for long periods of time. Mitigation options to fishers that cannot be easily implemented are unlikely to be adopted. Voluntary adoption of mitigation strategies will also never be achieved for measures that compromise the catch of target species. As such, some new innovations in mitigation may only be viable in well-regulated and high-value fisheries such as those targeting tunas and billfish. They will not be attractive options in most artisanal or other fisheries where product value is low.

Incentives are needed in fisheries dominated by a fishing fleet that is largely owner-operated and where immediate financial imperatives take priority over longer term indus-

Fig. 4.1. White-capped albatross, which nest on the Auckland Islands, are regularly caught in trawl fisheries (photo: B. Baker).

try sustainability. Measures will only be voluntarily adopted if it can be shown that they substantially improve fishing efficiency, such as through increased capture of target species, improvement to operational issues, such as easy stowing and deployment of gear, reduced labour costs, and improved fuel economy. Simply educating fishers in the use of mitigation measures in the absence of appropriate incentives and compliance is unlikely to change fishing behaviour.

Developing robust conclusions about the efficacy of mitigation measures requires rigorous monitoring and testing. Ideally, data would be sourced from experiments conducted at sea, where the mitigation measures in question would be contrasted with controls with no deterrent. Very few mitigation experiments are conducted in this way, which presents difficulties in drawing firm conclusions on effectiveness. Future efforts to develop and implement mitigation measures for all non-target taxa in fisheries should seek to embody these experimental principles.

Conclusion

The prospects for many seabird populations impacted by fishing will remain uncertain in the absence of dedicated attempts to solve bycatch issues. Only a thorough understanding of the ecology of a species and the nature of fisheries interactions, coupled with the widespread adoption of appropriate and effective mitigation measures, will give confidence in ensuring their long-term survival. The success achieved in Australian longline fisheries over the last 20 years provides encouragement and hope that dedicated attention to the problem, and establishment of collaboration between all stakeholders, can achieve outcomes that are sustainable for both seabirds and fisheries.

Acknowledgements

Progress in conservation of seabirds and development of mitigation in longline and trawl fisheries has benefitted over the years through the dedicated efforts of Jonathon Barrington, Nigel Brothers, John Cooper, John Croxall, Rosemary Gales, Ian Hay, Neil Klaer, Svein Løkkeborg, Andrew McNee, Malcolm McNeill, Ed Melvin, Janice Molloy, Narelle Montgomery, Tom Polacheck, Kim Rivera, Neville Smith, Ben Sullivan, Geoff Tuck and Sue Waugh. We also acknowledge the considerable contribution of the following members of the fishing industry for encouraging development and adoption of mitigation measures in Australasia: Phill Ashworth, Peter Ashworth, Dave Chaffey, Martin Exel, Tony Forster, Garry Heilmann, Hans Jusseit, Dave Kellian, Les Scott, Parvo Walker, Richard Wells and Nick Williams. Also influential in bringing about changes in fishing practices has been the efforts of Nicola Beynon, Alistair Graham and Alexia Wellbelove of Humane Society International, BirdLife International's Albatross Task Force, and the World Wide Fund for Nature through their Smart Gear Competition.

References

Agnew DJ, Black AD, Croxall JP, Parkes GB (2000) Experimental evaluation of the effectiveness of weighting regimes in reducing seabird by-catch in the longline toothfish fishery around South Georgia. *CCAMLR Science* **7**, 119–131.

Baker GB, Wise B (2005) The impact of pelagic longline fishing on the flesh-footed shearwater *Puffinus carneipes* in eastern Australia. *Biological Conservation* **126**, 306–316. doi:10.1016/j.biocon.2005.06.001

Baker GB, Gales RP, Hamilton S, Wilkinson V (2002) Albatrosses and petrels in Australia: a review of their conservation and management. *Emu* **102**, 71–96. doi:10.1071/MU01036

Baker GB, Double MC, Gales R, Tuck GN, Abbott CL, Ryan PG, *et al.* (2007) A global assessment of the impact of fisheries-related mortality on shy and white-capped albatrosses: conservation implications. *Biological Conservation* **137**, 319–333. doi:10.1016/j.biocon.2007.02.012

Brothers N (1991) Albatross mortality and associated bait loss in the Japanese longline fishery in the Southern Ocean. *Biological Conservation* **55**, 255–268. doi:10.1016/0006-3207(91)90031-4

Commonwealth of Australia (2014) 'Threat Abatement Plan 2014 for the incidental catch (or bycatch) of seabirds during oceanic longline fishing operations'. Department of the Environment, Canberra.

Cooper J, Baker GB, Double MC, Gales R, Papworth W, Tasker ML, *et al.* (2006) The Agreement on the Conservation of Albatrosses and Petrels: rationale, history, progress and the way forward. *Marine Ornithology* **34**, 1–5.

Croxall JP, Prince PA (1990) Recoveries of wandering albatrosses *Diomedea exulans* ringed at South Georgia 1958–1986. *Ringing & Migration* **11**, 43–51. doi:10.1080/03078698.1990.9673960

DSEWPC (2011) 'National recovery plan for threatened albatrosses and giant petrels 2011–2016'. Department of Sustainability, Environment, Water, Population and Communities, Hobart.

Gales R, Brothers N, Reid T (1998) Seabird mortality in the Japanese tuna longline fishery around Australia. *Biological Conservation* **86**, 37–56. doi:10.1016/S0006-3207(98)00011-1

Larcombe J, McLoughlin K (2007) 'Fisheries Status Reports 2006: status of fish stocks managed by the Australian Government'. Bureau of Rural Sciences, Canberra.

Melvin EF, Parrish JK (2001) 'Seabird bycatch – trends, roadblocks, and solutions'. University of Alaska Sea Grant, USA.

Phillips RA, Gales R, Baker GB, Double MC, Favero M, Quintana F, *et al.* (2016) A global assessment of the conservation status, threats and priorities for albatrosses and large petrels. *Biological Conservation* **201**, 169–183. doi:10.1016/j.biocon.2016.06.017

Robertson G, McNeill M, Smith N, Wienecke B, Candy S, Olivier F (2006) Fast sinking (integrated weight) longlines reduce mortality of white-chinned petrels (*Procellaria aequinoctialis*) and sooty shearwaters (*Puffinus griseus*) in demersal longline fisheries. *Biological Conservation* **132**, 458–471. doi:10.1016/j.biocon.2006.05.003

Robertson G, Ashworth P, Ashworth P, Carlyle I, Candy SG (2015) The development and operational testing of an underwater bait setting system to prevent the mortality of albatrosses and petrels in pelagic longline fisheries. *Open Journal of Marine Science* **5**, 1–12. doi:10.4236/ojms.2015.51001

Sullivan BJ, Kibel P, Robertson G, Kibel B, Goren M, Candy SG, *et al.* (2012) Safe Leads for safe heads: safer line weights for pelagic longline fisheries. *Fisheries Research* **134–136**, 125–132. doi:10.1016/j.fishres.2012.07.024

Trebilco R, Gales R, Lawrence E, Alderman R, Robertson G, Baker GB (2010) Seabird bycatch in the Eastern Australian Tuna and Billfish pelagic longline fishery: temporal, spatial and biological influences. *Aquatic Conservation* **20**, 531–542. doi:10.1002/aqc.1115

Waugh SM, Baker GB, Gales R, Croxall JP (2008) CCAMLR process of risk assessment to minimise the effects of longline fishing mortality on seabirds. *Marine Policy* **32**, 442–454. doi:10.1016/j.marpol.2007.08.011

Weimerskirch H, Jouventin P (1987) Population dynamics of the wandering albatross, *Diomedea exulans*, of the Crozet Islands: causes and consequences of the population decline. *Oikos* **49**, 315–322. doi:10.2307/3565767

Žydelis R, Small C, French G (2013) The incidental catch of seabirds in gillnet fisheries: a global review. *Biological Conservation* **162**, 76–88. doi:10.1016/j.biocon.2013.04.002

5

Mary's Famous Five: a story of connection, commitment and community in the recovery of threatened aquatic species in the Mary River catchment, Queensland

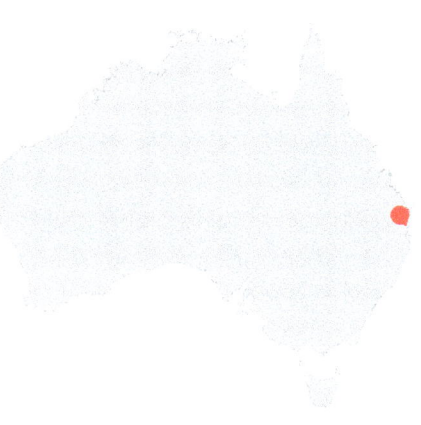

Tanzi Smith and Marilyn Connell

Summary

The problem

1. The Mary River catchment has been subject to more than 170 years of habitat fragmentation and modification, and direct threats at the species level such as overharvesting.
2. Recovery planning for a river, with its dynamic and complex connections – including between the river and floodplain, the freshwater and saltwater and upstream and downstream communities – involves some different challenges to terrestrial recovery planning.

Actions taken to tackle the problem

1. Species recovery efforts commenced in 1994 with recovery action for the Mary River cod (*Maccullochella mariensis*) that were inclusive of a diverse range of stakeholders.
2. A shift to a multispecies and river systems approach has helped conceptualise recovery of the river and its species, such as the Mary River turtle (*Elusor macrurus*), the giant barred frog (*Mixophyes iteratus*), the Australian lungfish (*Neoceratodus fosteri*) and the freshwater mullet (*Trachystoma petardi*) in a more holistic way.
3. Actions have included community engagement, riparian habitat restoration, protection of turtle nests, restocking of cod populations, surveys of threatened frogs and targeted field research.

Markers of success

1. Maintaining momentum on threatened species recovery for more than two decades and improving habitat through more than 800 individual on-ground projects and targeted conservation programs.

Reasons for success

1. Commitment and collaboration between multiple individuals and organisations.
2. High levels of community engagement and participation in the recovery effort.
3. Use of iconic threatened species and habitat to encourage connections between people and the river.
4. Partnerships with university researchers whose findings are building community knowledge and understanding of threatened species and their habitat.
5. More than 20 years of consistent and coordinated recovery effort has created a strong foundation for threatened species recovery.

Introduction

The 9600 km² Mary River catchment in south-eastern Queensland runs into the internationally recognised Great Sandy Strait Ramsar site adjacent to World Heritage listed Fraser Island (Fig. 5.1). It is the southernmost Great Barrier Reef Catchment. Throughout the past 170 years, the catchment has been subject to anthropogenic pressures such as gold mining, land clearing, sand and gravel extraction, flow modification through water extraction and construction of weirs and the introduction of exotic plants and animals (Mary River Threatened Species Recovery Team *pers. comm.*). Almost 60 species in the catchment and estuary are now listed as threatened under the *Environment Protection and Biodiversity Conservation Act 1999*. Management actions are vital to maximise the inherent resilience within the catchment and achieve recovery of these threatened species.

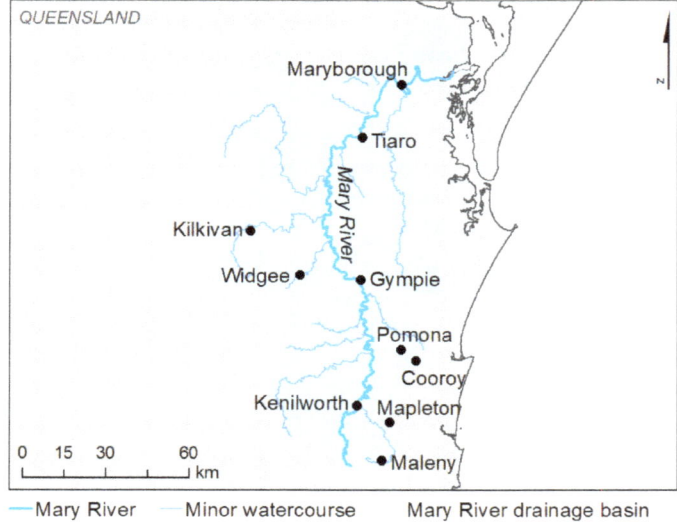

Fig. 5.1. The Mary River catchment in south-east Queensland, Australia.

This chapter provides an example of recovery planning for multiple species that was initiated by the local community in the 1990s with a single species, the Mary River cod. Recovery planning in the aquatic ecosystem context faces some unique challenges and opportunities. In contrast to terrestrial ecosystems, freshwater ecosystems exhibit diverse forms of connectivity that occur laterally, longitudinal and vertically within a river system and patterns of resilience that are dynamic in space and time (Barmuta *et al.* 2011). This connectivity links the main watercourse with the floodplain, the fresh water to salt water, upstream landholders with downstream landholders, upstream communities with downstream communities and upstream habitats and threats with downstream habitats and threats. The approach taken in the Mary River catchment provides an example of recovery planning at the catchment scale, incorporating the natural environment and the community in ways that are meaningful for the environment, local interest groups and individual landholders.

Conservation management

The Mary River cod is a large fish (up to 40 kg and 1.2 m in length, Fig. 5.2) endemic to the Mary River. By the mid-1990s, it was believed to be restricted to 30% of its original range (Simpson and Jackson 1999), and is listed nationally as Endangered. A Mary River Cod Recovery Team was formed in 1994 and developed a recovery plan in 1999 (Simpson and Jackson 1999). The plan focused on short-, medium- and long-term objectives across four integrated areas – production of cod fingerlings for release in priority areas, community engagement to establish the cod as a 'cultural icon', on-ground habitat restoration, and research and monitoring to fill knowledge gaps and assess progress. A wide range of stakeholders were involved in the recovery team, which was led initially by the Queensland Government Department of Environment and Heritage and then by the Queensland Department

Fig. 5.2. Mary River's Famous Five: (A) Mary River turtle, (B) Mary River cod, (C) freshwater mullet, (D) giant barred frog, and (E) Australian lungfish (photos: A–C, E, Gunther Schnida; D, Eva Ford).

of Primary Industries (Fisheries). Several different groups worked together in the implementation of the plan. Key organisations – the World Wildlife Fund, Barung Landcare and the Mary River Catchment Coordinating Committee – worked with other groups and landholders through forums such as the Hatchery Steering Committee and the Mary River Cod Recovery Network. Funding for the program was provided by World Wildlife Fund initially and later by the Commonwealth and local governments. State government staff provided significant scientific support and oversight of the program.

Implementation of recovery effort began in 1996, and actions to engage the community, involve landholders in rehabilitation and deepen understanding of the threats to the cod and to riverine health helped shape subsequent catchment rehabilitation planning (Stockwell 2001). At about the same time, the mystery of the native habitat of the 'pet-shop' turtle was solved, resulting in the naming of a new genus and species to describe the Mary River turtle (Cann and Legler 1994; Fig. 5.2). Also endemic to the Mary River, this species is threatened by loss of riparian and in-stream integrity. In 2001, the community group Tiaro and District Landcare initiated the Mary River turtle conservation project, which continues today (Flakus and Connell 2008). A key component of the program is to improve recruitment through an *in-situ* nest protection program. Since 2001, 672 nests have been protected, which has resulted in 4691 hatchlings. These turtles would probably have been consumed by predators in the nest were they not protected. In 2004, Tiaro and District Landcare established the Mary River Turtle Support Scholarship for postgraduates to fill gaps in knowledge about the turtle's population, biology and ecology. Two PhDs, two postdoctoral researchers and one Masters and one Honours student have been awarded the scholarship.

The Mary River catchment is also home to several threatened stream frogs. Concern for the fate of stream frogs was heightened by the extinction of the southern gastric brooding frog (*Rheobatrachus silus*) and the southern day frog (*Taudactylus diurnus*) in the Conondale and the Blackall ranges in the south of the catchment. Sightings of the Endangered giant barred frog (Fig. 5.2), and other threatened stream frogs – the cascade tree frog (*Litoria pearsoniana*) and tusked frog (*Adelotus brevis*) – outside of National Park areas in the early 2000s and the actions outlined in the Recovery Plan for Stream Frogs of South-East Queensland (Hines 2005) prompted the inception of the Living with Threatened Species Program in 2003. The World Wildlife Fund provided initial funding to focus on conservation of threatened species habitat on private land in the catchment. The Mary River Catchment Coordinating Committee took over the program, and continues to run it today through funding from Sunshine Coast and Noosa Councils. The giant barred frog has gained iconic status for rehabilitation (Ford *et al.* 2014). It occupies selected niches within 40 m of a stream and has large and long-lived tadpoles that require permanent water (Hines 2005).

In 2006, after more than 10 years of threatened species recovery work in the catchment, the Queensland Government proposed a dam on the main trunk of the river. The Federal Environment Minister, Peter Garrett, rejected the dam in 2009 and also recommended the development of a multi-species regional recovery plan as an appropriate planning response to guide recovery of the threatened species that were central to his decision.

Development of the Mary River Threatened Species Recovery Plan (Commonwealth of Australia *pers. comm.*) began in 2011, led by the Mary River Catchment Coordinating Committee in close collaboration with the Australian Government using a 'river systems approach' (Smith *et al.* 2012). The plan focused on recovery actions for the threatened aquatic species, the Mary River cod, the Mary River turtle, the giant barred frog and the vulnerable Australian lungfish. The freshwater mullet (Fig. 5.2) is also a focus of the plan due to its decline in the Mary River and neighbouring northern catchments. It is also locally significant to the Indigenous and non-Indigenous community and, as a diadromous species, symbolises the connectivity between the riverine and estuarine systems for river health. A formal assessment of the conservation status of the species is currently in process.

Extensive community engagement, including a specific Indigenous engagement program, informed the vision, objectives and actions of the Mary River Threatened Species Recovery Plan. A Recovery Team – including all levels of government, community groups, traditional owners, research institutions and water utilities – oversaw its development. The team will be convened again when the Australian Government indicates that the plan is ready for ministerial endorsement. In the meantime, members continue to collaborate and implement recovery actions through their own organisations. Implementation of the on-ground recovery actions has been given a significant boost by a 6-year A$2.4 million Clean Energy Futures Fund project titled 'Restoring Riparian Resilience: Implementing the Mary River Threatened Aquatic Species Recovery Plan' secured by the Mary River Catchment Coordinating Committee, which was completed in 2017.

Over the past 22 years, the Mary River Catchment Coordinating Committee has invested approximately A$6–7 million into actions to aid the recovery of threatened species. These funds have been sourced from direct grants from local government, the Australian Government, the Burnett Mary Regional Group and other sources. Inclusion of in-kind contributions would greatly boost the investment figures. This funding has supported more than 800 habitat restoration projects on private properties. Overall achievements include 400 km of riparian fencing installed, ~600 000 trees planted more than 4000 ha of weeds managed, many thousands of Mary River Cod fingerlings released, new areas surveyed for threatened frogs and other species and community participation in many dozens of threatened species-oriented events. During this time, a good understanding of what motivates and inspires landholders to be involved in these types of projects has been developed within the catchment through long-term field staff experience, as well as workshops and research (e.g. Smith 2016; Baldwin *et al.* 2017).

The cumulative benefit of all recovery actions on the threatened species is difficult to determine due to the lack of baseline data and subsequent monitoring of the threatened species populations. A review of the Cod Recovery Plan (Jackson 2008) found the goals of achieving self-sustaining populations, downlisting of the cod to Vulnerable and doubling the area of their confirmed range within the catchment had not yet been achieved, but the Mary River cod has recently been found in the Conondale, Kenilworth and Traveston reaches of the Mary River (DNRM 2016), indicating its range may be expanding. A baseline population is under way for the Mary River turtle and surveys for the giant barred frog, funded by the Living with Threatened Species Program, have expanded the known range within the catchment. Ongoing water quality monitoring and stream health assessments are occurring, and could be used, together with population studies, to build a catchment wide picture of the recovery process.

Success factors

Maintaining and growing the momentum for threatened species recovery actions for more than two decades is a mark of success. To identify what contributed to this success, the following key staff of the Mary River Cod Recovery Program were interviewed: Eve Whitney – facilitator of the Mary River Cod network, under the title 'Mary River Cod Mother' 1996–1997; Glenda Pickersgill – Mary River Cod Project Officer 1997–2000 and riparian landholder; and Bob Simpson – DPI Fisheries Scientist and Mary River Cod Recovery Plan author. The success factors that emerging from these interviews and from an analysis of community feedback generated during development of the Mary River Threatened Species Recovery Plan can be summarised as: connection, commitment and community.

Connection: use of iconic species and iconic habitat

> 'Once you started talking about the habitat for the cod their eyes went a little misty...' (Eve Whitney, describing the emotional reaction of long term landholders when talking about the Mary River cod)

The Mary River cod, Mary River turtle and giant barred frog have been powerful iconic species, which has helped people connect to the river and its tributaries and to act on these connections. During the development and implementation of the Mary River cod Recovery Plan, the strong sense of connection that many landholders in the catchment had to the Mary River cod became evident. This ranged from landholders for whom the cod triggered memories of their childhood and the way their creek used to be, to recreational fishers who were committed to restoring habitat for the cod. People were concerned because 'the numbers had dropped and their sizes had dropped' and because, in their lifetimes they had seen the loss of 'deep holes that were now filled with gravel' (Glenda Pickersgill). In the early days of the cod Recovery Plan, staff noticed the meaning people placed on the cod:

> 'It meant things like, if they had cod in their creek, they were being successful landholders. [they] didn't want to lose them. There was a sense of quest about it. A holy quest.' (Eve Whitney)

The Recovery Program staff interviewed speculated that the cod's uniqueness, its role as an apex predator and its impressive size may have contributed to this meaning. Its intrinsic value was also respected, with people wanting to:

> 'protect cod for its own value and ... its ... cultural value ... more so than its value as something to go and catch' (Bob Simpson).

Another important factor in building that sense of connection with the cod Recovery Plan was release of cod fingerlings at targeted locations:

> 'The fact you are putting fish in is right in front of your eyes, you can't miss it. Farmers down there ooh-ing and aah-ing and saying 'oh that one went under the log'. It piqued the interest of people more than talking about long-term habitat change.' (Bob Simpson)

The Mary River turtle nest protection program has played a similar role for the turtle, enabling people to relate to the turtle in direct ways. Thanks to Tiaro and District Landcare group, people can buy a chocolate turtle to help fund turtle nest protection work and they can visit the metre tall 'Mr T': a bronze turtle statue in Tiaro. Many community members have participated in field research on the turtle. Tiaro Landcare's actions have generated an international audience through presentations at four international symposiums and through the Mary River turtle Facebook page, which has attracted followers from over 40 countries.

Similarly the giant barred frog captivates people, potentially as a result of its large size, the large size of its tadpoles (>100 mm) and its beautiful appearance. The response when people

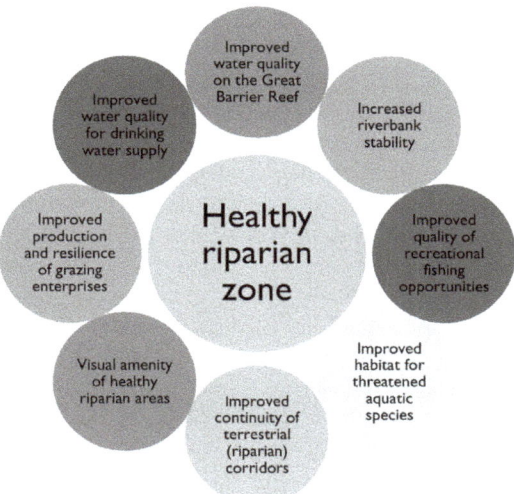

Fig. 5.3. Outcomes connected by a healthy riparian zone.

first see a giant barred frog during a frog survey is generally 'Wow!'; and a connection, once formed, helps foster an ethic of care for the frog's riparian habitat (Ford *et al.* 2014).

Activities have also focused on an iconic habitat – the riparian zone. One of the most significant threats to the priority species is loss of integrity of the riparian zone (Mary River Threatened Species Recovery Team *pers. comm.*). Although the reasons may differ, improving the riparian zone is a goal shared among a range of stakeholders. Figure 5.3 depicts how efforts to improve the health of the riparian zone connect with outcomes that are valued by a range of interests and stakeholders, such as landholders losing their land to erosion, water utilities wanting to protect water quality and recreational fishers wanting to enjoy a healthy ecosystem.

Commitment: role of key individuals and organisational longevity

> *'The key thing is the people who have the energy and passion to move things forward.'*
> *(Eve Whitney)*

A recurring theme in discussions about what enabled the cod recovery plan to be developed and implemented was the importance of key individuals who possessed the commitment, passion, patience and tenacity to overcome challenges. These individuals have included: government employees who negotiated funding to enable the plan to be completed, as well as cutting through internal bureaucracies that may otherwise have prevented progress; politicians who have championed the recovery plan; project officers who have had particular attributes and skills; and landholders who have led by example and became opinion leaders based on their 'knowledge, expertise and … local credibility' (Eve Whitney).

Involvement of engaged individuals from diverse organisations — local government, Queensland Government, Australian Government, universities, Landcare groups, catchment groups, large conservation groups, natural resource management organisations, recreational fishing groups, water utilities and industry bodies – has proved beneficial. Good working relationships became the foundation of long-term partnerships and collaboration. For example, government staff involved in developing the cod recovery plan found that using the community group networks to communicate information about cod habitat and cod restocking helped to 'bridge a gap' that sometimes exists between government and landowners.

Forums that facilitate connections and provide support and opportunities to share knowledge and resources renew and strengthen partnerships and can encourage the key individuals to keep pushing for recovery plan outcomes. In some cases, these connections may be formal, such as the delegates on the Mary River Catchment Coordinating Committee or the official membership of the recovery team. Informal connections that occur through years of working in the same catchment also prove invaluable.

Encouraging and supporting the commitment and passion of all involved during development and implementation of plans generates a range of positive outcomes. Without this commitment, 'Really in the end, a plan is meaningless if it is not implemented by people who have the passion to support it and see it through' (Eve Whitney).

Community: involvement has created better plans and better implementation

The community has played a fundamental role in recovery planning, contributing to – and, at times, leading – the various stages of the recovery planning process. Community knowledge of the Mary River cod underpinned the Cod Recovery Plan. Interviews with landholders were used to help put the scientific data collected during the development of the plan in a 'historical context' (Bob Simpson). Landholders also held positions on the Recovery Team. Catchment-wide consultation occurred in the early stages of the development of the draft Mary River Threatened Species Recovery Plan. As a result, the draft plan includes two objectives focused on people: one that specifies ways that traditional owners wish to contribute to the implementation of the plan and the other that identifies ways to increase involvement of the entire community.

Community groups can also play a leadership role in the development of a plan. The Mary River Catchment Coordinating Committee has done this with the draft Mary River Threatened Species Recovery Plan and has committed to oversee the Recovery Team and liaise with recovery team members as relevant issues arise.

The Cod Recovery Plan created a 'connection in terms of having a place and sense of belonging. It was an opportunity for people to connect with each other' (Eve Whitney). In the Mary River Catchment Coordinating Committee's work and through the Living with Threatened Species Program and the Restoring Riparian Resilience Project, implementation of actions to protect threatened species in partnership with landholders consistently deepen landholders' appreciation of their own properties and the role they can play in securing a species. Interaction with other landholders around these same goals further strengthens commitment.

The future

Future success relies upon embedding the threatened species and their recovery within the aspirations and perspectives of the catchment community. Although the source of future funding for recovery actions is uncertain, there is a significant and active group of people in the catchment who are committed to the conservation of these species. There are some major knowledge gaps regarding the species' status, but the main action needed to secure recovery of these species – namely riparian restoration – is well established and accepted throughout the catchment. The hope is that a multi-species approach, based on Mary's Famous Five, encourages stakeholders to conceptualise the river, the threats and recovery actions as a whole dynamic system and, in doing so, increases support for actions that have multiple benefits across the Mary River ecosystem and the human community.

Conclusion

Recovery planning in the Mary River catchment has a 22-year history. There are some recurring themes regarding the factors that have contributed to success in raising awareness of the threatened species and increasing adoption of practices that support their conservation. Three key aspects are evident: the use of iconic species and habitat to help captivate the community and form connections between people, and between people and the river; the commitment of key individuals from diverse organisations to progress the recovery process; and the invaluable role that community organisations and networks play in enabling translation of local knowledge into better planning and implementation of recovery actions.

Acknowledgements

This chapter would not have been possible without generous sharing of wisdom by Glenda Pickersgill, Bob Simpson and Eve Whitney. The Mary River Threatened Species Recovery Plan owes a particular debt to Tiaro and District Landcare and all of their volunteers and the Mary River Catchment Coordinating Committee's volunteer members and staff, particularly Eva Ford, Brad Wedlock and Deb Seal who have each supported the Committee for more than a decade. The contribution of all project officers and recovery team members who have applied their knowledge and passion over the last two decades is recognised. Most importantly the landholders of the Mary River catchment are acknowledged for their commitment to the river and to threatened species recovery.

References

Baldwin C, Smith T, Jacobson C (2017) Love of the land: social-ecological connectivity of rural landholders. *Journal of Rural Studies* **51**, 37–52. doi:10.1016/j.jrurstud.2017.01.012

Barmuta L, Linke S, Turak E (2011) Bridging the gap between "planning" and "doing" for biodiversity conservation in freshwaters. *Freshwater Biology* **56**, 180–195. doi:10.1111/j.1365-2427.2010.02514.x

Cann J, Legler JM (1994) The Mary River Tortoise: a new genus and species of short-necked chelid from Queensland, Australia (Testudines; Pleurodira). *Chelonian Conservation and Biology* **1**(2), 81–96.

DNRM (2016) 'Mary River cod – *Maccullochella mariensis* – an update on demography habitat movement ageing and genetics to inform effective management.' Department of Natural Resources and Mines, Brisbane.

Flakus S, Connell M (2008) 'Mary River Turtle, yesterday, today, tomorrow'. Tiaro and District Landcare Incorporated, Tiaro, Queensland.

Ford E, Wedlock B, Seal D, Watson D, Burgess S (2014) Frogs – Weapons of mass rehabilitation in the Mary River Catchment. In *Proceedings of the 7th Australian Stream Management Conference. Townsville, Queensland.* (Eds G Vietz, ID Rutherfurd, R Hughes) pp. 202–209. River Basin Management Society, Australia.

Hines H (2005) 'Recovery Plan for Stream Frogs of South-East Queensland 2001–2005.' State of Queensland, Environment Protection Agency, Brisbane.

Jackson P (2008) 'Appendix L Mary River Cod Review and Research Priorities.' Queensland Water Infrastructure, Brisbane.

Simpson R, Jackson P (1999) 'The Mary River Cod Research and Recovery Plan.' Queensland Department of Primary Industries – Fisheries Group, Brisbane.

Smith T (2016) Understanding the aspirations and perspectives of Rivercarers in the Mary River catchment, Qld. In *Proceedings of the 8th Australian Stream Management Conference*. 31 July–3 August 2016, Leura, New South Wales. (Eds G Vietz, A Flatley and I Rutherford) pp. 650–660. River Basin Management Society, Australia.

Smith T, Sim C, Ford E, Hogan G (2012) Lessons learned from developing the Mary River Threatened Species Recovery Plan – a first under the EPBC Act. In *6th Australian Stream Management Conference: Managing for Extremes*. 6–8 February, Canberra. (Eds JR Grove and ID Rutherford) pp. 89–97. River Basin Management Society, Australia.

Stockwell B (2001) 'Mary River Tributaries and Rehabilitation Plan – Implementation Edition.' Mary River Catchment Coordinating Committee, Gympie.

6

Spiny rice-flower: small, unassuming but with many friends

Vanessa Craigie, Debbie Reynolds, Neville Walsh, Steve Mueck, Liz James and Pauline Rudolph

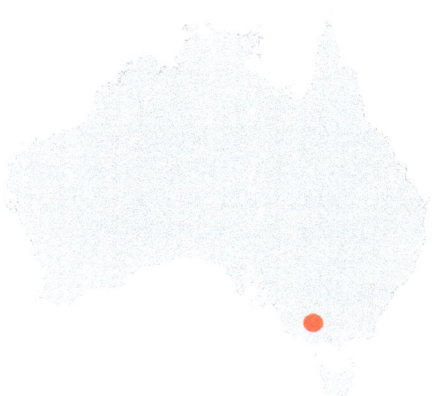

Summary

The problem
1. The grassland habitat of the spiny rice-flower has largely been cleared with remaining populations being small and highly fragmented.
2. These remnants are under threat from urban development, land clearing, weeds and inadequate removal of biomass.

Actions taken to manage the problem
1. Research on species biology and ecology to better inform management.
2. Dedicated reserves established and habitat managed with ecological burning and pest plant and animal control.
3. Education of land managers, support staff across many disciplines and the broader community to help protect vulnerable remnants.

Markers of success
1. More sites are being managed effectively to promote recruitment and some populations are being augmented.
2. The reasons for a lack of recruitment have been identified and are now effectively managed.
3. A recovery plan has been developed and is being implemented by an enduring and well-managed Recovery Team.
4. Community involvement and support has been fostered and is flourishing.

Reasons for success
1. An effective, collaborative, multi-disciplinary Recovery Team.
2. The species has become an iconic representative of much-depleted temperate grassland communities.

3. A dedicated Conservation Trust Fund, provides secure, long-term funding to support on-ground research, management and education.
4. Effective application of the adaptive management principle.

Introduction

The spiny rice-flower (*Pimelea spinescens*) (Fig. 6.1) occurs in native grasslands and grassy woodlands of the lowland plains of Victoria (Fig. 6.2) and has two subspecies: *spinescens* and *pubiflora* (Wimmera rice-flower). Subspecies *spinescens* occurs within the Victorian Volcanic Plain, Goldfields, Wimmera and Victorian Riverina bioregions. Wimmera Rice-flower is limited to three small populations in the Wimmera. The species is listed as threatened under the Victorian *Flora and Fauna Guarantee Act 1988* (FFG Act). The subspecies are listed separately as Critically Endangered under the federal *Environment Protection and Biodiversity Conservation Act 1999* (EPBC).

Most of the grassy ecosystems in which the two subspecies occur are listed under the EPBC Act as Critically Endangered (Natural Temperate Grassland of the Victorian Vol-

Fig. 6.1. The spiny rice-flower (*Pimelea spinescens*), a Critically Endangered plant of temperate native grasslands. (A) Flowers are about 2 mm across. (B) Flowers are produced in small clusters along branches. (C) These small shrubs rarely emerge above the surrounding grass sward (photos: D. Reynolds).

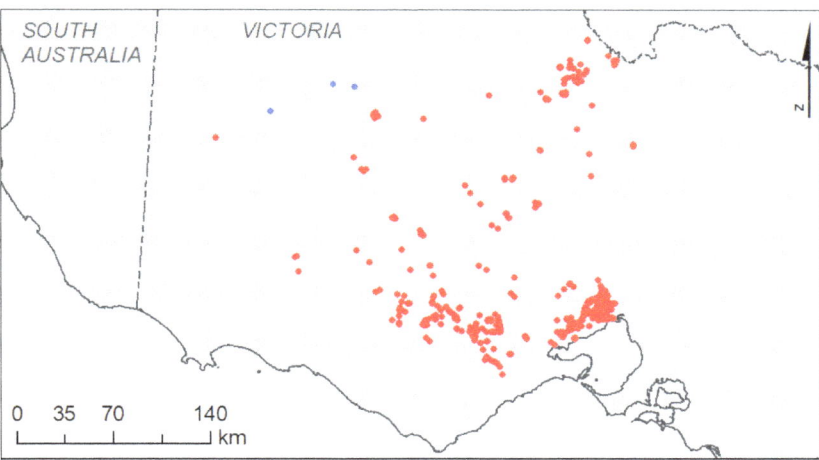

- *Pimelea spinescens* subsp. *spinescens*
- *Pimelea spinescens* subsp. *pubiflora*

Fig. 6.2. Distribution of the spiny rice flower (*Pimelea spinescens* subspecies *spinescens*) across western Victoria, Australia.

canic Plains, Grassy Eucalypt Woodland of the Victorian Volcanic Plain, and Natural Grasslands of the Murray Valley Plains). Because the relatively flat, fertile plains where they grow were quickly developed by European settlers for agriculture, most populations of spiny rice-flower tend to be on small parcels of public land such as roadsides, rail reserves, airports and cemeteries, or on private land that has not been ploughed (DEWHA 2009). The Recovery Team knows of 224 sites supporting subspecies *spinescens*, but 55% have fewer than 100 individuals, with 10% undocumented. Few of the known sites have accurate population counts or are monitored to detect persistence or population trends.

Nevertheless, the spiny rice-flower is enjoying an increased public profile as a result of being identified in the Australian Government's Threatened Species Strategy, and in the Melbourne Strategic Assessment (DEPI 2013), and through the efforts of a range of land managers and conservationists. Much has been learnt about its biology, life cycle, distribution and management requirements (Reynolds 2013).

Conservation management
Research, biology and ecology

Spiny rice-flowers are small slow-growing shrubs up to 50 cm high that re-sprout after fire from large taproots (up to a metre long) and are thought to live more than 100 years (Mueck 2000). They grow on heavy clay soils, are generally either male or female, flower from autumn to winter, and rely on small invertebrates for pollination (Reynolds 2013). Like many grassland species, spiny rice-flowers rely on open ground to allow seed germination (Gilfedder and Kirkpatrick 1994; Morgan 1997). This condition is maintained by regular fire, which removes build-up of smothering biomass (i.e. living or dead plant matter) (Lunt *et al.* 2012).

Seeds have a variable dormancy period (Reynolds 2013), so not all viable seed will germinate simultaneously, even with optimal conditions. Given favourable conditions, germination occurs in pulses, but survival seems to be irregular (Reynolds 2013). Consequently, populations of even-aged cohorts are vulnerable should plants die before the establishment of a significant soil seed bank.

Seedlings are most abundant in late winter, especially if there has been a fire in the preceding 6 months and rainfall has been above average, but few survive the hot, dry summers, presumably because their root systems are poorly developed. However, once a large taproot has developed, mature plants tolerate sustained drought or above-ground disturbance. Seedling development is enhanced by regular (ideally every 2–4 years) burning to remove smothering vegetation, mostly grasses (Reynolds 2013). Although plants propagate from cuttings, these individuals develop a fibrous root system rather than a taproot, and all four known translocations using this method have ultimately failed (B. Thomas *pers. comm.*).

Threats

Threats to the spiny rice-flower include clearing of grassland habitat, excess biomass accumulation from both native and introduced grasses, and competition from weeds, particularly introduced perennial grasses (Carter and Walsh 2006). Habitat loss and fragmentation have left many grassland species restricted to small, isolated populations, with little opportunity for genetic exchange, placing them at risk of inbreeding. Despite this, small areas of native grassland can resist weed invasion and retain high biodiversity values if there is no soil disturbance and biomass is regularly removed using suitable methods, particularly burning (Williams *et al.* 2006). In addition, most occurrences of the species are on land that is neither reserved for conservation nor actively managed. Plants within roadside and rail reserves are particularly susceptible to slashing, grazing, grading and soil compaction.

Planning and policy

The National Recovery Plan for the Spiny Rice-flower (Carter and Walsh 2006) sets clear, although broad, objectives, actions and performance criteria and is the key planning document guiding recovery for both subspecies. Since its publication, the Recovery Team estimates that over half of the performance criteria have been met, but actions to protect all populations would need additional funding. Other policies and plans providing guidance include the Action Statement for *Pimelea spinescens* (DSE 2008), which covers both subspecies, and the Policy Statement on Significant Impact Guidelines for subspecies *spinescens* (DEWHA 2009).

The Melbourne Strategic Assessment evaluated the impacts of the Victorian Government's urban development program for Melbourne. A Biodiversity Conservation Strategy for Melbourne's Growth Corridors, developed as part of the program, sets out the conservation measures required for matters of national environmental significance, including the spiny rice-flower (DEPI 2013).

The spiny rice-flower was identified as one of the 30 priority plant species in the Australian Government's Threatened Species Strategy that are targeted to have improved population trajectories by 2020 (Australian Government 2015).

The Recovery Team

In 2007, a *Pimelea spinescens* working group was formalised as the *Pimelea spinescens* Recovery Team. The current Team comprises representatives from state government (Royal Botanic Gardens Victoria, Department of Environment, Land, Water and Planning (DELWP), Parks Victoria and VicRoads), 22 local governments, six Catchment Management Authorities, Country Fire Authority, environmental consultants, Landcare groups

Fig. 6.3. The *Pimelea spinescens* Recovery Team Meeting, March 2015 at the spiny rice-flower sign on the Cressy-Shelford Rd (photo: D. Reynolds).

and universities. Meetings occur three times a year, alternating between Melbourne and regional centres (Fig. 6.3). Not all members attend every meeting, but a core group maintains regular attendance, and therefore consistency.

The Recovery Team oversees recovery and management of the species. Collaboration, cooperation and shared skills are the strengths of the Recovery Team, with its members having experience in on-ground management, policy, legislation and community engagement. The Team includes representatives from a range of public land managers and levels of government, so there is regular two-way engagement with government at the local, state and national level. It maintains a state-wide database of populations, prepares or commissions guidelines and protocols for subjects such as seed collection, monitoring and translocation, and assesses proposals for translocation. Although the activities of the Recovery Team are guided by the Recovery Plan, the Plan is now out of date, so the Team draws on the broad-ranging experience of members in conservation policy and practice.

The Team's approach is one of outcome-focused, scientific objectivity, and its members take every opportunity to share information and promote partnerships. Decisions are resolved through discussions, without conflict or self-interest. Members hold each other to account, and ensure that when members commit to participate in sub-groups or to comment on plans or projects, these commitments are delivered.

Money

A critical element of the program's success has been the ongoing funding provided by the Pimelea Conservation Trust Fund. This was established in 2005 as a result of an offset agreement between a developer and the federal Department of the Environment and Energy. The Trust for Nature was appointed as Trustee of the Fund, and in 2009, the Pimelea Conservation Trust Fund Committee was established to approve allocations of money from the Fund. The Committee comprises representatives from the Australian and Victorian Governments and Wyndham City Council.

The Fund has been of considerable importance to conservation management, because its existence means that conservation efforts are not tied to government budget cycles or short-term grant programs. Annual allocations provide for ongoing management of the two reserves purchased as part of the offset agreement. The Fund also supports recovery projects that aid the long-term conservation of the species. Each year, land managers, local communities, teaching institutions, private landholders or Landcare groups can apply for grants. After consideration by the Recovery Team and Trust Committee, funds are provided to projects that fulfil recovery plan objectives. Mid-project and annual reports are a condition of the funding.

DELWP has also received funding from federal and state sources over the last 10 years for implementation of on-ground conservation actions. This has included dedicated grants from state and federal threatened species initiatives, and 'in-kind' funding (e.g. salaries of government staff allocated to the project, including administrators and parks staff, and their operating costs).

Managing

The Trust funds the part-time Pimelea Conservation Officer, currently Dr Debbie Reynolds, who works one day per week to provide executive support for the Pimelea Conservation Trust and the Recovery Team, administer funding, support potential and successful applicants of Trust-funded projects, maintain population counts, keep the DELWP Victorian Biodiversity Atlas up to date and engage with the general community. The range and influence of the organisations represented on the Recovery Team has led to a high degree of exposure for such a small, unassuming plant, and prevented accidental or deliberate damage or loss to many populations of the species. Conspicuous signage featuring spiny rice-flowers has been erected on important roadsides to bring public attention to the species (Fig. 6.3).

Research

Among a range of minor projects, predominantly those on monitoring and population dynamics (e.g. Cropper 2009), there have been two major research efforts. A PhD project, part-funded through the Recovery Team, commenced in 2008 to investigate the persistence of spiny rice-flower populations (Reynolds 2013). It found that natural germination occurred *in situ* in greater numbers following biomass removal and declined with time since fire but longer term survival was low. Seed dormancy, rather than seed production and viability, was the reason for low annual recruitment. Dr Reynolds has also been researching how to establish new populations and under what conditions they survive. She has also established a seed orchard to capture genetic diversity across the Victorian Volcanic Plains populations and provide a seed source for future research and revegetation.

In 2011–12, the Trust funded the Royal Botanical Gardens Victoria to document how genetic variation is distributed through the range of the spiny rice-flower. The study found that spiny rice-flower is more genetically diverse than expected, probably because many plants are decades old and now exist in a very fragmented metapopulation. It also found that the northern populations of subspecies *spinescens* had diverged to some extent from the southern populations. In contrast, Wimmera rice-flower showed little variation, which is consistent with it having been very rare for many generations (James and Jordan 2014). An important management implication is that mixing genetic material within, but not between, northern and southern populations of subspecies *spinescens* should benefit the long-term viability of the species.

Translocation

Since 1998, land clearing associated with Melbourne's urban expansion has led to the loss of several populations of subspecies *spinescens*. This has triggered responses under state and federal legislation, such as mitigation and/or offset measures, or as a last resort, translocation. A review of 11 translocations of subspecies *spinescens* conducted between 1998 and 2012 found that all translocations retained some survivors (31%), with plants producing viable seed (Biosis 2014). Building on this, the Recovery Team produced translocation guidelines to cover seed collection (PsRT 2013a), monitoring (Reynolds 2014a, 2014b) and translocation (PsRT 2013b).

Other activities

The Trust has also funded projects on habitat condition and demographic structure, community workshops and production of a children's book and eBook, *Pimelea and Her Grassland Friends* (Fig. 6.4). On-ground projects have delivered regular ecological burning of many populations, fencing of a significant population on private land, annual pest plant and animal control, and best-practice management and monitoring of three populations in reserves to the west of Melbourne.

Learnings

Research has greatly increased knowledge of the species' biology, genetic variability and management requirements. We now know how to grow plants, how to move them successfully, how far genetic material should be dispersed and how best to manage and monitor populations.

The benefits and pitfalls of government policies and regulations have been illuminated. The survival of some spiny rice-flower populations has depended on strong regulation that has restricted clearing and development, but plants are still being lost. For example, since

Fig. 6.4. The children's book *Pimelea and Her Grassland Friends* is helping spread the story of the plant and its habitat, <https://friendsofiramoograsslands.wordpress.com/pimelea-and-her-grassland-friends/>.

2011 at least 400 plants have been destroyed through urban development in the Greater Melbourne area.

Controls relating to protected flora under the FFG Act do not operate on private land, and although native vegetation clearing controls apply, there are exemptions for permitted clearing, so some areas of native vegetation fall outside these planning controls. The EPBC Act has strong powers to regulate and mitigate impacts, but that too has its limitations. For example, in 2005, the Wimmera rice-flower, which at the time had an EPBC Act status of 'Extinct', was re-discovered. However, in 2008, when road maintenance works destroyed nearly 10% of the plants on one of only two known populations, its legislative status remained unchanged. It was not an offence to destroy something listed as 'Extinct' so there were no direct consequences under the Act. However, this was rectified by re-classifying Wimmera rice-flower as 'Critically Endangered' in 2009, and the local government responsible for the destruction voluntarily undertook remediation actions.

These and other learnings will be used to review and update the National Recovery Plan for the spiny rice-flower, and set better, clearer directions for future conservation.

Stakeholder support and involvement

Stakeholder and community support was variable when the Spiny Rice-flower Recovery Program began in 2005 (formalised as the Recovery Team in 2007). The species was regarded by some as unimportant and, to others, as an irritating impediment to land use change and development. Attitudes began to change when an EPBC Enforceable Undertaking was applied to a local council when roadworks destroyed plants, so many other local governments took steps to ensure they didn't make a similar error. Other EPBC Enforceable Undertakings entered into by agencies and individuals ensured that the spiny rice-flower became a high-profile species. Some developers were initially dismissive of the need to reduce losses of spiny rice-flower due to land clearing, but came to accept its importance, the controls that protected it, and the value of the Recovery Team's management advice.

At the community and individual level, support was inspired through the Trust Fund. Public workshops and community engagement were undertaken by the Pimelea Conservation Officer and other Recovery Team members to promote spiny rice-flower and native grassland management, and forge relationships with partner organisations. As a result, surveys of suitable habitat have been undertaken by keen local individuals and groups. At least 13 new discoveries of subspecies *spinescens* have been made, including 800 plants in far-western Victoria, thus extending the subspecies' known distribution by 100 km.

Involvement of the Country Fire Authority in well-publicised ecological burns, and the educational efforts of the Authority's Vegetation Management Officers, have led to a greater acceptance of ecological burning, and consequent improvements in management of spiny rice-flower and its native grassland habitats.

Setting priorities

Conservation management priorities have not been subject to rigorous cost–benefit analyses but the Trust Fund Committee makes decisions based on clear criteria associated with the National Recovery Plan. Projects that involve many or large populations (>300 individuals) are the most likely to receive funding, to provide the best 'bang for the buck'. Other priorities have been mostly reactive to date, because managing new threats as they arise tends to supplant more pro-active planning.

Monitoring

Few populations of spiny rice-flower are regularly monitored, making it difficult to determine whether the range or overall population size of the species have changed. Some populations continue to be monitored by either consultants or local governments. Monitoring of other populations tends to be somewhat unstructured and *ad hoc*. Promoting wider use of the Recovery Team's monitoring protocols aims to remedy these deficiencies.

Where translocation is undertaken as a requirement of a permit under the FFG Act, monitoring is a condition of the regulatory process. The Recovery Team must also be provided with an annual report of the progress of the translocations.

The future

The conservation of the spiny rice-flower continues to face challenges. Native grasslands and grassy woodlands are still the most threatened ecological communities in Victoria, and subject to threats from human population growth, agriculture intensification, weed invasion, urban and rural development. Climate change is a likely threat, but how it might affect the species is unknown. Wetter winters may provide conditions for better recruitment, but hotter, drier summers will reduce seedling survival. A key question is whether the long-term resilience of the species and its winter flowering habit will allow it to survive habitat changes. The high degree of genetic diversity within the species may permit adaptation to new conditions, and manipulating genetic mixing between populations may enhance their adaptive opportunities.

Nevertheless, future prospects for spiny rice-flowers have improved over the last decade – we know more, the species' profile has increased, and both government and non-government parties are better engaged. Ongoing research is investigating optimal establishment techniques, timing of management and seed production. Establishment of a seed orchard in the Western Grassland Reserve near Melbourne (DSE 2011) will provide an important future source of seed to help restore and augment secure populations in the future.

Victoria's new Biodiversity Plan, *Protecting Victoria's Environment – Biodiversity 2037*, which was released in April 2017, sets new directions for threatened species. These directions include moving away from managing threatened species 'one at a time', and focusing more on how ecosystems and ecological processes can be managed for the benefit of multiple species – such as spiny rice-flower and other species with similar habitat requirements and susceptibility to threats. It sets targets for broad-scale, cost-effective management actions, but also identifies where new types of management interventions might be needed under climate change, and has a strong focus on partnerships and collaboration.

Conclusion

The story of the spiny rice-flower is about how a small, relatively uncharismatic plant has become a flagship for Critically Endangered grasslands and grassy woodlands in Victoria. The spiny rice-flower is now well known and represents an example of how many and diverse partners can work together to deliver a result far greater than if each partner had worked in isolation. The species' Recovery Team includes people with diverse skills across multiple jurisdictions, and enjoys deep personal commitment from members and their organisations. The strength of the program has been the ongoing funding provided by the Pimelea Conservation Trust through its support for ongoing management and research, and employment of the Pimelea Conservation Officer – the essential human face of the program.

Acknowledgements

Many people and organisations have contributed to spiny rice-flower conservation, management and research. We thank the organisations who contribute by encouraging their staff to participate in the Recovery Team: Department of Environment Land, Water and Planning; Parks Victoria; VicRoads; Royal Botanical Gardens Victoria; Country Fire Authority; Trust for Nature; the Cities and Shires of Ararat, Brimbank, Campaspe, Central Goldfields, Colac Otway, Corangamite, Greater Bendigo, Golden Plains, Greater Geelong, Hobsons Bay, Loddon, Melton, Moonee Valley, Moorabool, Mt Alexander, Moyne, Northern Grampians, Pyrenees, Southern Grampians, Surf Coast, Wyndham, Yarriambiack; ABZECO; Aus Eco solutions; Biosis; Brett Lane & Associates; Ecology & Heritage Partners; Ecology Australia; Glenelg Hopkins Catchment Management Authority; Western Melbourne Catchments Network; and Dennis Family Corporation.

We also thank the people and groups who undertook or supported Trust Fund projects that contributed to ongoing conservation work: Enics Environmental; Mathews Ecological Services; Blue Devil Consulting; Country Fire Authority; Cairnlea Conservation Reserves Management Committee; Mt Rothwell Biodiversity Interpretation Centre; and Ullina Landcare Group.

References

Australian Government (2015) 'Threatened Species Strategy'. Australian Government, Canberra, <http://www.environment.gov.au/biodiversity/threatened/species/30-plants-by-2020/spiny-rice-flower>

Biosis (2014) 'Review of Spiny Rice-flower translocations in Victoria'. Report for *Pimelea spinescens* Recovery Team Project no. 15814. Biosis Pty Ltd, Melbourne.

Carter O, Walsh N (2006) 'National Recovery Plan for the Spiny Rice-flower *Pimelea spinescens* subsp. *spinescens*'. Department of Sustainability and Environment, Melbourne.

Cropper S (2009) 'Monitoring of *Pimelea spinescens* subsp. *spinescens* (Spiny Rice-flower) on Lake Borrie Spit in 2008 and a Discussion on the Appropriate Management of the Population'. Report prepared for Melbourne Water by Botanicus Australia Pty Ltd, Melbourne.

DEPI (2013) 'Biodiversity conservation strategy for Melbourne's growth corridors'. Department of Environment and Primary Industries, Melbourne.

DEWHA (2009) 'EPBC Act Policy Statement 3.11 – Significant impact guidelines for the Critically Endangered spiny rice-flower (*Pimelea spinescens* subsp. *spinescens*)'. Department of the Environment, Water, Heritage and the Arts, Canberra.

DSE (2008) 'Flora and Fauna Guarantee Act 1988 Action Statement No. 132 (Revised 2008): Spiny Rice-flower (*Pimelea spinescens* subsp. *spinescens*)'. Department of Sustainability and Environment, Melbourne.

DSE (2011) 'Western grassland reserves: grassland management targets and adaptive management'. Department of Sustainability and Environment, Melbourne.

Gilfedder L, Kirkpatrick J (1994) Genecological variation in the germination, growth and morphology of four populations of a Tasmanian endangered perennial daisy, *Leucochrysum albicans*. Australian Journal of Botany **42**, 431–440. doi:10.1071/BT9940431

James EA, Jordan R (2014) Limited structure and widespread diversity suggest potential buffers to genetic erosion in a threatened grassland shrub *Pimelea spinescens* (Thymelaeaceae). Conservation Genetics **15**, 305–317. doi:10.1007/s10592-013-0539-y

Lunt ID, Prober SM, Morgan JW (2012) How do fire regimes affect ecosystem structure, function and diversity in a grassland and grassy woodlands of southern Australia? In *Flammable Australia: Fire Regimes, Biodiversity and Ecosystems in a Changing World*. (Eds RA Bradstock, AM Gill and RJ Williams) pp. 253–270. CSIRO Publishing, Melbourne.

Morgan JW (1997) The effects of grassland gap size on establishment, growth and flowering of the endangered *Rutidosis leptorrhynchoides* (Asteraceae). *Journal of Applied Ecology* **34**, 566–576. doi:10.2307/2404907

Mueck SG (2000) Translocation of Plains Rice-flower (*Pimelea spinescens* ssp. *spinescens*), Laverton, Victoria. *Ecological Management & Restoration* **1**(2), 111–116. doi:10.1046/j.1442-8903.2000.00032.x

PsRT (2013a) Seed collection protocol. *Pimelea spinescens* Recovery Team 2013, Melbourne, <http://www.swifft.net.au/cb_pages/spiny_rice-flower.php>.

PsRT (2013b) Translocation protocol (March 2013). *Pimelea spinescens* Recovery Team, Melbourne, <http://www.swifft.net.au/cb_pages/spiny_rice-flower.php>.

Reynolds DM (2013) Factors affecting recruitment in populations of Spiny Rice-flower (*Pimelea spinescens* Rye subspecies *spinescens*) in Victoria's natural temperate grasslands: relationships with management practices, biological and ecological characteristics. PhD thesis. Victoria University, Australia.

Reynolds DM (2014a) 'Guidelines for monitoring *Pimelea spinescens* SOP No: 2.5_02.2014'. *Pimelea spinescens* Recovery Team, Melbourne.

Reynolds DM (2014b) 'Monitoring protocol for *Pimelea spinescens*. Version 1.2_01_2014'. *Pimelea spinescens* Recovery Team, Trust for Nature, Melbourne.

Williams NG, Morgan JW, McCarthy MA, McDonnell MJ (2006) Local extinction of grassland plants – the landscape matrix is more important than patch attributes. *Ecology* **87**, 3000–3006. doi:10.1890/0012-9658(2006)87[3000:LEOGPT]2.0.CO;2

7

Saving the pygmy bluetongue lizard

C. Michael Bull and Mark N. Hutchinson

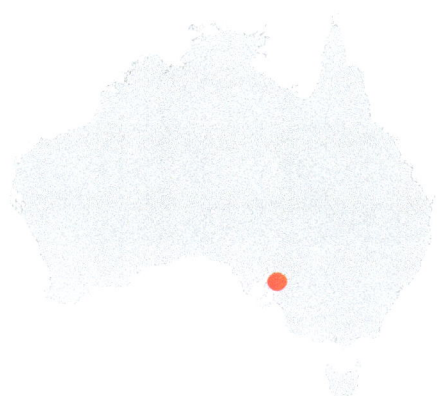

Summary

The problem
1. 'Rediscovered' in 1992, with little understanding of its biology and threats to guide conservation.
2. Habitat largely cleared with just over 30 small scattered populations remaining.

Actions taken to manage the problem
1. Research and monitoring program established to inform conservation.
2. Recovery plan developed, recovery team established.
3. Engagement of landholders encouraging compatible grazing regimes.
4. Community led awareness-raising activities.
5. Captive breeding.

Markers of success
1. Marked increase in knowledge of biology and status to inform management.
2. Key threats identified with management responses underway.
3. Collaborative recovery program guiding conservation effort.

Reasons for success
1. Building of ecological knowledge and its influence on conservation management.
2. Community awareness, support and involvement in recovery efforts.
3. A coordinated well-managed recovery process.

Introduction

This is the story of a lizard that we thought we had lost. The pygmy bluetongue lizard (*Tiliqua adelaidensis*), was described and named in 1863. From the start, it was recognised as an oddly small-sized bluetongue lizard, a group that otherwise includes several large, common and iconic Australian species. But the pygmy bluetongue lizard's status was

Fig. 7.1. (A) Volunteers surveying an area for pygmy bluetongue lizard burrows. (B) Pygmy bluetongue female. (C) Pygmy bluetongue site. (D) Pygmy bluetongue at burrow entrance (photos: Mark Hutchinson).

notably different to other bluetongue lizards: by the start of the 1990s, only 20 specimens were known, most collected in the late 19th century, and located in a few European and Australian museums. There was no information on the species' biology, and even its distribution was scantily documented, although known locations appeared to be scattered around the Adelaide region, hence the specific name, '*adelaidensis*'. The only records for the 20th century were specimens obtained near Burra in the 1940s and at Marion in 1959. The species was considered possibly extinct (Cogger 1992).

Then, on 14 October 1992, the story of the pygmy bluetongue lizard took a dramatic upswing when a specimen was discovered in the body of a road-killed eastern brown snake (*Pseudonaja textilis*) not far from Burra, 160 km north of Adelaide (Armstrong and Reid 1992). Discovery of the source population for the snake's meal a few weeks later led to the discovery of the unexpected and cryptic refuges in which the lizards live: inside the abandoned burrows of wolf spiders and trapdoor spiders in a patch of degraded native grassland. This knowledge led to the discovery of new populations, with just over 30 now known (Fig. 7.1).

The sites where the lizard occurs are scattered through areas of moderate rainfall (~400–600 mm annually) in an area bounded roughly by the towns of Kapunda, Peterborough, Jamestown and Kulpara. The species is listed as Endangered, both nationally under the *Environment Protection and Biodiversity Conservation Act 1999* (EPBC Act), and in South Australia under the *National Parks and Wildlife Act 1972*. Its status is based on it

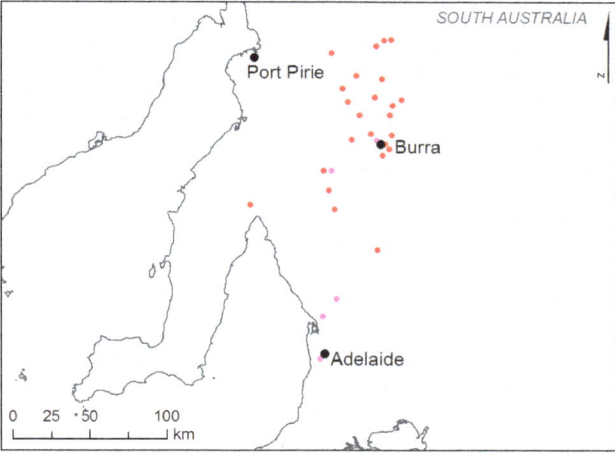

Fig. 7.2. Distribution of the pygmy bluetongue: sites discovered before and after 1992.

being known from a relatively small number of disjunct locations, many of which appear to support only small populations, and just a small number of which appear to be larger metapopulations – perhaps three or four such sites that probably harbour 1000 or more individuals.

The species now appears to occupy only the northern part of its former distribution, suggesting a loss of ~40% of its former range as well as a massive decline within the remaining area (Fig. 7.2). The locations occupied are all on privately owned land, and only a part of one site is currently a reserve for the species. All known sites are on land historically and currently used to graze sheep.

Because nothing at all was known about the species when rediscovered in 1992, there was initially no basis for any management actions for the species apart from minimising obvious harm, such as avoiding land clearance, or pesticide spraying. The early revelation that the species is a diurnal inhabitant of abandoned spider burrows meant that its life history and conservation could not be readily analogised from similar species: there were none. A Recovery Team was formed soon after the initial discovery, and it was clear that research on the ecology of the species had to be the priority.

From the time of the 1992 find, a research team from Flinders University and the South Australian Museum has focused on how to prevent us losing this lizard again. The work has had three major phases: (1) understanding the unusual ecology and behaviour of a spider-hole-living lizard; (2) developing advice for the management of the few known populations and getting the local communities to recognise, own and cherish their amazing biodiversity asset; and (3) planning how best to allow the lizard to overcome future threats to its persistence. The results of this work have been the basis for a Recovery Plan for the species, which first appeared in an early form in the early 2000s and has been revised and updated (Duffy *et al.* 2012).

The discovery phase has been very productive and we now believe we understand many of the important aspects of the lizard's ecology that will be necessary to conserve it. Our message is one of hope. We now understand to a large extent how the lizard lives and what it needs to survive. We have developed methods for finding and monitoring the species and have information on how to boost population sizes and also are close to having complete protocols for relocating lizards from hazardous areas to more secure locations. Active management of the species is as yet in its infancy, and more widespread application of the research findings, transformed into management directions, will need to be a priority.

Conservation management
The biology of the pygmy bluetongue lizard

The pygmy bluetongue's very particular shelter site has turned out to be the key to both studying the species and understanding the threats to it. Pygmy bluetongues prefer native grassland, but don't require pristine conditions – most of the known sites have native plant species accounting for little more than half the ground cover, with the remaining components being introduced plants and bare soil. The best sites have very open grass cover due to grazing. The lizards are extremely inconspicuous. Even at sites with high densities, they are seldom observed on the surface – the lizards appear to have very good vision and an approaching observer inevitably triggers a retreat into the spider hole long before the lizard can be easily spotted among the grass stems.

The optic fibrescope is an important piece of technology, used to probe down a likely spider hole to identify its occupant. The holes themselves are generally easy to miss, with openings no more than 20–25 mm in diameter. Thus locating a new population is a laborious process. Frequently, sites where the grassland seems generally suitable either have few searchable holes, or the holes contain no lizards. Even in locations where the lizards are found, they occur in local patches with extensive apparently lizard-less areas in between.

The pygmy bluetongue lizard relies entirely on spider burrows for refuge (Milne and Bull 2000) – in 24 years we have not recorded a lizard in any other home site and have seldom recorded them any other way than looking down a hole. This hole-owning behaviour means that once located, the same animal can be relocated and monitored at its 'home address': an unusual opportunity for researchers working with small vertebrates. Pygmy bluetongue lizards use the burrows as safe shelters from excessive temperatures, grass fires, and predators. During the day, they typically wait at their burrow entrance to bask and to ambush passing invertebrate prey, particularly grasshoppers (Fenner *et al.* 2007), always ready to instantly retreat from danger. As far as we know, the lizards do not dig their own burrows, nor have we observed them actively displacing resident spiders. Instead, we think they locate and move into vacated spider burrows. Once a lizard has a good burrow, it can stay there for the entire spring and summer 'activity' season, and even for more than one season, and it will aggressively defend that burrow from potential rival occupiers.

Known populations are confined to areas that have not been recently (20 years) ploughed, and where remnant populations of native grasses persist. They are often on gentle hill slopes, with lizard density highest where soil depth is more than 30 cm and soil structure has low clay content, allowing good drainage during rainy periods (Souter *et al.* 2007). All populations are on privately owned land, usually farms, although the Nature Foundation of South Australia has purchased a site with one population, and created the Tiliqua Reserve for the conservation of this lizard.

At several sites, we have monitored lizard numbers over several successive years. There is certainly variation in abundance across years, some of which correlated with rainfall. Population density usually (not in every case) declines following years of low rainfall, and increases after wet years, but there is some evidence that young and adults respond differently to the same climatic conditions. We do not yet know what influences the balance between recruitment and mortality in a given population. Certainly, immigration from adjacent sites has little influence on the population dynamics, because genetic analysis has shown no evidence of recent dispersal between populations as close as a kilometre apart (Smith *et al.* 2009; Schofield *et al.* 2012).

How to manage pygmy bluetongue lizard populations

These natural history observations have led us to three basic messages for the medium-term conservation of existing populations: (1) the lizards must be protected from major soil disturbance – ploughing or other intensive land uses are very likely to rapidly eliminate a local population; (2) spider holes are limiting – but human-made holes can act as temporary boosters to improve recruitment or allow releases; (3) grazing is essential, sheep are suitable grazers, and the levels of grazing needed are compatible with good agricultural land management. The current land use at each pygmy bluetongue site is, and has long been, medium impact grazing of sheep and, under this regime, the species has survived and in some cases thrived. Largely this translates as a message of 'if it ain't broke, don't fix it' to landowners.

Ploughing the soil is a lizard disaster: spider holes that persist in hard-packed, unploughed soil quickly erode in crumbling, ploughed soil. When native grassland and a ploughed field are adjacent to each other, the lizard population will abruptly stop at the dividing fence line, despite there being no other broad habitat change. Even when burrows are placed in the ploughed field right up to the fence, they are not used. This insight suggests the ploughing of much of the land area of the mid-north during the first 50 years of settlement is the likely reason for the decline in the pygmy bluetongue lizard's range and populations. We regard the surviving populations, limited to small, sloping and often stony areas (unsuitable for ploughing), as representing just the fringes of what must have been the main pygmy bluetongue habitat.

Burrows are essential for lizard survival. Lizards prefer burrows that are deep (>20 cm), presumably to allow them to escape both from digging predators and from the climatic extremes of the surface. They also select burrows with entrances that are just the right diameter for their heads, with bigger lizards preferring wider entranced burrows. Knowing these specifications, we have trialled providing extra holes, made from lengths of wooden dowel with the centre drilled out at the preferred diameter of 20 mm. Burrows provided in local patches within several populations were readily occupied by lizards, and used for living and reproduction (Souter *et al.* 2004). This in turn significantly increased recruitment success of juvenile lizards into the population. Our burrows had relatively short lives, the wooden dowelling decaying through rot and termite attack over 2 or 3 years. Spiders do a better job in providing a constant supply of new burrows, so conserving spiders ought to be an important strategy for conserving lizards, but artificial holes could be valuable in helping to boost survival in threatened populations, or even launching new populations.

Our third management essential is grazing. Grazing keeps the grassland very open, with scattered bare patches. The openness ensures that lizards have access to sunlight for basking and a good view from their burrows to allow them to see approaching prey or predators. In well-grazed habitats, lizards tend to sit at their burrow entrances for longer, and have more successful strikes at passing prey insects (Pettigrew and Bull 2014). Overgrazing, however, has a negative effect. When offered a choice, lizards avoided burrows where simulated hard grazing had removed all of the surrounding vegetation (Pettigrew and Bull 2012), and large numbers of sheep trample the surface and destroy burrows, with more sheep having a bigger impact (Clayton and Bull 2015). Farmers who manage their land well tend to avoid extremes of overgrazing, so good farming practices benefit both sheep and lizards.

Supporting groups and funding

In building our understanding of the pygmy bluetongue's conservation needs, the researchers have worked to varying degrees with other government and non-government groups and received key support at times from particular people in the community. The Recovery Team has continued to reflect those groups, namely the South Australian Government Environment Department, Flinders University, The South Australian Museum, Zoos SA, community conservation groups (membership has varied), local community members, including landowners with pygmy bluetongues on their property, and the consulting company EBS Ecology.

The South Australian Environment Department staff of both the central and regional offices provided overarching support for establishing the Recovery Team and for many of the actions that have begun to put the knowledge from research back into managing the species. State funding has supported groups who provided some of our first opportunities to recruit landowners who had pygmy bluetongues on their property to become part of the recovery process. Environment Department staff have coordinated volunteer groups directly supporting conservation efforts in their local areas, and the regional environment offices are vital conduits that help groups with overlapping interests in conservation to interact.

One such important group has been Greening Australia and its project to encourage less impactful use of native grasslands by stock, to simultaneously improve pasture quality and the diversity of the remaining native vegetation in stocked landscapes. Apart from establishing new lines of communication with landowners, the Greening Australia personnel have discovered new pygmy bluetongue colonies and their encouragement of rotational grazing looks likely to have long-term benefits for both landowners and lizards.

Although the Commonwealth (federal) Department of the Environment was an important early source of funding, the EPBC Act has been a crucial tool in managing the sometimes conflicting priorities of the lizards and a major economic development that has established in their habitat, namely wind farms. Wind farms place turbines on ridge tops, which are generally far too thin-soiled to support pygmy bluetongues, but the process of construction of both the turbines and the infrastructure can lead to the excavation of areas that have lizards, compaction of the soil by heavy earth-movers and soil runoff that could choke burrows. The strength of the EPBC Act has led to consultation and preplanning discussions between wind farm developers and Recovery Team members on several occasions, and each time has led to beneficial compromises regarding the placement of infrastructure and construction activities. During these processes, we have had the opportunity to train staff from the main environmental consulting company that has worked with these developers, EBS Ecology, with the result that EBS now has several staff who are expert in assessing landscapes for the potential presence of pygmy bluetongues as well as being skilled in locating the lizards; their efforts alone have discovered several new populations of the species. Currently, EBS is also overseeing a data gathering exercise in which the direct impact of nearby turbines on lizards is being assessed: this project paid for by a development company.

From the beginning, much of the research on pygmy bluetongues has come from grants in the Australian Research Council's (ARC) Linkage Grant Program. The pygmy bluetongue work has received a series of overlapping grants that have driven the immediate management related research questions and regularly involved as partners the other organisations South Australian Museum, Zoos SA, and state regional natural resources management boards.

Local community involvement

From the 1990s, meetings of the Pygmy Bluetongue Recovery Team were held in towns in the mid-north region of South Australia within or close to the lizard distribution. This was to keep open a line of communication with the local Department of Environment officers, and inform them about the research and recommended actions. Among other activities, we gave talks at local schools and addressed meetings of the local Goyder Council. In 2012, the Recovery Team decided to encourage local community members at Burra to form their own conservation group, and the Pygmy Bluetongue Conservation Association was born, thanks to ARC funding. That association is currently growing. Its current chair is a local farmer who has lizards on her property, and its membership includes other farmers, local high school students and residents from Burra and surrounding districts, with some research leaders to provide technical advice. It aims to increase awareness of this local lizard among the wider community, raising the profile of the lizard at country shows, supporting field days to inform locals and maintaining a visitor centre display in the old Burra railway station. It has been awarded several grants to promote its activities and has sponsored the production of a children's book about the lizard, *Pinkie and Pete*, which has been distributed to all the primary schools in the mid-north and to all of the primary schools in the Adelaide region.

The future

All of the climate change predictions are for hotter and drier conditions in South Australia. For pygmy bluetongue lizards, much of its current distribution will become much less suitable by 2100 (Fordham *et al.* 2012), so a strategy we need to consider is translocation: moving individuals to different sites, particularly to more southerly locations that are probably still within their historic range (Delean *et al.* 2013). Working with Zoos SA, we are discovering the factors that are likely to keep lizards at a release site and the best the time of release (Ebrahimi *et al.* 2015), with lizards less likely to move if released in late summer than in early spring. In early 2016, at Monarto Zoo, a small captive colony produced a batch of live young, and we have trials in place to determine how best to encourage this reproduction and to prepare the progeny for life in the wild.

The work done so far provides us with a manual for protecting pygmy bluetongues into the future. Some landowners are already committed to the minimal steps needed to guarantee the lizard's future. However, there is still work to do in improving community-level concern. The Pygmy Bluetongue Conservation Association has had a promising start as a community voice to reassure landowners and encourage a positive attitude to the lizards, but will need to expand its influence over time, and will need to continue to have a pool of expertise to refer to as issues arise. Recovery Team priorities will need to emphasise this supporting role for community conservation, to focus on gently improving the levels of interest and ownership within the community that harbours the last populations of this unique Australian lizard.

Acknowledgements

The conservation of the pygmy bluetongue lizard received a sad and unexpected blow late in 2016 with the death of Professor Mike Bull. His leadership and insights have driven the research that will underpin the management of the species into the future. Mike's PhD and

Honours students have been the ones at the coal face, discovering the information presented in this chapter – many of their names are listed as authors of the accompanying references. Major funding for research and conservation has come from the Australian Research Council, several programs for endangered species from the Commonwealth Department of the Environment, the South Australian Environment Department and the Nature Foundation SA. Zoos SA has supported the research and management of the species and helped to make the lizard more widely known in the community, and the staff of EBS Ecology have been vital in mitigating the effects of wind farm developments on pygmy bluetongue populations. Of the very many people who have helped in the work on the pygmies, we particularly wish to acknowledge Phil Ainsley, Dale Burzacott, Sylvia Clarke, Peter Copley, Angela Duffey, Ian Falkenberg, Mike Gardner, Dawn Hawthorn-Jackson, Paula Jones, Tim Milne, Terry Morley, Luke Price, Chris Reed, Ruth Robinson, Julie Schofield and Andy Sharp.

References

Armstrong G, Reid J (1992) The rediscovery of the Adelaide pygmy bluetongue *Tiliqua adelaidensis* (Peters, 1863). *Herpetofauna* **22**, 3–6.

Clayton J, Bull CM (2015) The impact of sheep grazing on burrows for pygmy bluetongue lizards and on burrow digging spiders. *Journal of Zoology* **297**, 44–53. doi:10.1111/jzo.12247

Cogger HG (1992) *Reptiles and Amphibians of Australia*. 4th edn. Reed Books Australia, Melbourne.

Delean S, Bull CM, Brook BW, Heard L, Fordham DA (2013) Using plant distributions to predict the current and future range of a rare lizard. *Diversity & Distributions* **19**, 1125–1137. doi:10.1111/ddi.12050

Duffy A, Pound L, How T (2012) 'Recovery Plan for the Pygmy Bluetongue Lizard *Tiliqua adelaidensis*'. Department of Environment and Natural Resources, Adelaide.

Ebrahimi M, Ebrahimie E, Bull CM (2015) Minimising the cost of translocation failure by using decision tree models to predict species behavioural response in translocation sites. *Conservation Biology* **29**, 1208–1216. doi:10.1111/cobi.12479

Fenner AL, Bull CM, Hutchinson MN (2007) Omnivorous diet of the endangered pygmy bluetongue lizard, *Tiliqua adelaidensis*. *Amphibia-Reptilia* **28**, 560–565. doi:10.1163/156853807782152462

Fordham DA, Watts MJ, Delean S, Brook BW, Heard L, Bull CM (2012) Managed relocation as an adaptation strategy for mitigating climate change threats to the persistence of an endangered lizard. *Global Change Biology* **18**, 2743–2755. doi:10.1111/j.1365-2486.2012.02742.x

Milne T, Bull CM (2000) Burrow choice by individuals of different sizes in the endangered pygmy blue tongue lizard *Tiliqua adelaidensis*. *Biological Conservation* **95**, 295–301. doi:10.1016/S0006-3207(00)00040-9

Pettigrew M, Bull CM (2012) The response of pygmy bluetongue lizards to simulated grazing in the field during three drought years. *Wildlife Research* **39**, 540–545. doi:10.1071/WR12086

Pettigrew M, Bull CM (2014) Prey capture behaviour by pygmy bluetongue lizards with simulated grazing. *New Zealand Journal of Ecology* **38**, 45–52.

Schofield JA, Fenner AL, Pelgrim K, Bull CM (2012) Male-biased movement in pygmy bluetongue lizards: implications for conservation. *Wildlife Research* **39**, 677–684. doi:10.1071/WR12098

Smith AL, Gardner MG, Fenner AL, Bull CM (2009) Restricted gene flow in the endangered pygmy bluetongue lizard, *Tiliqua adelaidensis*, in a fragmented agricultural landscape. *Wildlife Research* **36**, 466–478. doi:10.1071/WR08171

Souter NJ, Bull CM, Hutchinson MN (2004) Adding burrows to enhance a population of the endangered pygmy blue tongue lizard, *Tiliqua adelaidensis*. *Biological Conservation* **116**, 403–408. doi:10.1016/S0006-3207(03)00232-5

Souter NJ, Bull CM, Lethbridge M, Hutchinson MN (2007) Habitat requirements of the endangered pygmy bluetongue lizard, *Tiliqua adelaidensis*. *Biological Conservation* **135**, 33–45. doi:10.1016/j.biocon.2006.09.014

8

Malleefowl: answering the big questions that guide all malleefowl management

Sharon Gillam, Tim Burnard and Joe Benshemesh

Summary

The problem
1. Malleefowl habitat has been cleared and fragmented and widespread declines have occurred.
2. There are many threats, but not yet sure which is the most important.

Actions taken to manage the problem
1. Coordinated monitoring across the species' range.
2. Timely analysis of monitoring results.
3. Adaptive management trials building on existing dataset.
4. Respect and support for champions and volunteers.

Markers of success
1. Recovery Plan developed and implemented.
2. Recovery Team enduring and well managed.
3. Enhanced community involvement and support.
4. Statistically robust systems in place to track populations and evaluate management interventions.

Reasons for success
1. Dedicated champions and citizen scientist support.
2. A fascinating species.
3. A sophisticated monitoring system that engages volunteers at many levels.
4. A well-managed recovery process.

Introduction

Malleefowl (*Leipoa ocellata*) belong to one of the most unusual of bird families (Megapodiidae). Rather than sitting on eggs in a nest, as most birds do, the malleefowl are megapodes, which use compost, solar energy and sometimes volcanic heat to incubate their eggs. Unlike other megapodes, which are found in tropical and subtropical forests, the malleefowl is adapted to the drier environments of southern Australia – the species is typically found in semi-arid to arid shrublands and low woodlands, particularly those dominated by mallee and/or acacias. Malleefowl once occurred across the more arid parts of New South Wales, Victoria, South Australia, Western Australia and the southern Northern Territory. However, within the past century, much of the best habitat for malleefowl has been cleared for cropping or severely modified by grazing, causing contraction and fragmentation of the species' range and a decline in numbers.

As a consequence, malleefowl are listed nationally under the *Environment Protection and Biodiversity Conservation Act 1999* (EPBC Act) as Vulnerable, and as threatened in each state and territory within their range; the species may already be extinct in the Northern Territory. The malleefowl is one of 20 bird species identified in the Australian Government's Threatened Species Strategy for priority attention (Australian Government 2015).

Although there is evidence of ongoing continental-scale decline (Benshemesh *et al.* 2007), significant improvements in malleefowl conservation and understanding are under way. In particular, our ability to identify trends, respond to threats and assess the effectiveness of management actions aimed at benefiting malleefowl has increased enormously. Building on knowledge of trends gathered over decades, the Recovery Team and the many volunteers who have supported the bird have every hope of lifting the species' conservation status within the foreseeable future.

Conservation management
Biology and key threats
Malleefowl are generalist feeders on grain, vegetation and small animals. In winter, pairs construct mounds, usually renovating an old one, into which the female typically lays 15–25 eggs over spring and summer to be incubated by rotting vegetation and heat from the sun (Fig. 8.1). Initially, heat for incubation is provided by rotting compost, though this can be difficult to achieve if there has been too little winter rain to wet the compost. The male makes sure the mound is maintained at the right temperature by opening and closing the mound at different times. To lower the incubation temperature, he opens the mound very early and exposes the sand to the cool morning air; to raise the temperature, he opens the mound later and exposes the sand to the heat from the sun (Frith 1962). After hatching, the young receive no parental care, they make their own way to the surface, can fly within a day and lead solitary lives, but they are also subject to high levels of mortality from predators or starvation. Breeding begins when birds are 3–4 years old and continues for ~15 years (Benshemesh 2007).

Clearing of the mallee for wheat and sheep production has been the major factor in the decline of malleefowl in southern Australia. The fragmentation of malleefowl habitats limits the dispersal of birds among patches, amplifying other threatening processes and reducing the capacity of populations to recover. Such recovery can be slow: even in unfragmented landscapes, breeding can be reduced for at least 30 years after habitat is burnt. Other threats include inappropriate fire regimes, grazing by introduced herbivores, including sheep (*Ovis aries*), cattle (*Bos taurus*), rabbits (*Oryctolagus cuniculus*) and goats (*Capra hircus*), which may reduce food availability and change habitat structure, and predation by introduced foxes (*Vulpes vulpes*) and cats (*Felis catus*).

Without action, continuing declines are likely due to: the small size and isolation of many remaining populations; ongoing habitat clearance in some jurisdictions; the continued threat of introduced predators and competitors; the risk posed by recurring cata-

Fig. 8.1. Malleefowl incubate their eggs using heat from the sun and rotting vegetation (photo: S. Gillam).

strophic events (especially fire); and predicted declines in winter rainfall due to climate change (Benshemesh 2007).

Planning and policy

The current Recovery Plan (Benshemesh 2007) was preceded by a Research Phase Recovery Plan (Benshemesh 1992) and the first National Malleefowl Recovery Plan in 2000. Within these documents, goals, objectives and targets have undergone extensive review and refinement over time to reflect current research, management practices and knowledge. Reviews have been undertaken by the Recovery Team, which have been presented at National Malleefowl Forums (Copley 2012; Burnard 2016). Consultation on improvements was also sought from delegates at the 3rd and 4th National Forums, published as resolutions in the respective proceeding, and progress towards these were reviewed in subsequent proceedings. This demonstrates a willingness and flexibility within the Recovery Team to listen and adapt to concerns expressed by those with an interest in conserving the species, as well as continually striving to improve the malleefowl recovery process and effort.

People, agencies governance, and accountability

Fundamental to conservation of the malleefowl has been the National Malleefowl Recovery Team, formed in 1989. The main role of the team has been to coordinate, prioritise and promote the objectives and actions outlined in the Recovery Plan, with the aim of stopping the decline, supporting malleefowl recovery and securing existing populations across the species' range. The Recovery Team does this by providing advice and general guidance to state agencies, landholders and others on what actions are needed using the best and most reliable information available. Membership of the team comprises representatives from relevant state agencies and federal and stakeholders for industry and community groups. To communicate the national recovery effort to interested parties (including the general public), the Recovery Team has prepared advice to natural resource management organisations (J. Benshemesh unpublished), maintains a website, publishes a national newsletter twice a year and hosts national forums every 3–4 years. Although most land managers have been supportive, the Recovery Team is not in a position to prescribe or enforce management actions over the large range of the species.

The second key to the Malleefowl Recovery Program has been the commitment and dedication to the cause of malleefowl conservation by a handful of exceptional individuals, whose continuing drive, leadership and pursuit of excellence in science has inspired people to join the recovery process. Under the guidance of the Recovery Team, and with critical support from state agencies, community groups have embraced the objectives of the recovery process and raised the profile of the plight of malleefowl across its range. In recent years, successful grant applications have made it possible to employ trainers to travel the country and ensure training of volunteers is uniform, while also linking the various groups via face-to-face contact. We have also been able to fund meetings in Adelaide for several volunteer coordinators. The various groups, together with a large number of enthusiastic individuals, have grown to create one of the largest evidence-based citizen science programs currently running in Australia, with over 300 members involved in monitoring each year.

Monitoring

The best way to monitor malleefowl is to monitor their nest mounds. If a mound is active, the bird is most likely breeding – a good sign. If no mounds are active, there may be cause for concern, and there is a need to look for possible explanations. For example, declines may be caused by local failure of winter rains, by sudden habitat destruction due to fire, or a range of other more insidious factors such as predation, habitat change or a slowly changing climate. The need to monitor malleefowl to provide fundamental information on their trends and abundance was recognised as an essential tool in the late 1980s, and subsequently written into recovery plans and forum papers. As monitoring evolved over the last 25 years, it has become central to the malleefowl conservation effort, and continues to build momentum through the citizen-science community (Fig. 8.2).

Monitoring trends in malleefowl populations is essential for the evidence-based approach advocated in the Recovery Plan, but obtaining these basic data is not easy. This is because the species is sparsely distributed, shy and superbly camouflaged, making direct counts of populations unfeasible. Focusing on mounds overcomes this problem, but as breeding numbers in any year depend on environmental conditions, particularly rainfall, results are often highly variable from year to year, making trends difficult to determine. Moreover, monitoring mounds is labour intensive, and coupled with the enormous range over which malleefowl still occur, obtaining representative and reliable monitoring data has proved to be a huge undertaking that would not be possible without citizen scientists.

The first *National Malleefowl Monitoring Manual* was published in 2007 (National Heritage Trust National Malleefowl Monitoring Project 2007), describing in detail the standards, protocols and procedures for monitoring malleefowl and gathering data. These guidelines have been used in both Victoria and South Australia from the early 1990s, in Western Australia since 2004 and New South Wales since 2014, and have proven to be an invaluable reference and asset to all those involved in the monitoring process and beyond. The guidelines were updated in June 2016 (Tonkin 2016) to reflect current technologies, in conjunction with the development of a comprehensive *Coordinators Handbook* designed specifically to assist the data-handlers.

Fig. 8.2. Conservation of the malleefowl would not have been possible without an enormous contribution from volunteers (photos: Tim Burnard).

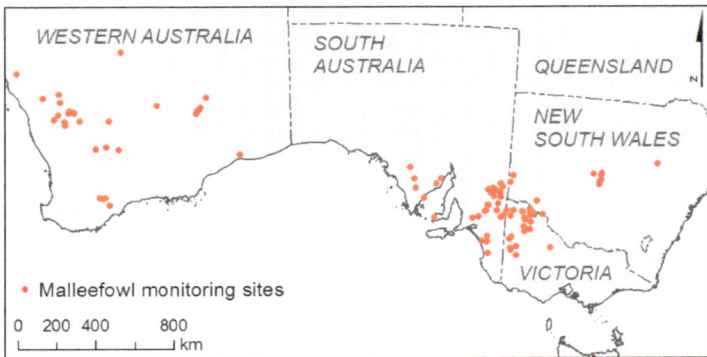

Fig. 8.3. Monitoring sites for malleefowl are scattered across their vast range demanding a high degree of coordination

With hundreds of volunteer citizen scientists involved in the collection of data from over 130 sites and up to 4000 mounds across Australia, there is an enormous challenge in managing the data collected and supporting the field work (Fig. 8.3). A high standard of incoming data must be maintained, as well as providing a well-organised experience for volunteers, to make the best use of their time. In the early days, there were issues with data quality control and storage, so, after the 2nd National Malleefowl Forum in 2004, a Multi-Regional Malleefowl Project was developed, which aimed to standardise and improve the monitoring at a national level, analyse the data collected to date and use the data to guide management. The Project, a massive undertaking, was extremely successful, largely due to the high level of collaboration among volunteer groups and agencies, driven by cooperation and enthusiasm in a renewed spirit of reaching for a common goal right across the country. Indeed, this Project created a momentum that was infectious in garnering further support and leveraging more funding for malleefowl recovery.

Through the Multi-Regional Project, the National Malleefowl Monitoring Database was initiated in 2007. With additional subsequent funding, it has evolved to provide a centralised storehouse for the monitoring data, and many of the processes previously done manually are now automated. It also provides an avenue to deliver feedback and reporting to state agencies, landholders and (other) volunteers, within a secure environment. The Database assists in the preparations before fieldwork, and the checking and validating of the incoming data post fieldwork, ensuring that a high standard is maintained and that the data are reliable and ready for analysis (Benshemesh *et al.* 2016).

Analysis of the monitoring data up to 2005 has indicated regional and national declines in the abundance of breeding malleefowl over the previous two decades, a decline estimated at ~2–3% per year nationally (Benshemesh *et al.* 2007). This worrying result was associated with declines in winter rainfall over this period, and more recent data suggests malleefowl numbers have generally improved since 2005. Whether this apparent improvement in trend is due to management actions or to environmental factors, such as improved winter rainfall, is being examined in an update of the trend analysis that is currently underway.

On-ground management

To date, on-ground management is undertaken by relevant agencies in each state. The Recovery Team seeks to provide guidance to actions by relevant agencies via the Recovery Plan and associated publications (J. Benshemesh unpublished). The success of this advocacy is difficult to measure, particularly because land managers rarely manage solely for malleefowl. The most targeted common management treatment has been fox control, based on studies showing the susceptibility of captive reared malleefowl to fox predation (Priddel and Wheeler 1994, 1997). However, although foxes (and cats) certainly kill malleefowl, especially young, whether reducing fox abundance leads to increases in malleefowl populations is not known. Analysis of data collected during over 20 000 mound inspections from 64 sites across Australia between the late 1980s and 2006, as well as information on fox control efforts, rainfall, fire history and landscape fragmentation, showed that winter rainfall profoundly affected breeding numbers in subsequent years (Benshemesh *et al.* 2007). Although fox baiting reduces fox abundance, there was no evidence that this was associated with increased malleefowl breeding numbers, even after several years of baiting (Benshemesh *et al.* 2007; Walsh *et al.* 2012).

These findings inspired a new evidence-based approach, the Adaptive Management Project, which draws on the re-invigorated, nationally coordinated monitoring program. Adaptive management is a logical process of 'learning by doing' that takes an experimental approach to management and evaluates the effectiveness of management actions through continuous monitoring. Its main aim is to reduce uncertainties in management while at the same time putting in place remedial action (Benshemesh and Bode 2012).

The Adaptive Management Project is operating through a team from Melbourne University in collaboration with the Recovery Team and Parks Victoria, and is assisted by land managers across all states. The Project is helping to identify and understand which of the threats to malleefowl are most important and which actions most effective. To this end, the team is establishing around 20 experimental sites of several thousand hectares each, with nearby control sites, across Australia, where a range of trials can be run. This Project constitutes one of the largest adaptive management experiments ever attempted in Australia. The Project has involved many stakeholders during both the research and planning stages. This process of inclusion (in workshops and forums) and feedback (through presentations and published papers) has been integral to the success of the Project so far (Benshemesh and Burnard 2016; Hauser *et al.* 2016). The first trial is focusing on the effect of fox and cat control on malleefowl populations, with the intention of resolving once and for all the question of whether controlling these predators is beneficial and cost effective.

Infrastructure

The basics of malleefowl surveying – volunteers moving through the bush on foot – has remained the same for decades. However, modern technologies such as hand-held electronic devices, GPS units and digital cameras to capture data in the field have been adopted, allowing for the replacement of reams of paper and compasses. As well as empowering volunteers to take control of the monitoring process, this led to substantial improvements in data accuracy and management. Attempts at using remote sensing – with the idea that warm mounds could be detected on cold mornings – proved to be expensive and of limited utility, as did helicopter searches, but recent analysis of 3-D photogrammetry and LiDAR imagery is showing great promise.

Money

The bedrock on which the Project has been built over many years is volunteer time – even when funding shortages occur for externally funded activities (as they often do), the volunteer network has been able to continue monitoring until additional funds were found.

Although each of the Recovery Plans was formally adopted by the Australian Government, funding was not automatically allocated for the plans or for running the Recovery Team. Nevertheless, the diligence and determination of a few dedicated parties has led to the successful acquisition of several grants, which have enabled a wide range of objectives and recovery activities to be instigated. For instance, the first Recovery Plan was funded by zoos in Adelaide, Melbourne, Perth and Sydney. The Natural Heritage Trust also provided substantial support that funded the national standardisation of the monitoring data, analyses and on-ground works. Local groups have used a wide range of approaches to obtain funding, from attracting funds from different Commonwealth (federal) and state government schemes, non-government organisations, through to commercial surveying of malleefowl mounds as part of state and Commonwealth development assessment and approvals processes. In the latter case, a contract was undertaken by the Recovery Team's incorporated body (National Malleefowl Recovery Group Inc.) to monitor 300 mounds on a Western Australian mining property. Profits from this effort have been used to carry out malleefowl works on nearby conservation reserves, as well as pay for coordinator training in Adelaide. Offset funds from mining industry have also been instrumental in developing the Database, analysing data, enabling several local and national initiatives, and leveraging for more substantial grants. For example, an Australian Research Council Linkage Grant was secured with backing from offset monies from Iluka Resources to fund the development of the Adaptive Management program at Melbourne University.

The future

We are optimistic that in the coming years there will be significant improvements in our approaches to malleefowl management, leading to improvement in the conservation status of malleefowl. Although uncertainty remains about the effectiveness of management actions in reversing the decline of malleefowl across Australia, the Adaptive Management Project will help demonstrate which actions work for malleefowl. However, predictions of higher temperatures and lower winter rainfall in southern Australia may lead to substantial reductions in malleefowl abundance and distribution (Parsons 2008), highlighting the need for both ongoing monitoring of malleefowl and continued adaptive management.

Although we are constantly looking to broaden our funding base, we only have some security for the next year or two. More staff are needed over the next 5 years so the results of the Adaptive Management Project can be shared with all relevant land managers. We are also working to upgrade our database to provide annual reports specific to Natural Resource Management Regions to assist in local management plans. We hope to use this as the basis for ongoing financial support.

Conclusion

We summarise our reasons for success as follows:

- Dedicated champions. Unfortunately it's hard to bottle this, but it is important that when champions are identified, we do what we can to support them.
- The fascinating biology of the species as the only arid zone mound-builder in the world. It engenders much publicity and general interest.
- A sophisticated monitoring system that engages volunteers at many levels: from camping and bushwalking to viewing thousands of motion camera images or programming apps to phones for data gathering. From one day a year to nearly full time jobs, malleefowl volunteering has something for everyone. We go to great

lengths to ensure our volunteers' time and all the data they gather are treated with respect.
- A well-managed Recovery Team that focuses on achieving Recovery Plan actions. The Recovery Team has been fortunate to have the involvement and leadership from some particularly talented and dedicated people. It is important to support these people when needed – it is a waste to have talented conservation specialists spending their time chasing funding.

Acknowledgements

The range of people involved in the malleefowl success story is many and varied. The efforts of all Recovery Team members over the years are acknowledged, particularly Peter Copley, Stephen Davies and the late Peter Sandell. Much of the momentum for progress has been provided by key community groups that have garnered volunteer support and made what seemed impossible possible: the Victorian Malleefowl Recovery Group (VMRG); North Central Malleefowl Recovery Group (NCMPG, WA); Malleefowl Preservation Group (MPG, WA – 1992–2014); and the WA Malleefowl Network, and all their members. Key players in coordinating and training volunteers include: Graeme Tonkin, Dave Setchell, Vicki Natt and Andrew Freeman (South Australia); Joy McGilvray, Carl Danzi, Gordon McNeil, Susanne Dennings and Sally Cail (Western Australia); Peter and the late Ann Stokie and Greg Davis (Victoria). The efforts of Peter and Ann Stokie, and Graeme Tonkin, have made particularly significant contributions to malleefowl conservation nationally. The Adaptive Management Team from Melbourne University have also proved to be a key component in the success story, particularly Michael Bode, Cindy Hauser, Jose Lahoz-Monfort, Brendan Wintle and, more recently, Darren Southwell.

References

Australian Government (2015) *Threatened Species Strategy*. Australian Government, Canberra.
Benshemesh J (1992) 'Recovery plan research phase for malleefowl'. Australian National Parks and Wildlife Service, Canberra.
Benshemesh J (2007) 'National Recovery Plan for Malleefowl'. Department for Environment and Heritage, South Australia.
Benshemesh J, Bode M (2012) Adaptive management of Malleefowl. In *Proceedings of the 4th National Malleefowl Forum 2011*. 29 July–1 August, Renmark, South Australia (Ed. SD Gillam) pp. 126–135. Colour Tech Digital Printing, Adelaide.
Benshemesh J, Burnard T (2016) Introduction to the National Forum. In *Proceedings of the 5th National Malleefowl Forum 2014*. 12–15 September, Dubbo, New South Wales. (Eds M Bannerman and SJJF Davies) pp. 3–7. Printak Pty Ltd, Adelaide.
Benshemesh J, Barker R, MacFarlane R. (2007) 'Trend analysis of malleefowl monitoring data. Revised 2007'. Milestone 3 report. Mallee CMA, Victorian Malleefowl Recovery Group, and multi-regional National Malleefowl Monitoring, Population Assessment and Conservation Action Project steering committee, Melbourne, Victoria.
Benshemesh J, Tonkin G, Stokie P (2016) Update on the National Malleefowl Monitoring Database: recent developments and new gear. In *Proceedings of the 5th National Malleefowl Forum 2014*. 12–15 September, Dubbo, New South Wales. (Eds M Bannerman, SJJF Davies) pp. 112–116. Printak Pty Ltd, Adelaide.

Burnard T (2016) How have we progressed as measured against the National Recovery Plan for malleefowl and resolutions from Renmark? In *Proceedings of the 5th National Malleefowl Forum 2014*. 12–15 September, Dubbo, New South Wales. (Eds M Bannerman, SJJF Davies) pp. 8–20. Printak Pty Ltd, Adelaide.

Copley P (2012) Performance evaluation of the National Recovery Plan for Malleefowl. In *Proceedings of the 4th National Malleefowl Forum 2011*. 29 July–1 August, Renmark, South Australia. (Ed. S Gillam) pp. 4–15. Department of Environment and Natural Resources, South Australia.

Frith HJ (1962) *The Mallee-fowl: The Bird that Builds an Incubator*. Angus and Robertson, Sydney.

Hauser CE, Bode M, Rumpff L, Lahoz-Monfort JJ, Benshemesh J, Burnard T, *et al.* (2016) Applying Adaptive Management principles to malleefowl conservation. In *Proceedings of the 5th National Malleefowl Forum 2014*. 12–15 September, Dubbo, New South Wales. (Eds M Bannerman, SJJF Davies) pp. 210–215. Printak Pty Ltd, Adelaide.

National Heritage Trust National Malleefowl Monitoring Project (2007) 'National manual for the malleefowl monitoring system'. Victorian Malleefowl Recovery Group and Mallee CMA, Melbourne.

Parsons B (2008) Malleefowl in the fragmented Western Australian wheatbelt: spatial and temporal analysis of a threatened species. PhD Thesis. School of Animal Biology. University of Western Australia, Perth, WA.

Priddel D, Wheeler R (1994) Mortality of captive-raised malleefowl *Leipoa ocellata*, released into a mallee remnant within the wheat-belt of New South Wales. *Wildlife Research* **21**, 543–552. doi:10.1071/WR9940543

Priddel D, Wheeler R (1997) Efficacy of fox control in reducing the mortality of released captive-reared malleefowl, *Leipoa ocellata*. *Wildlife Research* **24**, 469–482. doi:10.1071/WR96094

Tonkin G (Ed.) (2016) *National Malleefowl Monitoring Manual*. Edition: 2016–1. National Malleefowl Recovery Team, Australia.

Walsh JC, Wilson KA, Benshemesh J, Possingham HP (2012) Unexpected outcomes of invasive predator control: the importance of evaluating conservation management actions. *Animal Conservation* **15**, 319–328. doi:10.1111/j.1469-1795.2012.00537.x

9

From the brink of extinction: successful recovery of the glossy black-cockatoo on Kangaroo Island

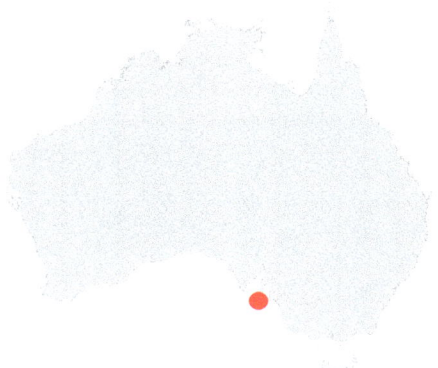

Karleah Berris, Michael Barth, Trish Mooney, David Paton, Martine Kinloch, Peter Copley, Anthony Maguire, Gabriel Crowley and Stephen T. Garnett

Summary

The problem
1. Glossy black-cockatoos are highly vulnerable to predation by common brush-tailed possums when nesting.
2. They also suffer from a shortage of nest hollows, with strong competition for nest hollows from other cockatoos and feral bees.

Actions taken to manage the problem
1. Ongoing protection of natural nest hollows from both predators and competitors.
2. Erection and maintenance of artificial nest hollows.
3. Extensive planting of food trees.
4. Ongoing efforts to maintain awareness of the cockatoos on the island.

Markers of success
1. The population trend was reversed with numbers steadily increasing for two decades.
2. The causes of decline have been identified and are now being managed effectively.
3. A recovery plan was developed and is being implemented by a well-managed, long-standing recovery team.
4. The level of community involvement and support has been sustained for a generation.

Reasons for success
1. Dedicated champions, both locally and among managers.
2. High quality research identified threats and how to manage them.
3. Strong community engagement and support, giving the species a strong local profile.
4. Sustained investment over two decades.

Introduction

The last remaining population of the South Australian subspecies of glossy black-cockatoo (*Calyptorhynchus lathami halmaturinus*, Fig. 9.1) has been confined to 4400 km² Kangaroo Island for ~40 years. The diet of this subspecies of glossy black-cockatoo is one of the most specialised among birds, comprising almost exclusively of the seeds contained within the hard, particularly large, cones of drooping she-oak (*Allocasuarina verticillata*). The bill of this subspecies is correspondingly larger than that of the eastern Australian subspecies, which feeds on smaller coned casuarina species (Schodde *et al.* 1993). As recently as 4000 years ago, *Allocasuarina* forests covered vast areas of south-eastern Australia (Bickford and Gell 2005) and the distribution of the cockatoos may have stretched from the hinter-

Fig. 9.1. The subspecies of glossy black-cockatoo on Kangaroo Island is characterised by having a particularly large bill, which it uses to extract seed from the cones of drooping she-oak, its only food (photo: Mike Barth).

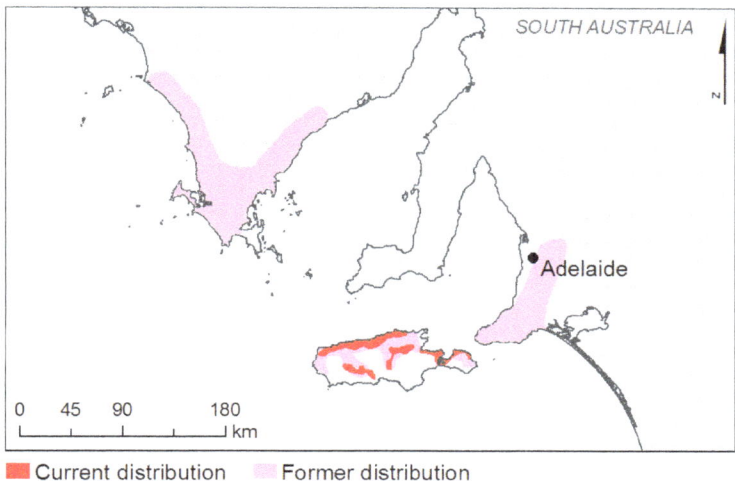

Fig. 9.2. The current and presumed historical range of the Kangaroo Island glossy black-cockatoo.

land from Eyre Peninsula in South Australia through to western Victoria (Schodde *et al.* 1993). Since then, much of the casuarina forest was replaced by eucalypts (Bickford and Gell 2005) so that the range of the cockatoos was already likely to have declined substantially before European settlement. However, the species still occurred throughout the Fleurieu Peninsula and southern Mt Lofty Ranges in the late 1800s (Baird 1986) and may have persisted on the lower Eyre Peninsula (Crowley *et al.* 1997; Fig. 9.2). Following colonisation, much of the mainland casuarina woodland was cut for firewood or felled to provide fodder for cattle and sheep (Stuart 2006). By 1995, it was estimated that only 161 ha of good quality drooping she-oak woodland remained on the entire Fleurieu Peninsula (Andrews 1995). The last confirmed mainland sighting of the South Australian subspecies of glossy black-cockatoo was in 1977 (Joseph 1989). As a result of this substantial decline, the subspecies is listed as Endangered at the national level under the *Environment Protection and Biodiversity Conservation Act 1999*.

Fortunately for the cockatoo, Kangaroo Island's separation from the mainland hindered agricultural development until after the Second World War. This meant that, although a soldier settlement scheme in the 1950s resulted in substantial land clearance and farm establishment (Argent 1997), much uncleared land remained when South Australia began limiting native vegetation clearance in the 1980s. As a result, an estimated 4800 ha of drooping she-oak woodland are distributed across Kangaroo Island (Crowley *et al.* 1998), which should provide ample food for around 600 cockatoos (Crowley *et al.* 1997; Chapman and Paton 2002).

Conservation management
Dealing with an unbalanced system
Concerns about the status of glossy black-cockatoos on Kangaroo Island surfaced during the 1980s and early 1990s when periodic population counts recorded fewer than 150 individuals. A key observation during these counts was the very low numbers of juveniles in cockatoo flocks, suggesting low reproductive success (Joseph 1982; Pepper 1997). In 1995, the population was deemed at risk of extinction due to low reproductive success, despite

significant areas of habitat still being available, so the Glossy Black-cockatoo Recovery Program was initiated.

Glossy black-cockatoos have a naturally low reproductive rate, with females laying only a single egg per clutch. If this egg is adequately cared for by the female, which is solely responsible for egg incubation, a small fluffy yellow chick will hatch after 30 days. The chick will remain in the nest for almost 3 months before fledging: the longest nest period of any cockatoo. This huge investment in rearing one chick – a process that binds the female to a nesting site for almost 4 months of the year – has a major disadvantage: each pair of cockatoos can rear a maximum of only one chick per year. Although some females will attempt to lay an egg again if their first nesting attempt fails in the early stages, in many cases the pair will wait after failure until the following year before attempting the laborious process again. In the 1990s, it was believed too few chicks were surviving each year to replace the adults being lost to natural mortality (Pepper 1997).

The Recovery Program began by monitoring nests to determine why so few chicks were surviving. Monitoring showed that nesting success rates were only 23% (Garnett *et al.* 1999) because common brushtail possums (*Trichosurus vulpecula*), a species native to Kangaroo Island, were entering nests at night and taking eggs and nestlings (Garnett *et al.* 1999). Possum densities are extremely high on Kangaroo Island (Garnett *et al.* 1999) due to the island's mosaic landscape of mixed farming and remnant vegetation. Studies elsewhere in Australia have indicated that habitat remnants in agricultural landscapes support much higher densities of brushtail possums than extensive native forest areas (Downes *et al.* 1997), and introduced pastures have become an important part of the possum diet (Green and Coleman 1986). It is likely that clearance for agriculture on Kangaroo Island, and subsequent fertilisation of pastures, has created a favourable landscape for brushtail possums, leading to increased densities of possums in cockatoo nesting habitat.

Clearance for agricultural production not only altered predation pressure but also resulted in a loss of nesting habitat. Clearance of large sugar gums (*Eucalyptus cladocalyx*) led to a shortage of nesting habitat in some parts of Kangaroo Island, and competition for remaining tree hollows was increased with the introduction of European honey bees (*Apis melifera*) and self-colonising galahs (*Eolophus roseicapillus*) and little corellas (*Cacatua sanguinea*). Changes in fire regime since European settlement on Kangaroo Island are also likely to have led to fewer old age hollow-bearing eucalypts in these areas.

With predation accounting for most nest failures, it was obvious that the high rate of possum predation should be tackled as a matter of urgency to prevent extinction of the black-cockatoos. Controlling possum densities on any meaningful scale would be an immense challenge. Instead, isolating individual nest trees from possums was considered the most feasible option. Once active nests were located, a corrugated iron guard was fixed to the base of the tree around the trunk to prevent possums from climbing them. In addition, overhanging canopy from surrounding trees was pruned to prevent possums accessing black-cockatoo nests by jumping across from adjacent trees. This ingenious low-cost solution proved to be highly effective at preventing possum predation, and nest success rates increased to 50% (Garnett *et al.* 1999).

Nesting habitat has also been increased with the installation of artificial nest boxes, constructed from PVC stormwater pipe. Feral bee hives are controlled within black-cockatoo nesting sites, and pest deterrents are installed in known nests to deter bees from invading hollows. Problem galahs and little corellas are also removed from nesting sites, reducing competition for hollows for glossy black-cockatoos, and other native species that use tree hollows.

The results of 20 years of continuous management

Since 1995, locating and protecting nest trees from possums has been a core priority of the Recovery Program. Installation of artificial nest boxes, and protection of the trees these boxes were placed in, has also provided the population with many more safe nesting options in the landscape (Fig. 9.3). Monitoring of nest success has continued annually, and has confirmed that this protection has been successful at maintaining reproductive success rates at 31–68% (in any one year), where nest success is measured as successful fledging (Berris and Barth 2016). Although highly variable, nest success in managed nests has consistently been above what was recorded in unprotected nests in the mid 1990s. Since 2007, between 50% and 60% of monitored breeding pairs nest in artificial nest boxes, and use of nest boxes is generally highest in areas of Kangaroo Island that have seen extensive native vegetation clearance (K. Berris unpublished).

What is most encouraging is that this improvement in reproductive success has translated into an overall increase in the population size. In the 2016 annual census of Kangaroo Island, a minimum of 373 individuals were counted, more than double the number of cockatoos in 1995 (Fig. 9.4) (Berris *et al.* 2017). This is a rewarding result for a species with such a low natural reproductive rate, and highlights how important this and other management actions have been for the recovery of this species.

As the population of black-cockatoos has increased, so too has the number of nest trees requiring protection. In 2016 at the time of writing, 295 nest trees are being managed by a Recovery Program team of two staff. Furthermore, the black-cockatoos have now been

Fig. 9.3. (A) Protection of nests from brush-tail possum predation has helped doubled the nesting success. (B) Glossy black-cockatoo chicks have rubbery bills that are no match against possums. (C) Volunteers have contributed enormously to conservation of the cockatoos, both by direct conservation and rehabilitation of habitat (photos: A, B, Rick Dawson; C, Colin Wilson).

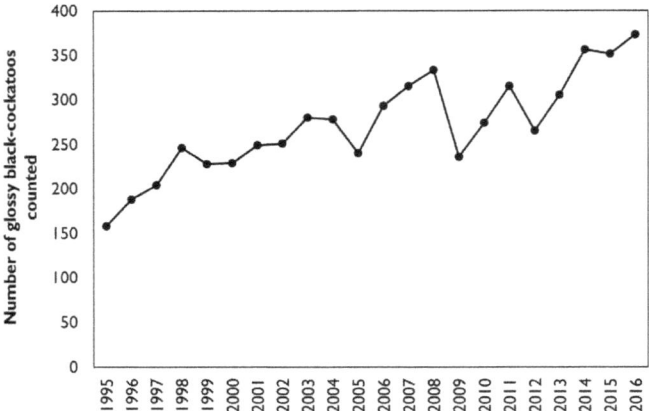

Fig. 9.4. Trends in the population of glossy black-cockatoos on Kangaroo Island, South Australia.

recorded nesting in some areas of Kangaroo Island from which they had previously disappeared. In 2015, the first nesting attempt on the Dudley Peninsula since the 1970s was recorded. Over the past 20 years the Recovery Program has recorded progressively larger flock sizes on the eastern end of the island. With nesting sites now scattered across Kangaroo Island, nest monitoring and maintenance of possum protection has become more time consuming and labour intensive.

Management actions have successfully mitigated the predation, competition and habitat loss threatening glossy black-cockatoos on Kangaroo Island. However, the threats themselves are still present. Canopy pruning of trees surrounding nest trees must be carried out at least every 3 years, and recent monitoring with motion activated remote cameras has shown that foliage growth from year to year can quickly make nest trees vulnerable to possum access. Artificial nest hollows also need regular maintenance. Some of these hollows are nearing 20 years old and are in need of upgrading. The number of little corellas is increasing on Kangaroo Island, which is heightening nest hollow competition in areas already suffering from a scarcity of large hollow-bearing eucalypts. Kangaroo Island glossy black-cockatoos depend upon continued management, and any cessation of current on-ground work would almost certainly result in reduced nesting success rates and hence a return to a trend of overall population decline.

Key elements to success

One of the key contributing factors to success of this Recovery Program has been the consistency with which possum exclusion management has been undertaken. The program is currently in its 22nd year of operation, and throughout this time protection of new nests and maintenance of existing possum exclusion barriers has been carried out annually. This has required ongoing funding: a luxury afforded to few recovery programs in Australia. A major factor in this Recovery Program's ability to consistently attract support and funding has undoubtedly been the hard work, dedication and long-term commitment of program staff and Recovery Team members. Another has likely been the realisation of all involved in the program that this species is management dependent, and that without continuation of on-ground works upward population trends will be reversed.

The recovery program has been 'blessed' with several field and program staff who have been employed for long periods. Each has brought different skills and views, and the benefits accrued have been enormous. Several previous staff members have continued involvement with the Recovery Team, contributing a diverse range of expertise. Continuity of participation is also a trend noted in the Recovery Team membership. Today the Recovery Team is a varied mix of ecological experts, government agency natural resource management staff and local community members. Some members have been involved with the program since it began in 1995, and continue to bring a wealth of experience to the Recovery Team. The benefits of long-term staff and Recovery Team membership are probably a key driver of the success of this Recovery Program

Continuity of staff and Recovery Team members has led to a logical continuity of applied research undertaken on the program. Early on, members recognised that a fundamental part of making good management decisions was gaining a better understanding of key aspects of the biology of the cockatoos and their principal food source, the drooping she-oak. Previous research projects have yielded information on everything from productivity of Kangaroo Island she-oaks to the ideal nest box size based on successful natural nests. Field staff often played a key role in undertaking or facilitating research, often on top of their workload associated with the Recovery Program objectives. Guidance in research direction has also benefitted from the knowledge of experts on the Recovery Team. This history of commitment to research means good science now underpins the actions carried out by the Recovery Program, and has resulted in an ongoing refinement and improvement of management actions and monitoring. The Recovery Team still seeks to fill key knowledge gaps on the ecology of the species and refinement of management actions. Recent research has identified that climate change, under a high CO_2 emissions scenario, may pose a significant threat to the population (Harris *et al.* 2012). Increasing wildfires under a predicted climate change scenario were identified as a key future threat to the population through this research (Harris *et al.* 2012), hence further research is being undertaken to determine whether glossy black-cockatoos have a preference for a particular age class of she-oaks, with the aim of improving fire management strategies to reduce the chances of large-scale loss of key feeding areas.

A high level of cooperation between program staff and Recovery Team members has enabled findings from research to be translated rapidly into effective management actions. Regular meetings have provided members with an opportunity to discuss strategies freely and undertake long-term planning, while staff have been essential in testing and providing feedback on the feasibility of management actions. The result of this collaboration has been a sequence of three highly effective Recovery Plans and the current Strategic Plan for this species. These plans have focused on the long-term goals of increasing the black-cockatoo population to a level that should allow downlisting of their status from Endangered to Vulnerable, through ensuring that a viable breeding population persists on Kangaroo Island and that it returns to the Fleurieu Peninsula. Quality research and adequate planning has not only assisted in the effective implementation of management actions for this species, but has also likely been the reason the Recovery Program has had much success attracting funding over the years.

Another benefit of this continuity has been the long-term relationships forged between staff, Recovery Team members and the community. This continuity of personnel on the Recovery Program has no doubt enhanced volunteer commitments to and support for the program. An important role of staff has been to raise awareness of the Recovery Program and actively recruit volunteers to assist with implementation of recovery actions. Thanks to regular media articles, radio interviews and one-on-one conversations with local residents, there are few people on Kangaroo Island today who are not aware, and supportive, of the Recovery Program. The 1997 annual black-cockatoo population census involved 160 volunteers participating, and community involvement has continued to be high. A core group of 10–15 local Kangaroo Islanders generally contributes around half (~350 h) of all volunteer hours recorded per year. Events such as the annual community planting day, where drooping she-oak seedlings are planted to re-establish woodlands, attract around 40–50 volunteers per day, with high levels of participation from young families. These high levels of volunteer participation on the Recovery Program have probably been in part due to the good rapport that program and field staff have built with the local community over time, in addition to the iconic appeal of this species.

Through the extensive community engagement activities undertaken by staff on the Recovery Program, Kangaroo Island landholders are generally well informed as to how they can help preserve their glossy black-cockatoo population. Landholders lucky enough to have black-cockatoo feeding and/or nesting habitat often feel fortunate to have this remarkable bird on their property. Throughout the life of the Recovery Program, staff have spent many hours on properties while undertaking population censuses, nest monitoring and nest protection works. With around 65% of current known nest trees occurring on private property, gaining continued access to key habitat areas to implement management actions has been vital. Face-to-face contact and cooperation between program field staff and landholders has been pivotal to the program's success.

The future

Despite the success achieved so far, there is uncertainty about the future of this Recovery Program. The largest problem at present is one that faces almost all long-term natural resource management projects – that of securing continued funding. Government funding cycles have meant that the Recovery Program has gone through a series of peaks and troughs in the level of funding received, primarily through a succession of Commonwealth (federal) and South Australian Government Natural Resources Management programs. At present, delivery of the recovery plan strategies is being funded through a Biodiversity Fund grant that was obtained in 2012 and will expire in 2017. Funding from the Australian Government's 20 Million Trees Programme is also currently supporting revegetation of 170 ha of glossy black-cockatoo habitat on Kangaroo Island and the Fleurieu Peninsula from 2016 to 2018. No funding beyond 2017 has been secured for implementing the Recovery Plan strategies.

Given how dependent this species is on active conservation management, with predation of eggs and young by possums continuing to be managed with laborious site-based exclusion from black-cockatoo nest sites, it is highly likely that its population will again decline if funding ceases. Although, ideally, a landscape-scale reduction in possum populations would reduce the need for such intensive management, for many reasons, this challenge is likely to remain unresolved for the foreseeable future.

Conclusion

For the past 20 years, the last remaining population of glossy black-cockatoo in South Australia has benefitted from a continuous and active Recovery Program characterised by skilled long-term staff, a competent and engaged Recovery Team, high-quality applied research, effective strategic planning, and tremendous support from the local community. The cumulative effect has been to enable this long-lived species with a naturally low reproductive rate to more than double its population size in the last two decades. The Recovery

Team and Recovery Program staff are dedicated to maintaining the gains made so far, in the hope that this charismatic cockatoo will one day soon make a re-appearance on mainland South Australia. But to do this, ongoing financial support to implement the recovery plan and strategic plan are essential.

Acknowledgements

The authors would like to acknowledge all past and present members of the Recovery Team, the hundreds of volunteers who have contributed countless hours to the Recovery Program, and all former staff, many of whom are still contributing to the program in a voluntary capacity. In particular, the Recovery Program would like to mention the following people for their ongoing or long-term support: Lynn Pedler, Leo Joseph, Terry Dennis, John Pepper, Wally Meakins, Philippa Kneebone, Anne Morrison, Eleanor Sobey, Dai Morgan, Bryon Buick, Chris Denman and Bill Prime.

References

Andrews M (1995) 'Assessment of the availability of potential nesting and foraging habitats for the reintroduction of the South Australian glossy black-cockatoo, Southern Fleurieu Peninsula, South Australia.' BSc (Hons) thesis, University of South Australia, Adelaide.

Argent N (1997) Rural crises and local farm/non farm business linkages: a case study of Kangaroo Island. *South Australian Geographical Journal* **96**, 3–19.

Baird RF (1986) Historical records of the glossy black-cockatoo *Calyptorhynchus lathami* and red-tailed black cockatoo *C. magnificus* in south-eastern Australia. *South Australian Ornithologist* **30**, 38–45.

Berris KK, Barth M (2016) 'Glossy black-cockatoo Recovery Program, 2015 Annual Report'. Natural Resources Kangaroo Island, Department of Environment, Water and Natural Resources, Kingscote, South Australia.

Berris KK, Barth M, Kinloch M (2017) 'Glossy black-cockatoo Recovery Program, 2016 Annual Report'. Natural Resources Kangaroo Island, Department of Environment, Water and Natural Resources, Kingscote, South Australia.

Bickford S, Gell P (2005) Holocene vegetation change, Aboriginal wetland use and the impact of European settlement on the Fleurieu Peninsula, South Australia. *The Holocene* **15**, 200–215. doi:10.1191/0959683605hl800rp

Chapman TF, Paton DC (2002) 'Factors influencing the production of seeds by *Allocasuarina verticillata* and the foraging behaviour of glossy black-cockatoos on Kangaroo Island'. Wildlife Conservation Fund, Canberra.

Crowley G, Garnett S, Peder L (1997) 'Assessment of the role of captive breeding and translocation in the recovery of the South Australian subspecies of the glossy black-cockatoo *Calyptorhynchus lathami halmaturinus*'. Report 5. Birds Australia, Melbourne.

Crowley G, Garnett S, Carruthers S (1998) 'Mapping and spatial analysis of existing and potential glossy black-cockatoo habitat on Kangaroo Island'. Department for Environment and Heritage, Adelaide.

Downes ST, Handasyde KA, Elgar MA (1997) The use of corridors by mammals in fragmented Australian eucalypt forests. *Conservation Biology* **11**, 718–726. doi:10.1046/j.1523-1739.1997.96094.x

Garnett ST, Pedler LP, Crowley GM (1999) The breeding biology of the glossy black-cockatoo *Calyptorhynchus lathami* on Kangaroo Island, South Australia. *Emu* **99**, 262–279. doi:10.1071/MU99032

Green WQ, Coleman JD (1986) Movement of possums (*Trichosurus vulpecula*) between forest and pasture in Westland, New Zealand: implications for bovine tuberculosis transmission. *New Zealand Journal of Ecology* **9**, 58–69.

Harris JBC, Fordham DA, Mooney PA, Pedler LP, Araujo MB, Paton DC, *et al.* (2012) Managing long-term persistence of a rare cockatoo under climate change. *Journal of Applied Ecology* **49**, 785–794. doi:10.1111/j.1365-2664.2012.02163.x

Joseph L (1982) The glossy black-cockatoo on Kangaroo Island. *Emu* **82**, 46–49. doi:10.1071/MU9820046

Joseph L (1989) The glossy black-cockatoo in the south Mount Lofty Ranges. *South Australian Ornithologist* **30**, 202–204.

Pepper JW (1997) A survey of the South Australian glossy black-cockatoo (*Calyptorhynchus lathami halmaturinus*) and its habitat. *Wildlife Research* **24**, 209–223. doi:10.1071/WR94063

Schodde R, Mason IJ, Wood JT (1993) Geographical differentiation in the glossy black-cockatoo *Calyptorhynchus lathami* (Temminck) and its history. *Emu* **93**, 156–166. doi:10.1071/MU9930156

Stuart E (2006) Cultural landscape change in the Willunga Basin from European settlement to the present. In *Valleys of Stone: The Archaeology and History of Adelaide's Hills Face.* (Eds PA Smith, FD Pate and R Martin) pp. 113–130. Kopi Books, Adelaide.

10

Science, community and commitment underpin the road to recovery for the red-tailed black-cockatoo

Vicki-Jo Russell, Richard Hill, Tim Burnard, Bronwyn Perryman, Peter Copley, Kerry Gilkes, Martine Maron, David Baker-Gabb, Rachel Pritchard and Paul Koch

Summary

The problem

1. Most of the feeding and breeding habitat of the south-eastern red-tailed black-cockatoo has been cleared for agriculture.
2. The two eucalypt species on which it feeds do not return to full seed production capacity for a decade after intense fires while the third feed tree, buloke, takes a century to mature.
3. The quantity and quality of both feeding and nesting habitat continues to decline with continuous burning and loss of paddock trees.

Actions taken to manage the problem

1. Research to understand the key ecological requirements of the South-eastern red-tailed black-cockatoo.
2. Habitat protected and, in strategic areas, expanded.
3. A strong relationship built with the community to value the birds.
4. Fire management protocols developed to reduce the impacts of fires on habitat quality.

Markers of success

1. The major causes of decline have been identified and establishing policy change to achieve management of most is now the priority.
2. A recovery plan has been developed and is being implemented and revised regularly under guidance from a long-standing Recovery Team.
3. There is strong community commitment to the taxon with substantial ongoing volunteer support.
4. Recovery has not occurred yet – but much worse declines have been avoided.

Reasons for success
1. The effectiveness of red-tailed black-cockatoos as a flagship species.
2. The consistent application of science to change practice and policy.
3. The adoption of a landscape vision for recovery combined with a strong respectful relationship with the local community.
4. An effective, representative Recovery Team.

Introduction

The south-eastern subspecies of red-tailed black-cockatoo (*Calyptorhynchus banksii graptogyne*, Fig. 10.1) is confined to an area of ~18 000 km^2 in south-eastern South Australia and adjacent far south-western Victoria. A single population of ~1400 individuals persists, despite reduced feeding and nesting habitat. A decline in the subspecies was first recognised in 1982 (Joseph 1982) and then confirmed by the first recovery plan written in 1995–96 (Garnett and Crowley 1996). The taxon is currently listed as Endangered under the Commonwealth *Environment Protection and Biodiversity Conservation (EPBC) Act 1999*.

This cockatoo is an unusually selective feeder, relying almost exclusively on the seeds of just three tree species: brown stringybark (*Eucalyptus baxteri*), desert stringybark (*Eucalyptus arenacea*) and buloke (*Allocasuarina luehmannii*). Both stringybark species hold

Fig. 10.1. The proportion of male (left) to females or immature (right) red-tailed black-cockatoos has been used to indicate breeding success and recruitment (photo: Bob McPherson).

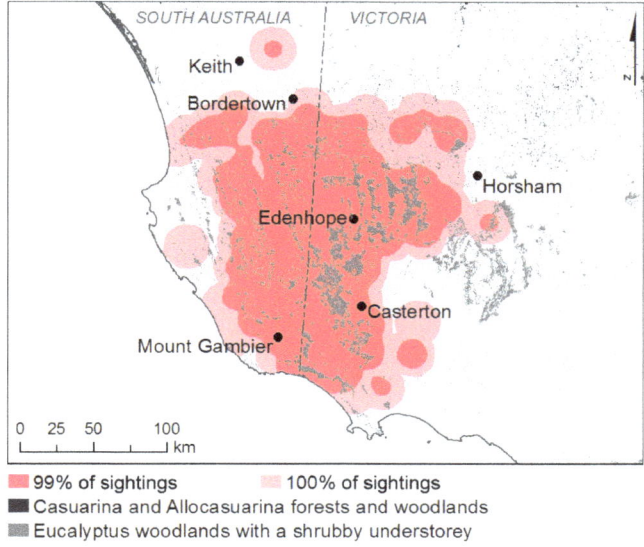

Fig. 10.2. Distribution of the red-tailed black-cockatoo in south-east South Australia and south-west Victoria and of their remaining feeding habitat.

seeds for 2 or more years and these are available year round. In contrast, buloke holds its seeds for a few months in summer and autumn. When feeding, red-tailed black-cockatoos target stringybark and buloke trees with large seed crops and high seed densities to maximise foraging success. Feeding habitat quality of stringybark varies considerably from year to year, with highly productive years interspersed with years of low seed production.

Since the 1840s, ~54% of stringybark and 97% of buloke woodlands within the red-tailed black-cockatoo's historic feeding range have been cleared for agriculture and forestry, with limited subsequent natural recruitment (Fig. 10.2). Furthermore, stringybarks need to be at least 10 years old, and bulokes at least 100 years old, to produce adequate seed crops for cockatoos to feed upon. Most feeding in bulokes has been observed in trees estimated to be greater than 200 years old. Paddock trees are particularly productive and valuable as feeding habitat, but continue to be lost (Maron 2005).

Fire has a major impact on food quantity. After crown scorch, stringybark trees take, on average, 10 years to produce seed crops similar in size to long-unburnt trees (Koch 2003). Increased bushfires and planned burns in Victoria have seen an almost doubling of crown scorch in the past 10 years.

Nesting habitat also may become limited in the future. The large hollows required as nest sites take centuries to form and the cockatoos prefer these in dead trees (81% of known nests) (Joseph *et al.* 1991). An estimated 4–7% of such dead trees in paddocks are being lost through collapse each year (Hill and Burnard 2001) suggesting that a serious shortfall in suitable hollow-bearing trees is imminent.

Suitability of breeding sites and habitat is also influenced by proximity to feeding habitat: most known nests are within 5 km of blocks of stringybark vegetation, 5 ha or more in size (Hill and Burnard 2001). Having nesting and feeding habitats adjacent to one another in the landscape is therefore important and provides a focus for conservation management.

Remaining feeding and nesting habitat within the birds' range is at risk through ongoing land clearance, often linked to intensive agricultural practices such as centre-pivot irrigation, and to tree senescence, wild and planned fires, invasive woody weeds, and the impacts of pathogens and pests on food trees. Climate change is also likely to increase wildfire intensity and stress on habitat.

Conservation management

Conservation attention has been focused on the south-eastern subspecies of red-tailed black-cockatoo since the early 1980s. Research between 1982 and 1995 clarified its range, population size and its declining trend, and described feeding, nesting and roosting habitats (Garnett and Crowley 1996). In 1993, a Victorian Flora and Fauna Guarantee Action Statement provided a list of suitable management actions, and a meeting at that time of representatives of Victorian and South Australian state agencies highlighted the plight of the bird and the need to take action (Venn and Fisher 1993).

The taxon's first National Recovery Plan (Garnett and Crowley 1996), implemented through a national Recovery Team, aimed to: understand the bird's key ecological requirements; locate and protect known nest sites; retain and expand feeding and nesting habitat; and work with the community to build knowledge and deliver recovery activities. Over time, the Recovery Team's inclusive approach has been just as important as its strategy and, for this reason, there has long been a strong focus on community representation in recommending action and community engagement for the plan's delivery. A revised recovery plan was signed-off under the EPBC Act in 2007 (Baker-Gabb 2007), and there is a new draft plan awaiting final community consultation and sign-off (July 2017).

Among many recovery activities, the Recovery Team has overseen research projects that have provided robust evidence on the decline and recruitment of buloke woodlands, factors affecting food availability (including the effects of fire on stringybark seed production) and the factors influencing breeding success. The team has also identified priority areas for revegetation with stringybark and buloke and with gum eucalypts.

Building on this knowledge, the program has worked with the Victorian and South Australian Governments and local governments and other authorities to improve awareness, legislation and planning tools to better protect nesting and foraging habitat. These initiatives have included statutory protection in both states for standing dead trees with large hollows and specific habitat protection controls in two planning schemes. Also, fire management guidelines have been developed to reduce the risk of canopy-scorch in stringybark feeding habitat. The guidelines aim to maintain at least 85% of stringybark habitat unburnt across the bird's range over a given 10-year period. This objective has been built in to fire management strategies (e.g. DEWNR 2014). Before controlled burning operations, assessments are used to identify sites carrying the greatest seed loads and burns are deferred at these sites, if possible.

Red-tailed black-cockatoos can nest in any suitable habitat within their 18 000 km^2 range and nests are very hard to find. Since 2011, a nest reporting incentive program has provided a modest payment to community members for information leading to the location of new nests so that they can be protected from predators and competitors. About 10% of known nests have been protected by landholders reticent to report nests because they want to avoid disturbance at the site, but who have then implemented suitable protection measures themselves.

Since 1996, BirdLife Australia and the Recovery Team have coordinated a range-wide annual cockatoo count by volunteers, most of whom are local residents. The annual count raises awareness of the specialised needs of the cockatoo, offers volunteers a chance to get involved and helps the team to locate flocks for follow-up sex-ratio counts. In 2016, the program celebrated its 20th annual count, with some local volunteers having participated in every single count.

A significant research and synthesis exercise was used to identify priority locations and targets for future habitat regeneration (Maron *et al.* 2008). The model signalled a significant shift in the program's approach from protecting sites and general awareness of threats to working towards landscape change at large scales.

Building on this, a market-based incentive program was developed that offered habitat stewardship payments on private land through an open tender process. The program paid incentives to landholders to implement 5-year management plans to improve the quality of red-tailed black-cockatoo habitat in the Wimmera Catchment Management Authority region. Among numerous other landscape restoration projects are the innovative 'Cockies helping Cockies helping Cockies' and Kowree Tree Farm projects. These projects are driven by local farmers (known colloquially as 'cockies') supporting other farmers to achieve conservation outcomes on their land or in their local area with a focus on cockatoo feeding habitat. Local schools have joined forces with these initiatives through the 'Schools helping Cockies helping Cockies' project and now provide additional stringybark seedlings to participating landholders each year. Such social programs have been greatly aided by targeted social surveys to understand local landholder perspectives on the importance and requirements of the cockatoos and barriers to protecting and re-establishing critical habitat on their own land.

Meanwhile, awareness among the broader public was greatly increased when the cockatoo became official mascot of the Melbourne 2006 Commonwealth Games. Subsequent sale of number plates featuring the mascot (known as 'Karak') contributed to red-tailed black-cockatoo conservation. A newsletter and website regularly communicate the latest news to an audience of 1500 including best practice advice to landholders and fire managers.

Over the last decade, the Recovery Team has met four times a year, twice by phone and twice in face-to-face meetings. The team rotates venues across the region for each of its face-to-face meetings. Since 1996, the team has always worked to a recovery plan even in the absence of Commonwealth (federal) and/or state endorsements. Two detailed recovery plan reviews (in 2002 and 2012, Pritchard and Burnard 2012) formed the basis for ongoing refinement for subsequent plans. Team partners report every 2 years on their own contributions to the plan and a strategic workshop is held every second year to test the relevance and priority of the plan's objectives and strategies, ensuring it is still fit for purpose.

Key features in success

The recovery effort operates across the full range of the taxon crossing the boundary of two states, not just as a series of sites within an area. We argue that it has succeeded in two key ways: first, by influencing legislation, regional planning and landholder decision making to the extent that the cockatoo's requirements are now embedded, at least in part, in standard practice; second, by moving from a science-led program with strong community support to a community-led program supported by strong science.

Science

One of the strong planks of the program has been its consistent application of research to practice and policy. For example, among other insights, a PhD undertaken by Paul Koch in 2003 demonstrated the long period required for scorched stringybark to return to their previous level of seed production (Koch 2003). This fundamentally changed the way the team and, as a consequence, regional land managers viewed paddock trees and the impact of planned burning on the food resource, leading to a transformation of burning practices.

One of the significant turning points for the program occurred at a meeting in 2003 when the team asked itself what the landscape in the range would need to look like if the cockatoos were going to have a long-term future. This shift to a landscape vision has proved very useful and prompted the development and application of a future habitat regeneration model (Maron *et al.* 2008) that, in turn, inspired landscape restoration projects such as those done by the Kowree Tree Farm Group in Victoria and 'Cockies helping Cockies helping Cockies' project in South Australia, which have protected and/or re-established hundreds of hectares of feeding habitat. The model was not just a guide to strategic investment but articulated a vision that the team could communicate and aspire to. This vision allowed the Recovery Team and associated projects to inspire landholders with a plan based on what those landholders could do, rather than what they couldn't.

People

The Recovery Team has been operating for 20 years and, fortunately, the membership has remained consistent and passionate over much of that time. Half of the current membership has been associated with the team for more than a decade and one quarter for longer. Some stalwarts have been involved continuously in recovery efforts, even from before the team's formation. A consequence of this longevity is that the team's corporate memory is unbroken and the trust between members around the table and between members and the community is outstanding. This trust has been helped by having long-term staff associated with the program: of the eight people employed over that time, six are still employed or associated with the team in some capacity. Leadership has also been strong and relatively constant, with just four chairs: David Baker-Gabb (1996–2010), co-chairs Dr Martine Maron and Vicki-Jo Russell (2010–2013), Tim Burnard (2013–2015) and Vicki-Jo Russell (2015–present). All chairs have been independent of government. The Recovery Team has always had strong relationships with relevant Victorian and South Australian agencies, and the independent chair has provided an avenue to work across boundaries and hierarchies in a way that would have been difficult for many government employees to manage.

Landholder surveys in 2003 and 2009 were fundamental to shifting the Recovery Team's relationship with the local community. The surveys prompted development of a communication strategy that tailored relevant requests, messages and channels to key partners and stakeholders. This led to a blossoming of partnerships and many of the successful communication products are available currently on the website. The Kowree Tree Farm Group and 'Cockies helping Cockies helping Cockies' project continue to redefine our partnership with landholders within the cockatoo's range and today many of the most influential team members are local farmers.

Governance

The recovery program's de-centralised governance might also be considered one of its strengths. The program operates under the auspices of BirdLife Australia who contract the project officer and chair and administer funds. The chair supports the project officer and is responsible for managing the Recovery Team. BirdLife is an active member of the Recovery Team and in its national advocacy role can make representations on issues facing the cockatoos, such as the management of fire and vegetation clearance, independently of the team. The independent chair is not bound by any one group's participation or agenda and can focus on ensuring the team works effectively. This can be challenging, with potential conflicts between what is best for the cockatoos or for stakeholder interests, including those of agencies from two states. In particular fire risk has been hotly debated. Two specialist sub-committees meet out of session to discuss dilemmas or proposals: the Science Working Group and the Finance and Management (Governance) Group. Many recovery team partners operate complementary projects independent of BirdLife, but which contribute to delivery of the collective strategy defined in the recovery plan. This kind of open arrangement allows partners to play to their strengths and spreads the risk should funding be reduced.

Money

Funding for a project officer has been available since 1996. In the early years, modest funds were split between a part-time science officer and part-time community engagement officer who also supported the Recovery Team. Since 2011, a single part-time project officer has supported delivery of an annual work plan guided by the Recovery Plan, and that officer also seeks and manages grants, manages partnership requests, coordinates communication or volunteer activities and responds to sightings, nest reports or other information or requests. Base level funding for this professional support has been critical to the program's longevity and success and helped leverage partner contributions many times over in value. Other funding sources have included government and research grants, agency and university in-kind contributions, private donations, philanthropic organisations, environmental NGO contributions, local industry sponsors (particularly local wineries), landholder in-kind contributions, fundraising events, minor retail lines and a range of pro-bono and volunteer contributions. Voluntary contributions have also been vital for the program's annual count, nest incentive and protection program, and many other public engagement activities.

Partners

The Recovery Program has had support from a diverse range of partner organisations, the majority of them over many years. These partners have been key to success of the program, both through the extent of their commitment and their durability. Recovery of such a long-lived species in such an altered environment is going to take many decades and will only be achieved through partnerships with organisations that extend beyond the commitments of individuals.

The future

In a decade, the program partners will have protected and replanted thousands of hectares of feed trees for the south-eastern red-tailed black-cockatoo within priority areas in the range. However, an unprecedented drop in the ratio of females and young birds to males in the annual flock counts suggests that breeding success rates have been decreasing in recent years, probably due to poor seed production observed over the last few years. Based on this evidence, our strategies may require another review.

If climate or pests reduce the capacity of existing or new trees to sustain viable populations of black-cockatoos, the team may need to explore measures to increase the amount of food that trees can produce, especially in dry years. In the meantime, the recovery priorities are for: protection and restoration of buloke and stringybark habitats; protection of nest sites and nest trees (and their replacement as attrition occurs); maintaining existing monitoring; and continuing to strengthen the team's relationships. However, the task of maintaining these ongoing recovery ingredients may be challenged further given that, in a more constrained funding environment, no support for the project officer has been secured beyond June 2018. Maintaining the team and a plan will remain a high priority. The team is fortunate that most of its current membership are not due to retire in the next decade and that it still receives regular requests by program partners and stakeholders to join, so succession planning is not an immediate concern. Losing the chair and/or project officer role are both significant potential threats to the team's effectiveness.

Conclusion

The south-eastern red-tailed black-cockatoo is a genuine flagship species, charismatic enough to engage the public's imagination or empathy. That empathy catalyses community support, which is required for the protection of large hollow-bearing trees and foraging habitat, and implementation of fire management strategies, with such management responses likely to assist a large number of other native species. The species is not yet secured, but the recovery effort has provided a remarkable story of a community, bonded by a shared passion for the bird, working together over a long period to change both the physical and policy landscape. The strategies adopted mirror what are commonly phases of long-term recovery programs, in particular, to: confirm a species status and threats (understand); work to halt those threats (protect); expand habitat and/or spread the risks across the range (restore); work with stakeholders and partners to better align management practices and policy (embed change); and then monitor and refine as we learn more (continuously improve). It could be argued that these last few steps are critical to make the shift from saving to recovering a species.

Given the longevity of this species and its slow reproductive rate, and the great extent of historic loss and fragmentation of its habitat, the recovery effort must work at decadal periods and landscape scales. The program's long-term objective (50 years) is to have sufficient buloke, stringybark and eucalypt woodland habitats to sustain a stable, viable population of the cockatoos. However, to achieve this in a drying and variable climate may be the program's greatest challenge yet.

Acknowledgements

Project partners have included BirdLife Australia; Conservation Volunteers Australia; Greening Australia; Victorian Department of Environment, Water, Land and Planning; Parks Victoria; South Australian Department of Environment, Water and Natural Resources; Forestry SA; Glenelg-Hopkins Catchment Management Authority; Wimmera Catchment Management Authority; Natural Resources South East; Kowree Tree Farm Group; Nature Glenelg Trust; The University of Queensland; The University of Adelaide; Monash University; University of Ballarat; Trust for Nature; Zoos SA; WWF-Australia; Timberlands Pacific; Conservation SA; West Wimmera, Glenelg and Horsham Shire Councils (Victoria); and Grant, Tattyara, Naracoorte-Lucindale and Wattle Range District Councils (South Australia). The authors also wish to acknowledge: all Recovery Team members, past and present; all research partners and local experts who have contributed knowledge to the program, especially those who forged the way for the red-tail's plight to be recognised – Leo Joseph, Bill Emison, Wayne Caldow and David Venn; all funding bodies and partners who provided financial or in-kind resources; all who reported a red-tail or nest sighting and/or joined in the annual count; every landholder who protected or planted red-tail habitat on their property; the artists and story-tellers who have shared skills and passion for red-tails; the local makers of fine wine and caterers who have regularly supported the program; local horticulturalist Ralph Sheels and the schools and students who so lovingly grow trees for landholders to plant.

There are too many individuals to name but important contributions have been made by Wendy Beumer, Tania Rajic, Dave Williams, David Paton, Evan Roberts, Jim McGuire, Julie Kirkwood, Mark Bachmann, Andrew and Ros Bradey, Oisin Sweeney, Bryan Hayward, Stephen Garnett, Gabriel Crowley, Briony Jarmyn, Mick Fendley, Troy Horn, Bill Wallace and Dick Cooper. Thanks to Jennifer Howe and David Edey for assistance with the writing process.

References

Baker-Gabb D (2007) 'National Recovery Plan for the south-eastern red-tailed black-cockatoo *Calyptorhynchus banksii graptogyne*'. Australian Government, Department of the Environment and Water Resources, Canberra, <http://www.environment.gov.au/resource/national-recovery-plan-south-eastern-red-tailed-black-cockatoo-calyptorhynchus-banksii>.

DEWNR (2014)' Ecological fire management strategy, south-eastern red-tailed black-cockatoo *(Calyptorhynchus banksii graptogyne)*'. Department of the Environment, Water and Natural Resources, Adelaide, <http://www.environment.sa.gov.au/files/903a1b6a-0be8-4602-9e4f-9fff0100196e/fm-gen-redtailedblackcockatoofirestrategy.pdf>.

Garnett S, Crowley G (1996) 'Red-tailed Black-Cockatoo Recovery Plan *Calyptorhynchus banksii graptogyne*'. Environment Australia, Canberra and Birds Australia, Melbourne, <http://www.redtail.com.au/uploads/RECPLAN%201996.pdf>.

Hill R, Burnard T (2001) 'A Draft Habitat Management Plan for the south-eastern red-tailed black-cockatoo'. Red-tailed Black-Cockatoo Recovery Team, Melbourne, <http://www.redtail.com.au/uploads/Management%20Plan%20Final%20Version%202001.pdf>.

Joseph L (1982) The red-tailed black-cockatoo in south-eastern Australia. *Emu* **82**, 42–45. doi:10.1071/MU9820042

Joseph L, Emison WB, Bren WM (1991) Critical assessment of the conservation status of the Red-tailed Black-Cockatoo in south-eastern Australia with special reference to nesting requirements. *Emu* **91**, 46–50. doi:10.1071/MU9910046

Koch P (2003) 'Factors influencing food availability for the endangered south-eastern red-tailed black-cockatoo *Calyptorhynchus banksii graptogyne* in remnant stringybark woodland, and implications for management.' PhD thesis, The University of Adelaide, Australia.

Maron M (2005) Agricultural change and paddock tree loss: implications for an endangered subspecies of Red-tailed Black-Cockatoo. *Ecological Management & Restoration* **6**, 206–211. doi:10.1111/j.1442-8903.2005.00238.x

Maron M, Koch P, Freeman J, Schultz S, Dunn P, Apan A (2008) Modelling and planning to increase future habitat of the red-tailed black-cockatoo. Wimmera Catchment Management Authority, Horsham, Victoria, <http://www.redtail.com.au/uploads/file/RTBC%20Habitat%20Modelling%20Report22_12_08.pdf>.

Pritchard R, Burnard T (2012) South-eastern Red-tailed Black-Cockatoo Recovery Plan Review. Department of Sustainability and Environment, Carlton, Victoria, <http://www.redtail.com.au/uploads/RtBC_RP_Review_20120723.pdf>.

Venn DR, Fisher J (1993) 'Red-tailed Black-Cockatoo *Calyptorhynchus banksii graptogyne*'. Flora and Fauna Guarantee Action Statement No. 37. Department of Conservation and Natural Resources, Melbourne, <https://www.environment.vic.gov.au/__data/assets/pdf_file/0024/32883/Red-tailed_Black-Cockatoo_Calyptorhynchus_banksii-graptogyne.pdf>.

11

Collaborative commitment to a shared vision: recovery efforts for noisy scrub-birds and western ground parrots

Allan Burbidge, Sarah Comer and Alan Danks

Summary

The problem
1. These two bird taxa are now confined to tiny areas of highly fire-prone habitat.
2. Even with successful fire management, introduced predators pose an ongoing threat.

Actions taken to manage the problem
1. Research has been conducted over many decades to refine understanding of habitat requirements.
2. Noisy scrub-birds have been translocated successfully to additional sites, including an offshore, predator-free island.
3. Intensive programs to reduce fire risk and control foxes and feral cats are being applied to habitat of both species.
4. The community has become closely involved in all aspects of the Recovery Program.

Markers of success
1. The risk of extinction of both species has been reduced, with the population of the scrub-bird substantially larger than when rediscovered.
2. A long-standing recovery team has effectively managed all aspects of the recovery process.
3. There is a strong commitment from the community to prevent extinction.

Reasons for success
1. A solid understanding of the ecology of each species.
2. Development of, and commitment to, a shared vision for south coast threatened birds, and a recovery team working collaboratively within that vision.

3. The capacity of individuals involved to think in the long term and act collaboratively to engage others and ensure optimal outcomes.
4. Significant support from Commonwealth (federal) and state agencies over a period of about five decades.
5. Comparable management requirements and conservation foci for several broadly co-occurring species allows for some efficiency in management and higher priorities set for this area.

Introduction

Following discovery of the noisy scrub-bird (*Atrichornis clamosus*) by John Gilbert in south-western Australia in 1842, clearing and altered fire regimes as a result of European colonisation caused severe decline of the species to such an extent that it was thought extinct (Abbott 2008; Danks *et al.* 2011). Because of the historical decline, rediscovery of the species at Two Peoples Bay on the south coast of Western Australia in December 1961 after a 72-year absence from the official record was a thrilling event in conservation history and marked the beginning of a classic threatened species recovery program. Initially focusing on the scrub-bird, the recovery effort came to encompass other threatened birds, including the western ground parrot (*Pezoporus flaviventris*).

Practical recovery efforts for threatened birds on the south coast of Western Australia began in the early 1960s. Formal recovery teams were formed in the 1990s and subsequently, the Western Australian South Coast Threatened Birds Recovery Team was formed in 1996 by amalgamating former teams established for the noisy scrub-bird, western bristlebird (*Dasyornis longirostris*) and western ground parrot. This conservation effort also encompasses conservation management actions for the western whipbird (*Psophodes nigrogularis nigrogularis* and *P. n. oberon*) and the western subspecies of the rufous bristlebird (*Dasyornis broadbenti litoralis*) (DPaW 2014). The emphasis of recovery efforts has largely been around the noisy scrub-bird and western ground parrot, because of the enigmatic and unusual nature of these two species and their sometimes perilous conservation status. As noted by Libby Robin (2001), 'The Noisy Scrub-bird is unique in having gone from 'extinct' to 'critically endangered' to 'vulnerable' in *that* order over just a few decades.' More recently, it has been reclassified as Endangered, due to multiple large fires causing a decline in population numbers. The western ground parrot is now classified as Critically Endangered following an ongoing and dramatic reduction in range and abundance over the last two decades.

Conservation management

Noisy scrub-bird conservation commenced following the announcement of its rediscovery in 1961. At the time, the area around Two Peoples Bay was earmarked for development as a town site, which would almost certainly have spelled the end for the scrub-birds. People soon rallied to the cause, and many visited Two Peoples Bay helping to conduct searches to locate and define the population, which turned out to be very small. Action was needed to protect and manage the bird's habitat and the best way to do this was to create a fauna reserve, but there was considerable resistance to this proposal. Conservationists kept the scrub-bird's plight in the public eye through newspaper articles and public lectures here and overseas, and eventually the government was convinced not to develop the area and in 1967 established Two Peoples Bay Fauna Reserve/Wildlife Sanctuary.

The 1971 management plan for the reserve identified fire control and visitor management as the critical management issues. The then WA Fisheries and Fauna Department initially provided part-time wardens and, in the early 1970s, a full-time on-site reserve officer was appointed.

Virtually nothing was known about the noisy scrub-bird's biology or habitat requirements. Early research efforts focused on identifying population size, geographic extent and determining breeding biology. A captive breeding program was trialled from 1975 to 1981 but was ultimately deemed unsuccessful. Due to concern about the limited geographic range of the species, and the risk of unplanned fire, trials in 1983 tested the feasibility of translocation and 16 birds were moved to nearby Mt Manypeaks. The success of this trial showed that translocation was a viable management tool to expand the range and reduce the risk of the threat of wildfire.

Habitat management and translocations formed the backbone of the 1986 recovery plan (Burbidge *et al.* 1986) and this approach has continued since then. In addition to the Mt Manypeaks translocations (1983 and 1985), scrub-birds have been translocated successfully to Bald Island, East Waychinicup and the Mermaid area near Cheynes Beach, increasing the geographic range and size of the population (Fig. 11.1). Despite the failures of attempts to reintroduce the species at four sites outside the greater Albany area, translocations have been a key strategy to buffering the scrub-bird against the impact of fire (Comer *et al.* 2010), and at the last full count, nearly 70% of the 2011 population was in sites established through translocations.

As a result of the capture and release activities, much was learned about scrub-bird ecology and management (e.g. Danks 1997; Comer *et al.* 2010). The 1996 recovery plan

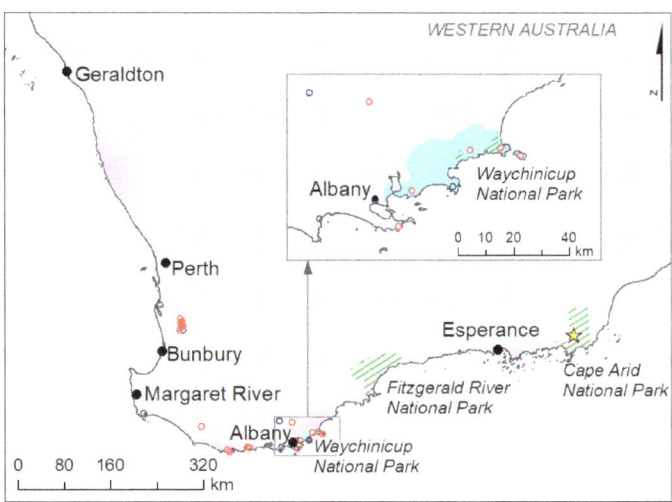

Fig. 11.1. Historical and current distributions of the Noisy Scrub-bird and the Western Ground Parrot.

(Danks *et al.* 1996) detailed recovery actions related to monitoring, preparation and implementation of area management plans, awareness raising, and research focused on increasing our understanding of scrub-bird biology and ecology (e.g. in relation to genetic variation in the source and translocated populations). Research also has been designed to answer specific challenges, such as in the scrub-bird translocation program where improving understanding of factors influencing the success of this important conservation tool has been crucial to maximising effectiveness and optimising use of limited resources (Danks *et al.* 2011; Comer *et al.* 2015).

As with the scrub-bird, attempts to secure some of the most important western ground parrot habitat have been challenging. In the mid-1980s, the Western Australian Government was intent on releasing land for agriculture along the then northern margin of the Fitzgerald River National Park. It was only through the efforts of local people working with non-government organisations and the Department of Fisheries and Wildlife, that the importance of the area to western ground parrots was recognised. Eventually, the decision to release the land was rescinded.

Both species occur in geographically restricted areas and in a region where climatic conditions often create a high risk of fire, and in habitats that are greatly changed by fire; recently burnt vegetation is generally unsuitable for these species (Fig. 11.2). Fire has always been a major management concern and responses have included actions to stop or slow the progress of bushfires (traditional breaks, slashed breaks and wind-driven burns to create buffers of lower fuel load). The success of scrub-bird translocations inevitably led to suggestions that translocation also would be appropriate for western ground parrots, and so it was planned in 2004 to re-introduce some birds from Fitzgerald River National Park to places within its historical range near Walpole. However, an unexpected and large reduction in the size of the proposed source population meant that this proposal has been shelved until the cause of the decline is determined and threats mitigated. Flexibility is a

Fig. 11.2. Recovery Team inspecting fire damage at Mt Manypeaks Nature Reserve in 2005 (photo: Sarah Comer).

valuable attribute within recovery teams. Another key activity concerns the raising of awareness among managers, neighbours and other interested persons around the impact of fires on these and other fire-sensitive species (e.g. Comer et al. 2015, 2016).

However, evaluation of monitoring data revealed that recent declines in parrot distribution and abundance have not been due simply to inappropriate burning regimes, as previously thought, leading to the hypothesis that feral cat predation is a major factor limiting parrot numbers. Although fox baiting under the Department's Western Shield program commenced in Fitzgerald River National Park in 1996, this program did not control feral cats, and there was no robust framework in place to test its effectiveness. This led to the Recovery Team initiating, in 2009, an ambitious project to test the predation hypothesis in an adaptive management framework (Comer et al. 2013; Burbidge et al. 2016).

The western ground parrot population was in a perilous situation by this time, and the Recovery Team was not convinced that impacts of predation could be adequately understood and mitigated in time to save the species from extinction. A small captive population was therefore established in 2009, first to learn how to maintain the species in captivity but with the ultimate aim of being able to breed for release once predation pressures were successfully controlled. To date, the captive program has not achieved that long-term aim.

Following devastating fires in late 2015 that burnt 90% of the known western ground parrot habitat, the Recovery Team initiated a workshop in March 2016, at which 39 people from 19 organisations gathered to help create a future for this rare Australian bird and for other species that share its habitat. Discussion and development of recommendations focused around protection of extant populations, establishing additional populations, securing long-term support for recovery efforts and optimising the value of the captive program (Burbidge et al. 2016).

Biology

Apart from their distinctive calls, noisy scrub-birds and western ground parrots are highly cryptic in nature, so monitoring of both species relies on auditory methods. The territorial songs of male noisy scrub-birds (Fig. 11.3) are loud and far-reaching and regular counts of singing males have provided an index to population trends since the 1960s, allowing the progress of recovery to be tracked over time (Fig. 11.4). Such counts also allow monitoring of birds at translocation sites. The ability to monitor the population is very important in recovery management. However, breeding females cannot be counted in this way and the relationship between the number of singing males and the total population remains unknown.

Ground parrots (Fig. 11.5) are less strongly territorial, making them harder to count. Particularly challenging is the fact that reliable calling is confined to the hour before sunrise and the hour after sunset, creating logistical challenges in getting observers into, or out of, listening points in the dark. Since 2010, our use of autonomous field recording units has assisted greatly in monitoring, because they can be deployed for long periods, sometimes in sites that are remote from vehicular access, without the need for regular visits. Nevertheless, going through the many hours of audio files in the laboratory is time-consuming, and the software to automate this process requires considerable development to allow reliable separation of western ground parrot calls from those of tawny-crowned honeyeaters (*Glyciphila melanops*).

Infrastructure

The most important infrastructure has been the protected lands on which the birds persist – for both species, it has been essential as a constraint on agricultural development. From the early years of the translocation program, a research station was established at Two

Fig. 11.3. Although noisy scrub-birds are rarely seen, males have an exceptionally loud call that can be used to monitor abundance (photo: Alan Danks).

Peoples Bay to provide the infrastructure and a base for noisy scrub-bird survey and translocation work, as well as accommodation for staff and volunteers participating in these activities. In contrast, the relative remoteness of Cape Arid National Park, with limited infrastructure, means extra financial and time costs in getting there, and research and monitoring work needs to be camping-based. Technical equipment available for conservation management has increased greatly, particularly recently, with miniaturisation of radio-transmitters and automation of call monitoring.

People

None of the successes that have been achieved would have been possible without the contribution of a wide range of people from diverse backgrounds. For example, Harley Webster, whose audacious announcement of the rediscovery of the noisy scrub-bird in *The West Australian* newspaper on Christmas Day 1961, ensured that the species received the attention it deserved and helped reverse the decision to establish the area as a town site. Don Merton, a New Zealander with a strong reputation for saving highly threatened bird

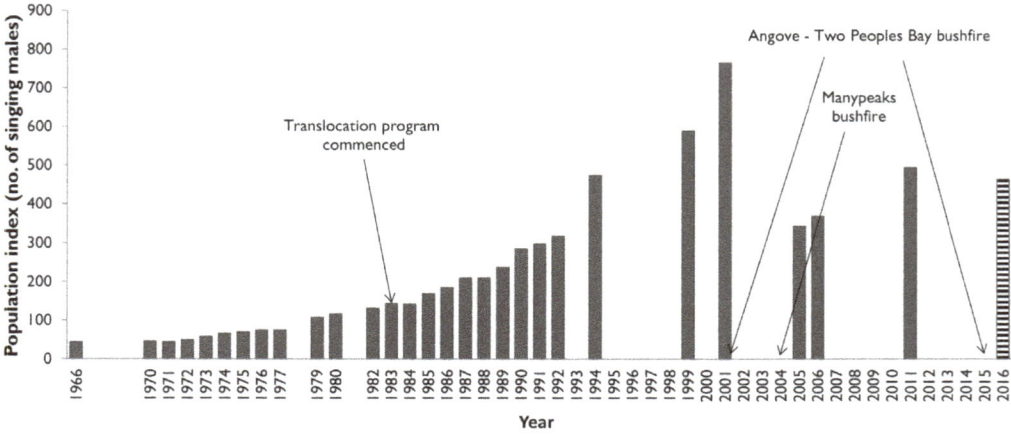

Fig. 11.4. Trends in singing male noisy scrub-birds over time, showing major discontinuities as a result of fire. All data are derived from a full census except final hatched column which is an estimate.

Fig. 11.5. Western ground parrots are diurnal, consuming thousands of tiny seeds each day (photo: Alan Danks).

populations, was invited to assist with the pilot translocation project in 1983. Don promoted the need to run a long-term program of translocations. This provided the basis for successive recovery plans, resulting in very effective recovery efforts.

Early on, key people in the state agency recognised two important prerequisites for recovery efforts: (1) good science to underpin informed management action; and (2) collaborative action to achieve optimal outcomes. This philosophy has continued to underpin recovery efforts, and ensured communication of new knowledge and understanding to operational staff responsible for the effective implementation of on-ground recovery actions and involvement of volunteers and community through direct participation or popular media.

Similarly, contributions from particular individuals have been crucial to successes in western ground parrot recovery. Key individuals who had been involved in earlier western ground parrot conservation efforts became members of the Recovery Team, and a similar situation occurred within the Friends of the Western Ground Parrot, an organisation that has grown to have considerable success in raising awareness about the plight of the species, and attracting significant funding for the broader project. Independent members of the Recovery Team also continue to ensure that the hard questions are asked and properly debated in the team. Although there have been changes in the composition of the South Coast Threatened Birds Recovery Team, many members have committed over long timeframes. This continuity has assisted in maintaining a longer term vision and a collaborative commitment to practical management and recovery efforts.

Balance is the key to every team and diversity of backgrounds and experience within the membership of the Recovery Team has been a major factor in success. The team functions through different people bringing different questions and different perspectives, followed by robust, but respectful debate, ensuring that the team has been able to progress with an agreed approach to challenges, confident that we have made sound decisions and recommendations, even in the face of sometimes considerable uncertainty and risk. Sometimes, there are high stakes, and potentially high emotions around some questions (e.g. 'Should we take more ground parrots from the wild for the captive program, given the small numbers in the wild?'). The Recovery Team is strongly participatory: it tackles emotionally and politically risky issues and promotes active participation in decision making, rather than simply operating as a mechanism to inform interested parties.

Successive project officers for both species have spent countless hours in the field. Survey and monitoring has relied heavily on volunteers to provide the necessary observer effort, because the only practical way to survey western ground parrots is by listening for calls. Such data have contributed to species' distribution modelling (Gibson *et al.* 2007), as well as to evaluation of the impacts of fires and predator control. With the commencement of the ground parrot captive program, other skills were needed – experts in husbandry, social interactions and veterinary assessments. Because the ground parrot recovery focus broadened to include attempting feral cat control, multi-disciplinary skills were required in the field team to integrate feral cat control with western ground parrot monitoring.

Social support and governance

One of the key factors in recovery has been the active involvement of private individuals, non-government organisations and government-based researchers and managers in debating the content of recovery plans and assisting to implement them. None of the plans we have used have been written by a single individual. One of the reasons is that, early on, the precedent was set that members of the Recovery Team should be contributors to the team effort, rather than simply representatives of a stakeholder organisation.

Before 2003, planning dealt with either area management (e.g. Two Peoples Bay Management Plan 1994) or the species of concern (e.g. Burbidge *et al.* 1986; Cale and Burbidge 1993), and were overseen by a scrub-bird recovery team (formed in 1992), bristlebird team (formed in 1994) or ground parrot team (1996). There was high overlap in composition of these separate recovery teams, and the birds' geographic distributions, and therefore combining the individual teams to form the South Coast Threatened Birds Recovery Team was a logical step. Additionally, the threats and relevant management actions and responsibilities for each were similar. In the same year, increasing concern about ground parrots led to the writing of an interim recovery plan for the species (Burbidge *et al.* 1997). In 2003, the Recovery Team initiated an ambitious project to write a multi-species recovery plan that also considered multiple management areas in the same document. This was completed in 2006, and a more traditional but still multiple species plan was endorsed by the Australian Government in 2014 (DPaW 2014).

In the early days, with few people involved, there was little need for a recovery team, given that senior staff in the Department were committed to action in relation to both research and management. However, there are now many more people involved in the delivery of both research and management actions, and managers are faced with an increasing array of pressures, often within short-term political contexts. In such an environment, the Recovery Team provides a longer term perspective based on input from a range of specialists from various backgrounds, each of whom has strong connections with different sectors with an interest in environmental management. Together, the Team provides informed decisions and recommendations that otherwise would be difficult to achieve.

Money

Over more than 50 years, funding for noisy scrub-bird and western ground parrot recovery has come from a variety of sources. CSIRO supported early research on the scrub-bird. The Western Australian Government has provided funding for research staff and ongoing reserve management actions such as fire management, introduced predator control and the early scrub-bird translocation work. Funding received directly from the Australian Government and through South Coast Natural Resource Management Inc. has supported various aspects of recovery efforts. Special allocations of Parks and Wildlife funds, supported by funds raised by the Friends of the Western Ground Parrot, allowed for the establishment of a ground parrot captive facility and commencement of feral cat control efforts. From 2012, state natural resource management and Commonwealth (Biodiversity Fund) funding supported the western ground parrot recovery project and the trials of feral cat control across over 500 000 ha of habitat for south coast birds between Two Peoples Bay and Cape Arid–Nuytsland Nature Reserve (Burbidge et al. 2016; Comer et al. 2016). The Friends of the Western Ground Parrot have lobbied for funds since ~2009 and, in addition to supporting the captive management program, have provided financial support for field surveys and purchases of field equipment. There has also been massive time input from volunteers, mostly through survey and monitoring work.

The future

Although much remains unknown about the biology and ecology of both scrub-birds and ground parrots, the major threats have been identified, and relevant actions for recovery are being implemented, but it is unclear as to how this might be sustained into the future. A lack of ongoing and reliable funding is a major challenge to maintaining momentum for critical aspects of management such as integrated predator control.

In addition, as the climate of south-western Australia continues to get warmer and drier, the incidence and extent of uncontrolled bushfires appears to be increasing, resulting in direct and indirect threats to habitats and food resources. Combating these changes will depend on developing innovative solutions and continued engagement of managers and the general public in the recovery of these two unusual and enigmatic species.

Conclusion

Substantial successes have been achieved in scrub-bird and ground parrot recovery, but the key point is that it has been about people: participation in the development of a clear shared vision, commitment to that vision, and working collaboratively within it. Across the entire recovery period, implementation of outcome-focused research, practical management action followed by monitoring and evaluation, and subsequent communication of findings in both popular and scientific literature, have been essential. The contribution of volunteer naturalists and postgraduate students has been enormous, and careful fostering of collaborative interaction between such people and professional managers and scientists has been critically important.

The Recovery Team continues to build on the work of those who first committed to saving scrub-birds and ground parrots. A culture of good two-way communication, trust and respect has allowed robust debate and frank discussions about future directions and collective confidence in implementing agreed decisions, even in the face of great uncertainty and the potentially confronting outcomes of failure. It is almost certain that these two birds would have become extinct were it not for coordinated and dedicated research and management effort, and community engagement, extending now over 50 years. However, despite this progress, there are still many challenges to overcome and the fate of these two species may never be entirely secure.

Acknowledgements

Many people have contributed to the success of the recovery actions for noisy scrub-birds and western ground parrots. They include past and present Recovery Team members, project officers, technical support staff, scientists from within and beyond government, Perth Zoo staff, students and, especially, volunteers. The number of people who have made significant contributions over the last half-century has been huge, so we apologise for not mentioning people individually. But to all of you, we acknowledge your outstanding enthusiasm, commitment and support for implementing recovery actions for these wonderful birds.

References

Abbott I (2008) Historical perspectives of the ecology of some conspicuous vertebrate species in south-west Western Australia. *Conservation Science Western Australia* **6**(3), 1–214.

Burbidge AA, Folley GL, Smith GT (1986) 'The noisy scrub-bird'. CALM WA Wildlife Management Program No. 2, Perth, WA.

Burbidge AH, Blyth J, Danks A, Gillen K, Newbey B (1997) 'Western Ground Parrot Interim Recovery Plan 1996–1999'. Interim Recovery Plan No. 6, Department of Conservation and Land Management, , Perth, WA.

Burbidge AH, Comer S, Lees C, Page M, Stanley F (Eds) (2016) 'Creating a future for the western ground parrot: workshop report'. Department of Parks and Wildlife, Perth, WA.

Cale P, Burbidge AH (1993) 'Research Plan for the Western Ground Parrot, Western Whipbird and Western Bristlebird'. Unpublished report from WA Dept of Conservation and Land Management to ANPWS Endangered Species Unit, Canberra, <https://library.dpaw.wa.gov.au/static/FullTextFiles/060178.pdf>.

Comer S, Danks A, Burbidge AH, Tiller C (2010) The history and success of Noisy Scrub-bird re-introductions in Western Australia: 1983–2005. In *Global Re-Introduction Perspectives: Additional Case-Studies from Around the Globe*. (Ed. PS Soorae) pp. 187–192. IUCN/SSC Re-Introduction Specialist Group, Abu Dhabi, United Arab Emirates.

Comer S, Burbidge AH, Algar D, Berryman A, Bondin A (2013) Kyloring, cats and conservation: the race to save the western ground parrot. *Landscope* **29**(2), 40–45.

Comer S, Berryman A, Burbidge AH, Broomhall G, Moar S (2015) 'Fire Management Guideline No. S13. Western Ground Parrot'. (Revised from 2009 version). Fire Management Services, Department of Parks and Wildlife, Perth, WA, <https://fms.dec.wa.gov.au/documents/fire-management-guidelines-principles/fire-management-guidelines>

Comer S, Burbidge AH, Algar D, Clausen L, Berryman A, Pinder J, *et al.* (2016) From the ashes. Creating a future for Western Ground Parrots. *Landscope* **31**(4), 11–15.

Danks A (1997) Conservation of the Noisy Scrub-bird: a review of 35 years of research and management. *Pacific Conservation Biology* **3**, 341–349. doi:10.1071/PC980341

Danks A, Burbidge AA, Burbidge AH, Smith GT (1996) 'Noisy Scrub-bird Recovery Plan'. Dept of CALM, Wildlife Management Program No. 12. Department of Conservation and Land Management, Perth, WA.

Danks A, Comer S, Burbidge AH (2011) Back from the brink: 50 years of conservation at Two Peoples Bay. *Landscope* **27**(1), 32–38.

DPaW (2014) 'South Coast Threatened Birds Recovery Plan'. Western Australian Wildlife Management Program No. 44. Department of Parks and Wildlife, Perth, WA.

Gibson L, Barrett B, Burbidge AH (2007) Dealing with uncertain absences in habitat modelling: a case study of a rare ground-dwelling parrot. *Diversity & Distributions* **13**, 704–713. doi:10.1111/j.1472-4642.2007.00365.x

Robin L (2001) *The Flight of the Emu. A Hundred Years of Australian Ornithology*. Melbourne University Press, Melbourne.

12

Back from the brink – again: the decline and recovery of the Norfolk Island green parrot

Luis Ortiz-Catedral, Raymond Nias, James Fitzsimons, Samantha Vine and Margaret Christian

Summary

The problem
1. Shortage of nest sites for Norfolk Island green parrots safe from cat and rat predation or competition with introduced birds.
2. Lack of intensive management of nest sites or monitoring to understand parrot trends.
3. Nest boxes and protection of natural nest sites had been effective before in recovering parrot numbers but the ongoing implementation of such measures was neglected once the parrot had recovered from near extinction in 1988.

Actions taken to manage the problem
1. Based on an Action Plan developed for the Norfolk Island National Park in early 2013, and the appointment of a Natural Resource Manager, predator control was enhanced including increased trapping of cats near parrot nests, revised rodent baiting practices and culling introduced avian nest-competitors.
2. A population survey in mid-2013 confirmed the very small size of the population. The survey was commissioned by non-government organisations (NGOs) following concerns about the population being raised by local residents in 2012.
3. Following the release of the 2013 survey results, the Australian Government implemented an action plan that included expert training for national park rangers to ensure reproductive output was maximised and plans for establishment of an insurance population on predator-free Phillip Island.

Markers of success
1. Increasing parrot numbers.

2. Renewed and ongoing commitment to maintenance of nest site maintenance and protection.
3. Creation of a partnership between Parks Australia, the local community and the NGO sector.
4. A forward-looking plan that identifies a durable solution.

Reasons for success
1. Resources available to rapidly commission population surveys.
2. Strong community interest in the parrot that meant that declines were identified despite a lack of formal monitoring.
3. Existing knowledge of the actions needed to manage the parrot, and the parrot's high fecundity once adequate protection was provided.
4. Effective action by Norfolk Island National Park staff once the problem was identified, working in partnership with external organisations.
5. An inclusive approach and targeted capacity development that resulted in national park staff believing in the outcomes, having the skills to implement recovery actions and able to take pride in their achievements.

Introduction

The Norfolk Island green parrot (*Cyanoramphus cookii*), also known as the Tasman parakeet, is endemic to Norfolk Island, off eastern Australia. There have been two waves of decline, near extinction, and managed recovery for this species: (1) decline and near extinction from an early loss of habitat after settlement, combined with predation and competition from introduced species (Hicks and Greenwood 1989; Hill 2002); (2) recovery after 1988 following predator control and nest-protection; (3) decline again during the period before 2013; and (4) recovery again, in 2014, after emergency intervention including the predator proofing of suitable nesting sites.

The Norfolk Island green parrot is listed as Endangered under the federal *Environment Protection and Biodiversity Conservation Act 1999* (EPBC Act). Garnett *et al.* (2011) reviewed the conservation status of the Norfolk Island green parrot and considered it to be Critically Endangered. The Norfolk Island green parrot is one of 20 bird species identified in the Australian Government's Threatened Species Strategy for priority attention (Australian Government 2015).

A recovery plan for the species was released in 2002 (Hill 2002). The overall objective of that plan was to shift the conservation status of the species from Endangered to Conservation Dependent within 10 years (Director of National Parks 2010). The single species recovery plan was superseded in 2010 by the Norfolk Island Region Threatened Species Recovery Plan (Director of National Parks 2010).

Conservation efforts for the Norfolk Island green parrot (Fig. 12.1A) provide numerous lessons for threatened species recovery, ranging from the engagement of community, the importance of monitoring, the danger of shifting emphasis from actions that were successful, and the benefits of targeted surveys. They also highlight the importance of building capacity within conservation staff in a remote location and of clear governance arrangements to ensure good decision making and accountability.

Conservation management

The Norfolk Island green parrot was a common forest bird at the time of the discovery of Norfolk Island in the late 1700s, but by the late 1970s the species had declined to fewer than

Fig. 12.1. (A) This male Norfolk Island green parrot fledged from the first active predator-proofed nest located in July 2013 and had already paired with an adult female by the time of the photograph 4 months later. (B) The parrots are well camouflaged while foraging in red guava (*Psidium cattleianum*), highlighting the importance of targeted surveys (photos: Luis Ortiz-Catedral).

50 individuals (Silva 1989; Forshaw 2002). The main causes of decline included shooting by early settlers, egg and chick depredation by introduced rats and cats and shortage of safe nesting sites (hollows and cavities) owing to felling of large trees since the early 1800s (Hermes 1985; Hicks and Greenwood 1989). Further pressures on declining parrots included competition for nesting sites by the introduced crimson rosella (*Platycercus elegans*) and common starling (*Sturnus vulgaris*) (Forshaw 2002). Efforts to reverse the decline of the species commenced in 1983 when the Government Conservator oversaw the construction of a captive breeding aviary. The captive breeding program failed to fledge sufficient chicks to aid in the recovery program, but succeeded in building local engagement and nurturing a sense of stewardship in the local community.

In 1987–1988, the staff from Norfolk Island National Park officially began rat and cat control operations on Mount Pitt (cat trapping occurred unofficially before the park was declared) and provisioning of purpose-built nesting sites and regular monitoring (Hicks and Greenwood 1989). The Norfolk Island green parrot nests in tree hollows and these can be modified to prevent rats from gaining access to the nest. From 1987 to the early 2000s, ~250 chicks fledged successfully as a result of the nest-provisioning and predator-control programs. Sightings of green parrots in the National Park and nearby orchards and gardens became more common. However, after that time changes in the management of the

Norfolk Island National Park resulted in a shift in emphasis away from specific management of the parrot population to habitat rehabilitation activities that were stated to be of benefit to all native species.

Concern over the status of the green parrot, was raised again in 2009–2012 by naturalists on the Island (e.g. in Garnett *et al.* 2011). A survey of forest birds in 2009, funded by BirdLife International, detected parrots at a few survey locations and estimated a population of ~240 birds (Dutson 2013). The study also highlighted the need for a targeted survey of green parrots due to the challenge of estimating actual numbers with an acceptable confidence interval (Fig. 12.1B). Despite these concerns and those of local naturalists, the National Park Manager and Canberra-based officials at the time stated in late 2012 there was no evidence for, nor any reason to suspect, a recent decline. In response, the Norfolk Island Flora and Fauna Society, Island Conservation, BirdLife Australia and The Nature Conservancy commissioned an expert in the *Cyanoramphus* genus (Luis Ortiz-Catedral from Massey University) to undertake a targeted population survey (Fig. 12.2A).

A new Natural Resource Manager was appointed in April 2013 and this bought considerable additional expertise to the National Park. An interim Action Plan was developed for the parrot and a small grant to re-construct some of the old nest sites was secured in May 2013.

The first targeted survey for Norfolk Island green parrots since the 1990s was completed over 25 days in July–August 2013. Mapping of all visual and acoustic records obtained during 220 h of searches along 282 km of transects by two observers with 5–10 years' experience of *Cyanoramphus* parrots, revealed that only between 42 and 96 individuals survived, with only 10 confirmed adult females in ~300 ha of remnant forest (Ortiz-Catedral 2013). Although it is unclear when population declines started, it appears it was linked to nest deterioration and possibly a lack of intense rat and cat control around active nesting sites and throughout the National Park. Out of the 72 nests available by the late 1990s, only 12 were deemed suitable in late 2013. This prompted and enabled an emergency program targeting the shortage of safe nesting sites.

In November 2013, the combined NGOs wrote to the Federal Minister for the Environment, the Hon. Greg Hunt MP, with the findings of this survey and a list of urgent recommendations. This was followed by a series of meetings with ministerial staff and department officials highlighting the urgency of the situation and the feasibility of solutions. In December 2013, a response was received indicating that the Minister had accepted the findings and agreed to implement a comprehensive recovery effort. Within 12 months, 78 rat and cat-proof nests were built and rat and cat control intensified in the immediate vicinity of these nests. These actions resulted in 13 active nests, which successfully fledged chicks by May 2014 (Milman 2014). Regular monitoring of the green parrot population revealed an increase in sightings by June 2015, when the population had increased to ~220 individuals as estimated by DISTANCE surveys (Skirrow and Ortiz-Catedral unpublished). This increase is better explained as pulses of juvenile birds resulting from the assisted nesting program, rather than steady recruitment. In support of this view, periodic declines in population estimates are still detected. For instance, in March 2015, a population of 144 was estimated using the same methodology. The measurable increase in the population of the species is supported by a significant six-fold increase in detection rates: from 0.74 detections per hour in July 2013, to 4.22 detections per hour in June 2015. In addition, mapping of records indicates the presence of breeding pairs in previously unoccupied areas. However, the proportion of juveniles recruiting into the breeding population is still under investigation (Ortiz-Catedral unpublished).

Fig. 12.2. (A) Luis Ortiz-Catedral measuring a fledgling Norfolk Island green parrot. Males and females can be distinguished before fledging by the dimensions of their bills. (B) Joel Christian inspecting a nest cavity protected from climbing rats and cats by the smooth cement surface surrounding the cavity entrance (photos: A, Mark Delaney, B, Luis Ortiz-Catedral).

On average, nest success (i.e. nests where at least one chick fledged; Fig. 12.2A) is 70% by the 2016 breeding season. In contrast, during most of its management, green parrot nest success has rarely exceeded 40%. Thus, with the current management, more nests are successful and more chicks produced per nest and per year as a result of providing new or safer nests (Fig. 12.2B) and improved ranger capacity to monitor nests without disturbance.

Research, biology and identification of key threats

The Norfolk Island green parrot, like other *Cyanoramphus* species, has high reproductive potential, laying large clutches of up to eight eggs (Forshaw 2002). The species also reaches sexual maturity at a relatively early age, making it possible for juveniles to begin nesting within months of fledging. This makes the species a useful indicator of the effectiveness of management actions, because a positive effect would be reflected in parrot numbers and breeding activities in a relatively short time and over a small area: Norfolk Island green parrots inhabit a core forest remnant of ~300 ha (Fig. 12.3).

The key threats to the species are well known (predation and competition for nest hollows). Thus providing safe nesting sites proved to be an effective way to boost the population and resulted in a detectable increase in breeding pairs from 10 confirmed breeding pairs in 2013 to ~22 pairs in 2014. At least four breeding pairs observed in 2014 involved juvenile females, most likely hatched in the preceding 8 months.

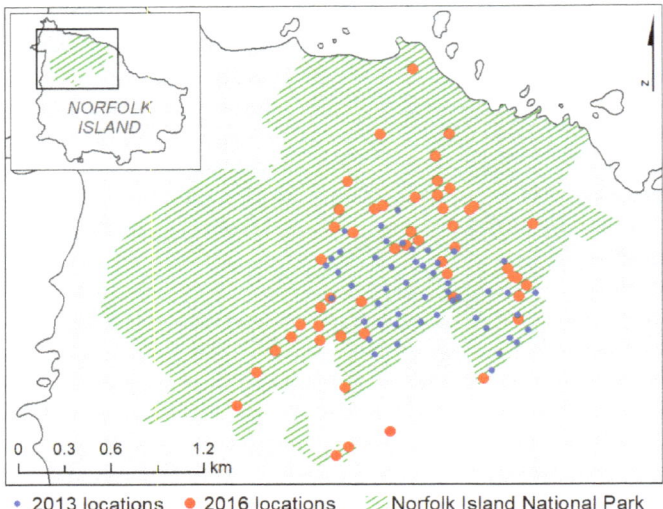

Fig. 12.3. Distribution of Norfolk Island green parrots on Norfolk Island in Oct–Nov 2013 and Aug–Sept 2016.

Planning and policy

A Norfolk Island Green Parrot Recovery Plan was in place and resourced from 2002 (Hill 2002) but replaced in 2010 by the multispecies Norfolk Island Region Threatened Species Recovery Plan (Director of National Parks 2010). Despite a new plan being in place, the species' population continued to decline and key components of the previous 2002 Recovery Plan, which were known to be successful, including maintenance of 'rodent-proof' nesting sites had ceased by 2013. When the recovery efforts re-commenced in earnest in 2014, it again focused on these proven successful strategies.

People, agencies, governance, and accountability

Before the Australian Government's involvement in Norfolk Island's environmental issues from 1979, the late Owen Evans monitored a range of bird species, the island's flora and other terrestrial biota, raising the alarm and taking remedial action. The island's early Government Conservators took ecological advice from Evans and he influenced the direction for management plans. Neil Hermes, the second Conservator, instigated the actions that began recovery of several bird species, including the green parrot. Locally appointed staff carried a strong sense of responsibility and ownership, with the community remaining involved and interested, as they had been encouraged to be over the years by Evans. Despite changes in the Norfolk Island National Park managers, with their differing approaches, the consistency that local staff and a strong, passionate, hands-on Advisory Committee brought, provided continuity and purposeful, outcome-oriented management.

Policy changes saw a formal shift in emphasis from single species to multi-species recovery approaches and broad-scale threatening processes. As good news stories had been written about the success of the green parrot program, there seemed to be no need to protect an already 'saved' species, and the budget became consumed by competing priorities. Local naturalists raised concerns regarding the population size and trajectory with the Australian Government in 2010 (reflected in Garnett *et al.* 2011) and with Australian non-government conservation groups in 2012. By March 2013, in a meeting with the concerned NGOs, the Department acknowledged the value of developing a robust, repeatable methodology for determining a reliable figure for the population of green parrots in the Park.

The independently NGO-commissioned 2013 surveys allowed much flexibility with how the findings could be communicated both to government and the general public. Articles in birding magazines on the dire state of the species alerted bird interest groups while media was used by the partners to communicate the message to the general public. We also ensured the results and straightforward solutions were rapidly communicated to the Australian Government. Because the majority of the green parrot's habitat occurred in Norfolk Island National Park (a park managed by the Australian Government), there was a more direct responsibility here for the Australian Government than for most other Australian threatened species. We were conscious that the most recent vertebrate extinction to that date (the Christmas Island pipistrelle) had occurred in another Australian Territory, and that the imminent loss of another species from a Commonwealth-managed territory would want to be avoided by the Australian Government.

Managing

Management actions were influenced through a combination of expertise, local knowledge and public support and the ability to access money to commission new science relatively quickly. In our discussion with the management agencies in early 2014, we came to agreement that the management actions that had worked in the past were the ones again that needed to be implemented immediately, or significantly enhanced (such as predator-proofing of nest-sites; Fig. 12.2B). Other actions, such as translocating an insurance population to nearby Phillip Island following methods developed in other *Cyanoramphus* species in New Zealand (Ortiz-Catedral and Brunton 2010) could be undertaken once the immediate extinction crisis had been averted. The historical range of green parrot most likely included Phillip Island, because previous accounts of that site describe a similar composition of plant species to areas near Mount Pitt (Coyne 2009) and because Phillip Island is within the flying capabilities of *Cyanoramphus* parrots (Ortiz-Catedral 2010).

An outcome of this engagement, and a shared understanding of the issues and appropriate management responses, has seen a strong and productive relationship formed between various stakeholders. Not only do the recovery actions continue to proceed, an advisory group has been formed with stakeholder representatives, new partners (e.g. WildMob) have come to the scene with additional resources and new actions have been developed that may not have otherwise been possible. One example of this has been the development by the local community and Norfolk Island National Park of an integrated rodent-control program covering private properties in the Anson Bay areas, the adjacent Forestry Zone and the National Park. The aim is to increase the area of Norfolk Island under effective predator control, creating additional safe habitat for the expanding green parrot population.

Monitoring

Surveys of Norfolk Island green parrots have continued every 4–6 months by volunteers and staff from the Norfolk Island National Park to determine population trends. Over the last 3 years, two monitoring approaches have been used: visual and acoustic detections along visitor tracks and DISTANCE sampling on 60–80 random locations. The latter, although more labour intensive, has yielded the most consistent results.

Infrastructure and technology

A key aspect of the recovery of the Norfolk Island green parrot has been provisioning of safe nesting sites combined with intense rat and cat control around active nests. The development of a protocol for nest checking without disturbing nesting females (which might lower hatching success and hence reproductive output) has also been important. To prevent nest desertion or altering the duration of incubating stints, the contents of active nests are checked weekly and only after males call females out of the nest to feed them. These incubating or brooding 'breaks' occur roughly every 2–4 h, and last for 10–25 min. This behaviour is stereotypical of *Cyanoramphus* parrots and stops when chicks are ~2–3 weeks old (Ortiz-Catedral 2006).

Money

Seed funding for first surveys and diagnosis of the problem came from The Nature Conservancy's Ecological Science Program (supported by The Thomas Foundation) and through grants to Island Conservation from the Norman Wettenhall Foundation and Packard Foundation, the Mohamed bin Zayed Species Conservation Fund, WildMob, Massey University, The World Parrot Trust and Parrot Society UK. This equated to ~A$90 000. Funding of A$50 000 to the Norfolk Island Flora & Fauna Society from the Australian Government's Caring for our Country Program in 2013–2014 also allowed the society to highlight several issues relating to the species during 'Invasive Species Month'.

Grants of A$27 500 to the Norfolk Island National Park from the Foundation for National Parks and Wildlife and the Parrot Society of Australia allowed construction of 78 rat and cat-safe nests and kick-start a pilot study on the movements and survival of fledglings.

In-kind support provided by volunteers from WildMob has an additional estimated value of around A$50 000 at late 2016.

The operating cost for the Norfolk Island National Park and Botanic Garden was A$1.183 million in 2014–2015 (Director of National Parks 2015), although it is not specified how much was allocated to actions that directly targeted to the Norfolk Island green parrot. The Threatened Species Commissioner (2015) committed A$300 000 of Australian Government funding in 2015 for rodent control on Norfolk Island to benefit several species, including the green parrot.

The future

Emergency actions started in 2013 have made it possible to increase the numbers of green parrots within the Norfolk Island National Park and strengthen the capacity of the park staff to manage the species. As long as rodents exist on Norfolk Island, there will be an ongoing need to maintain a vigilant effort to control their numbers and to provide sufficient predator-proof nest-sites for an expanding population of parrots. Maintaining adequate and ongoing funding to essential management activities, such as rat and cat control and nest hollow maintenance, will be critical into the future.

However, the species is still restricted to a single breeding population. Experience with other *Cyanoramphus* parrots in New Zealand show that translocation to predator-free sites is an effective measure to increase the number of populations and individuals (Ortiz-Catedral and Brunton 2010). Thus, the next milestone in the conservation of the species is to establish a population on rodent-free Phillip Island that should provide a valuable insurance population in case of extreme events such as disease outbreaks or major habitat loss. Significant resources will need to be sourced for this proposal and in 2017, more than A$86 000 was raised for the translocation through a crowdfunding campaign (Jeffery 2017).

No formal recovery team existed at the time of writing, but a recovery team with independent representatives would provide an important governance structure that can weather the vagrancies of changing government priorities and provide a mechanism that can respond to changes in the parrot's status if necessary. The Recovery Plan is due for review and revision to guide future recovery effort.

Conclusions

The ongoing success of recovery for the Norfolk Island green parrot will build upon and rely on the ongoing collaboration between the key players and the management structures established by the Norfolk Island National Park. The ultimate recovery of the Norfolk Island green parrot will always remain at risk so long as invasive predators and competitors remain in large numbers on Norfolk Island. For the foreseeable future, the fate of the species will require careful monitoring and active management to reduce threats from invasive species.

Acknowledgements

The success in the conservation of the Norfolk Island green parrot has been possible thanks to the logistical and financial support of: Massey University; Mohamed bin Zayed Species Conservation Fund; World Parrot Trust; The Parrot Society, UK; Norfolk Island Flora & Fauna Society; WildMob; Island Conservation; The Packard Foundation; The Nature Conservancy's Ecological Science Program generously funded by The Thomas Foundation; BirdLife Australia; Auckland Zoo Conservation Fund; Australian Government's Department of the Environment and Energy; Foundation for Parks and Wildlife Australia; The Parrot Society of Australia. Alan Saunders, Guy Dutson and Derek Ball provided expert input. Special thanks to the staff from the Norfolk Island National Park: Craig Doolan, Joel Christian, Ken Christian, Dids Evans, Rossco Quintal, Abigail Smith, Matt King and Cassandra Jones. Volunteers Liz Whitwell, Matt Upton, Daniel Waldmann, Luke Martin, Emma Wells, Nat Sullivan, Tansy Bliss, Rebecca Hamner, John Steemson, Amy Waldmann, Mike Skirrow and Jessica Barr provided assistance in the field. Derek Greenwood, Beryl Evans, Dave South and Rob Ward shared their knowledge about green parrots, and people of Norfolk Island have welcomed us in their community. Aaron Harmer developed the basis for the map. Craig Doolan, Judy West, Peter Latch and Stephen Garnett commented on an earlier draft of the chapter.

References

Australian Government (2015) 'Threatened species strategy'. Australian Government, Canberra.
Coyne P (2009) *Incredible! The Amazing Story of the Birth and Rebirth of a Natural Treasure Phillip Island, South Pacific*. Petaurus Press, Canberra.
Director of National Parks (2010) 'Norfolk Island Region Threatened Species Recovery Plan'. Department of the Environment, Water, Heritage and the Arts, Canberra.
Director of National Parks (2015) 'Director of National Parks Annual Report 2014–15'. Director of National Parks, Canberra.
Dutson G (2013) Population densities and conservation status of Norfolk Island forest birds. *Bird Conservation International* **23**, 271–282. doi:10.1017/S0959270912000081
Forshaw J (2002) *Australian Parrots*. 3rd edn. Alexander Editions, Robina, Queensland.
Garnett S, Szabo J, Dutson G (2011) *The Action Plan for Australian Birds 2010*. CSIRO Publishing, Melbourne.

Hermes N (1985) *Birds of Norfolk Island*. Wonderland Publications, Norfolk Island.

Hicks J, Greenwood D (1989) Rescuing Norfolk Island's parrot. *Birds International* **1**, 34–47.

Hill R (2002) 'Recovery Plan for the Norfolk Island green parrot *Cyanoramphus novaezelandiae cookii*'. Environment Australia, Canberra.

Jeffery S (2017) Crowdfunding secures Norfolk Island green parrot 'insurance colony'. *Sydney Morning Herald*, <http://www.smh.com.au/environment/crowdfunding-secures-norfolk-island-green-parrot-insurance-colony-20170328-gv8p7r.html>.

Milman O (2014) Norfolk Island green parrot rescued from the brink of extinction. *The Guardian*, <https://www.theguardian.com/world/2014/sep/22/norfolk-island-green-parrot-extinction>.

Ortiz-Catedral L (2006) Breeding ecology of a translocated population of red-crowned kakariki (*Cyanoramphus novaezelandiae*) on Tiritiri Matangi Island, New Zealand. MSc thesis, Massey University, Auckland, New Zealand.

Ortiz-Catedral L (2010) Homing of a red-crowned parakeet (*Cyanoramphus novaezelandiae*) from Motuihe Island to Little Barrier Island, New Zealand. *Notornis* **57**, 48–49.

Ortiz-Catedral L (2013) 'The population and status of green parrot (Tasman parakeet) *Cyanoramphus cookii* on Norfolk Island 2013'. Institute of Natural and Mathematical Sciences, Massey University, Auckland, New Zealand.

Ortiz-Catedral L, Brunton DH (2010) Success of translocations of red-fronted parakeets *Cyanoramphus novaezelandiae novaezelandiae* from Little Barrier Island (Hauturu) to Motuihe Island, Auckland, New Zealand. *Conservation Evidence* **7**, 21–26.

Silva T (1989) *A Monograph of Endangered Parrots*. Silvio Mattacchione and Co., Pickering, Ontario, Canada.

Threatened Species Commissioner (2015) 'Threatened Species Commissioner Report to the Minister for the Environment'. Department of the Environment, Canberra.

13

Progress in the conservation of populations of the eastern bristlebird from central coastal New South Wales and Jervis Bay Territory

David B. Lindenmayer, Chris MacGregor and Nick Dexter

Summary

The problem
1. The Eastern bristlebird has disappeared from many parts of its range.
2. Fire and feral predators have negative impacts on the species, especially in combination, and invasive weeds can replace suitable habitat.

Actions taken to manage the problem
1. Intensive control of feral predators.
2. Effective fire management means that fragments of suitable habitat remain and birds survive in those patches, even after fire.
3. Restoration of weed-infested habitat with native vegetation.
4. Successful translocation to previously empty habitat.

Markers of success
1. Population increasing.
2. Causes of decline identified and now effectively managed.
3. Enhanced community involvement and support.

Reasons for success
1. Dedicated, science-literate staff within key management agencies.
2. Close relationship between researchers and managers.
3. Continuity of funding to maintain monitoring and in turn assess the effectiveness of management actions.
4. An apparently simple relationship between the birds and the threats to them, lending to rapid recovery once threats are removed.

Fig. 13.1. Eastern bristlebirds are easy to hear, but difficult to see in the dense vegetation they prefer (photo: Graeme Chapman).

Introduction

The nationally Endangered eastern bristlebird (*Dasyornis brachypterus*) is a small bird (30–50 g, Fig. 13.1) that lives primarily in dense heathy vegetation, and feeds and breeds close to the ground. Its diet is primarily seeds, small fruits and invertebrates and it lives singly, in pairs or small groups (Higgins and Peter 2002). The species occurs in three geographically separate populations all of which have contracted since European settlement (Garnett *et al.* 2011). These three populations are south-eastern Queensland and north-eastern New South Wales, central New South Wales including the Jervis Bay and Beecroft areas (which are the focus of this chapter), and far south coast of New South Wales and north-eastern Victoria on the border between these two states (Fig. 13.2). This chapter briefly outlines some of the successful conservation outcomes for the coastal central New South Wales population and outlines 10 factors that have contributed to these positive outcomes.

Conservation management

Study areas and long-term monitoring programs

The eastern bristlebird is one of a suite of bird species that we have studied in detail at Booderee National Park, managed by Parks Australia, and at Beecroft Weapons Range, managed by the Department of Defence over the past 5–14 years. Several major investigations have been completed including those of the species' response to fire (Lindenmayer *et al.* 2008, 2009, 2016a), targeted weed management programs aimed at controlling the invasive bitou bush (*Chrysanthemoides monilfera* ssp. *rotundata*) (Lindenmayer *et al.* 2015) and the impacts of weapons use and associated military training (Lindenmayer *et al.* 2016b).

Previous research, mostly of relatively short duration, provided some broad generalisations about the ecology of this species and indicated some likely threats; this evidence

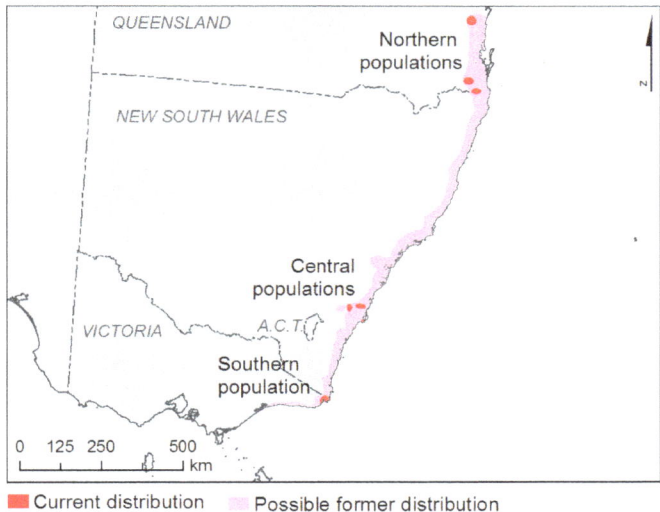

Fig. 13.2. Distribution of the eastern bristlebird across three geographically separate locations.

provided the basis for some general management directions. The more enduring and substantial studies described here have produced much more detailed and robust evidence, which has allowed for more focused and nuanced management response. For example, these studies have revealed some important findings about the response of the eastern bristlebird to disturbance, including post-disturbance recovery. First, repeated surveys at Booderee National Park have shown that the eastern bristlebird inhabits a wider range of vegetation types than previously recognised (e.g. Pyke *et al.* 1995) and occupies not only heathlands and sedgelands but also shrublands and woodlands. Perhaps most importantly, recent analyses of our data suggest that the eastern bristlebird now occupies more of the 134 long-term field sites at Booderee National Park than it did when intensive monitoring commenced in 2002 (D. Lindenmayer *et al.* unpublished).

Second, studies of the effects of fire have revealed that the eastern bristlebird either remains on burned sites or recovers to pre-disturbance levels of occurrence within 2–3 years after fire (Lindenmayer *et al.* 2009). However, follow-up monitoring and research at Booderee National Park has indicated that the Eastern bristlebird is more likely to occur on long-unburned sites (Lindenmayer *et al.* 2016a). In earlier work, we suggested that the eastern bristlebird can persist in burned areas (or recover quickly in them) because poison-baiting of feral predators reduces the risk of the birds being predated after vegetation cover is removed in a fire (Lindenmayer *et al.* 2009). Similarly, populations of the species have persisted at Beecroft Weapons Range, despite repeated fires in that area, probably because intensive baiting has also been a key component of management activities. Our results are broadly consistent with suggestions by Baker (2000) that the eastern bristlebird is cover-dependent and therefore at risk (e.g. of predation) when vegetation is removed.

A third major finding has been that actions to remove bitou bush generally have few negative impacts on native vertebrates, including the eastern bristlebird. There is a significant negative relationship between the occurrence of the eastern bristlebird and the number of bitou bush plants. However, the eastern bristlebird responded positively to increased native vegetation cover following bitou bush removal (Lindenmayer *et al.* 2017). This is a welcome result given the extent of bitou bush infestations in Booderee National Park, in part due to widespread deliberate past establishment of this invasive plant, and the intense program of spraying, burning and follow-up spraying that is now used in controlling this weed species (Lindenmayer *et al.* 2015).

Fourth, a dedicated translocation program has resulted in the successful establishment of populations of the eastern bristlebird at Beecroft Weapons Range on the northern peninsula of Jervis Bay (Baker *et al.* 2012). Indeed, repeated surveys conducted by The Australian National University across the Beecroft Weapons Range have revealed that the eastern bristlebird now occurs at all 40 long-term sites established in the area. Part of this work also explored the impacts of fire and demonstrated that the species was more likely to be recorded on long-unburned sites (Lindenmayer *et al.* 2016b) – a result consistent with studies in Booderee National Park (Lindenmayer *et al.* 2016a). Notably, the entire bird assemblage has been monitored at Beecroft Weapons Range, and our data indicate there has been a temporal decline in overall bird species richness, with the declines of 12 individual species being most prominent in the area subject to repeated bombing (Lindenmayer *et al.* 2016b). Notably, the eastern bristlebird has not been subject to temporal declines and patterns of occurrence do not differ significantly between areas where bombing and other weapons use has occurred regularly for the past century relative to sites outside this high impact zone where weapons have not been tested for the past 25 years (Lindenmayer *et al.* 2016b).

Key features in success

Flexibility in the biology of the target species

Part of the success of conservation efforts for the eastern bristlebird lies with the ecology of the species itself. As a generalist ground-foraging species that appears to have a relatively limited ability to fly, the eastern bristlebird is likely to be vulnerable to predation by exotic predators such as the red fox (*Vulpes vulpes*) and feral cat (*Felis catus*). Therefore, intensive feral predator-control programs are likely to contribute significantly to the maintenance of populations of eastern bristlebird. Low numbers of predators due to sustained control programs is probably the key reason why the species is widespread throughout a range of vegetation types in Booderee National Park and Beecroft Weapons Range. Flexibility in habitat use, coupled with the rarity of feral predators, also may explain the high level of success of the translocation program at Beecroft Weapons Range.

Maintenance of key management activities

As outlined above, a surprising result from the research and monitoring program has been the rapid recovery of the species following fire (Lindenmayer *et al.* 2009). The maintenance of an intensive feral animal baiting program within Booderee National Park and Beecroft Weapons Range appears to be a key reason for the apparent success of conservation efforts for the eastern bristlebird. In both places, ongoing hard work by dedicated field staff has been essential to maintain effective programs for poison baiting of feral predators. The recurrent nature of this program has led to sustained marked reduction in predator levels; indeed, in the case of Booderee National Park, levels of take of poison baits has declined to almost zero and patterns of predator activity in sand plots are similarly minimal (Parks Australia *pers. comm.*).

Continuity of funding

Continuity of funding is often crucial to success of conservation efforts. Work on the eastern bristlebird has been part of broader monitoring programs supported by Parks Australia and the Department of Defence, which have focused on quantifying temporal changes in a range of biotic groups including birds, mammals, reptiles, amphibians and vascular plants (Lindenmayer *et al*. 2014).

Work on the eastern bristlebird and other biota has been funded since 2002, although multiple sources of funding have been required at all times to ensure that the work can continue, creating an enormous challenge for the project leader (David Lindenmayer). Creative funding strategies have been required such as subsidising the work with book royalties and the use of money from prizes and awards.

The long-term nature of the work has meant that, since 2011, it has qualified for support under the Australian Government's National Collaborative Research Infrastructure Scheme (NCRIS), which funds the Terrestrial Ecosystem Research Network (TERN) of which the Long-term Ecological Network (LTERN) is a part. The network of long-term sites at Jervis Bay is one of 12 plot networks within LTERN. Unfortunately, LTERN will be axed from the end of 2017 and the future of all plot networks, including those at Booderee National Park remains uncertain, but with most of them likely to collapse.

Continuity of on-ground presence of field staff

The ability to determine whether conservation programs for the eastern bristlebird have been successful is due, in large part, to the skills and dedication of a field staff member from The Australian National University (Chris MacGregor) who has worked full-time within Booderee National Park for more than a decade. He has been responsible for the maintenance of ongoing, long-term biodiversity monitoring, including monitoring of the eastern bristlebird. He also works closely with members of the area's traditional land-owners (the Wreck Bay Aboriginal Community), and has been responsible for training and mentoring of Indigenous people in biodiversity survey and other related skills. He has also worked closely with key staff from the Department of Defence.

The monitoring work on the eastern bristlebird is part of a much larger ongoing monitoring effort within Booderee National Park and Beecroft Weapons Range. The total funding for all monitoring work amounts to A$40 000 per annum from Parks Australia (2003–2017) and A$20 000 from the Department of Defence (2008–2015), as well as A$50 000 per annum (2010–2016) from the LTERN. Hence, the total investment in monitoring since 2003 is ~A$1 million, of which an estimated 20% of effort or A$200 000 could be crudely assigned to work on the eastern bristlebird. In addition, investments of ~A$30 000 per annum at Booderee National Park are estimated to be associated with intensive fox baiting (labour and time to set out poison baits as bait stations) across the reserve. A smaller cost of ~A$20 000 per annum is estimated for efforts to remove bitou bush from the subset of habitats and associated areas within Booderee National Park where the eastern bristlebird is known to occur (Lindenmayer *et al*. 2015). Thus, in total, about A$700 000 in management costs plus A$200 000 in monitoring has been expended over the 14-year duration of the program as direct grants from government.

Robust design of studies and associated expert statistical support and analysis

A robust design for monitoring programs has a significant influence on the ability to determine whether a species is (or is not) responding to a given management intervention. Our studies of the eastern bristlebird have been underpinned by robust designs following detailed consultation with a panel of expert professional statistical scientists. Statisticians have been involved not only in the design of field studies but also the analysis of datasets and the interpretation of results. They have also been key co-authors in scientific articles reporting the outcomes of various studies of the eastern bristlebird and other elements of the biota (e.g. Lindenmayer et al. 2009, 2016a, 2017).

Simple site-level infrastructure that is easy to maintain

Site infrastructure underpinning the work on the eastern bristlebird is simple (comprising permanent steel posts as markers) and is easy to maintain. These site markers are easy to locate, including by volunteer observers, which facilitates field data collection. They are also inexpensive, which means they are not targets for theft – a potential issue in places such as Booderee National Park, which has more than 450 000 visitors annually. However, site markers cannot be used at Beecroft Weapons Range because of the risks posed by driving steel posts into soil containing unexploded ordnance.

High-quality field management staff and volunteer support

Conservation management programs are rarely successful without high-quality field teams. This includes dedicated staff from management agencies responsible for key conservation actions such as the control of feral animals and the control of invasive plant species. At Booderee National Park, these vital tasks are undertaken by national park staff. In the case of the management of Beecroft Weapons Range, the Department of Defence undertook an extensive clearance program to remove unexploded ordnance to enable field surveys of the eastern bristlebird to proceed.

Strong scientifically literate engagement between partner organisations

The key partner organisations involved in work on the eastern bristlebird have been extremely supportive of monitoring and research, not only on the target species but also other elements of the biota (Lindenmayer et al. 2013). Indeed, ongoing work would not be possible with this support. We suggest that high levels of scientific literacy among the staff at Parks Australia and Department of Defence have meant that there is a strong understanding of the value of monitoring and research, including that on rare and threatened species. Moreover, agency staff have been invited to contribute to the monitoring and research and are often co-authors on peer-reviewed scientific articles based on the work. Most importantly, the results of research and monitoring have been made available to management staff and are often used to guide on-the-ground conservation actions. For example, the results of studies of fire indicated that the patchiness of burning had a significant effect on the persistence of the eastern bristlebird on burned sites (Lindenmayer et al. 2009). Managers of Booderee National Park then made it clear that post-fire 'black-out burning', in which green areas within the boundaries of a fire are burned (see Backer et al. 2004), would not occur in the reserve. Similarly, because overall bird species richness at sites within Booderee National Park is reduced with each additional fire at those sites (Lindenmayer et al. 2008), prescribed burning policies would avoid successive re-burning of already repeatedly burned areas.

Connection to higher level education programs, including contributions from postdoctoral researchers

The involvement of postgraduate and postdoctoral researchers has increased the scientific and management value of work on the eastern bristlebird, leading to new discoveries that underscore the success of conservation efforts for the species. For example, recent studies have quantified patterns of eastern bristlebird distribution across a range of habitats within Booderee National Park. Moreover, the species has been shown to have a key role as an indicator of high species richness in the remainder of the bird community, thereby acting as a robust surrogate for bird biodiversity (A. Tulloch *pers. comm.*). These discoveries could not have been made without the considerable efforts of outstanding postgraduate and postdoctoral researchers.

Serendipity and agility

Serendipity plays a role in the success of many projects, and this case was no exception. In late 2003, a major fire burned approximately half of Booderee National Park, including more than half of our sites that had been established just 6 months earlier. Fortunately the spatial extent of the fire, coupled with the completion of pre-fire surveys, meant that it was possible to quickly redesign the original monitoring program with new objectives focused on quantifying post-fire recovery. Moreover, through support of Parks Australia, it was possible to obtain insurance funding from damaged infrastructure to re-establish each of the burned sites and recommence surveys soon after the 2003 conflagration. Indeed, this change in project focus has arguably delivered more insights into the efficacy of management actions, including those for the eastern bristlebird, than might otherwise have been the case (Lindenmayer *et al.* 2014).

The future

There is reason to be cautiously optimistic about the conservation of central populations of the eastern bristlebird. However, there is clearly a need to maintain the intensive predator-baiting program in key areas such as Booderee National Park and Beecroft Weapons Range, because current evidence suggests that this is benefitting the eastern bristlebird, although there do appear to be some unexpected perverse effects on other species (see Lindenmayer *et al.* 2011; Dexter *et al.* 2012). Nevertheless, it is not clear for how long intensive baiting programs will remain effective because such programs typically select against non-responding individuals (i.e. predators that are bait shy). Other factors argue against complacency in conservation, including huge upheavals in the governance of Booderee National Park and the potential loss of expertise and skills in maintaining effective management programs for biodiversity conservation. The development of enhanced biodiversity survey and management skills within the local Indigenous community is critical because it takes on the responsibility for sole management of Booderee National Park. To date, funding and opportunities for such enhanced on-the-ground training have been limited. Finally, the funding landscape to maintain long-term monitoring programs and associated researcher–manager partnerships is increasingly tenuous; the collapse of any set of studies – no matter how prolonged or successful, or how iconic the target species – may be only a grant proposal rejection away. Indeed, the body of work at Booderee National Park is now at severe risk of collapse as a consequence of the collapse of Australia's Long-term Ecological Research network, funding for which will cease as at the end of 2017.

Conclusion

The eastern bristlebird may be easier to conserve than many species because it appears to be relatively adaptable. The key management intervention is control of feral predators, although the species may also benefit from weed removal and fire management. Fire on its own, however, appears not to be as great a threat as previously thought. These insights, and their implications for management of the species, have only been possible through extended partnerships between managers and researchers.

The prospects for the central New South Wales populations of eastern bristlebird are reasonably positive if the baiting and its effectiveness can be maintained. It is likely that these same factors affect the other populations of this species that are more threatened.

Acknowledgements

Many people and organisations have contributed to the success of conservation efforts for the eastern bristlebird. Parks Australia, the Australian Research Council and the Department of Defence have provided funding for the work at Booderee National Park, HMAS Creswell and Beecroft Weapons Range for more than a decade. Our work has been championed by key individuals within Parks Australia and the Department of Defence, but particularly Martin Fortescue and Fred Ford. Since 2010, the long-term sites in Booderee National Park have been part of a plot network funded by the LTERN within the TERN. We are indebted to the people of the Wreck Bay Aboriginal Community for continued access to their land in Booderee National Park. Volunteers from the Canberra Ornithologists Group have assisted in annual bird counts, including those of the eastern bristlebird since 2002. In particular, we thank Bruce Lindenmayer, Jenny Bounds, Martyn Moffat, Terry Munro, Noel Luff and Peter Roberts for their help with field work. Many staff from The Australian National University have been part of the bird counts at Booderee National Park. These include Mason Crane, Damian Michael, Sachiko Okada, Dan Florance and Thea O'Loughlin. Statistical support for our work has been crucial, particularly the major contributions by Ross Cunningham, Jeff Wood, Wade Blanchard and Steve Candy. Claire Shepherd and Tabitha Boyer assisted in manuscript production.

References

Backer DM, Jensen SE, McPherson GR (2004) Impacts of fire suppression activities on natural communities. *Conservation Biology* **18**, 937–946. doi:10.1111/j.1523-1739.2004.494_1.x

Baker JR (2000) The eastern bristlebird: cover dependent and fire sensitive. *Emu* **100**, 286–298. doi:10.1071/MU9845

Baker J, Bain D, Clarke J, French K (2012) Translocation of the eastern bristlebird 2: applying principles to two case studies. *Ecological Management & Restoration* **13**, 159–165. doi:10.1111/j.1442-8903.2012.00640.x

Dexter N, Ramsay DSL, MacGregor C, Lindenmayer DB (2012) Predicting ecosystem wide impacts of wallaby management using a fuzzy cognitive map. *Ecosystems* **15**, 1363–1379. doi:10.1007/s10021-012-9590-7

Garnett ST, Szabo JK, Dutson G (2011) *The Action Plan for Australian Birds 2010*. CSIRO Publishing, Melbourne.

Higgins PJ, Peter JM (2002) *Handbook of Australian, New Zealand and Antarctic Birds: Pardalotes to Shrike-thrushes*. Oxford University Press, Melbourne.

Lindenmayer DB, Wood JT, Cunningham RB, MacGregor C, Crane M, Michael D, et al. (2008) Testing hypotheses associated with bird responses to wildfire. *Ecological Applications* **18**, 1967–1983. doi:10.1890/07-1943.1

Lindenmayer DB, MacGregor C, Wood JT, Cunningham RB, Crane M, Michael D, et al. (2009) What factors influence rapid post-fire site re-occupancy? A case study of the endangered Eastern bristlebird in eastern Australia. *International Journal of Wildland Fire* **18**, 84–95. doi:10.1071/WF07048

Lindenmayer DB, Wood JT, McBurney L, MacGregor C, Youngentob K, Banks SC (2011) How to make a common species rare: a case against conservation complacency. *Biological Conservation* **144**, 1663–1672. doi:10.1016/j.biocon.2011.02.022

Lindenmayer DB, MacGregor C, Dexter N, Fortescue M, Cochrane P (2013) Booderee National Park management: connecting science and management. *Ecological Management & Restoration* **14**, 2–10. doi:10.1111/emr.12027

Lindenmayer DB, MacGregor C, Dexter N, Fortescue M (2014) *Booderee National Park: The Jewel of Jervis Bay.* CSIRO Publishing, Melbourne.

Lindenmayer DB, Wood J, MacGregor C, Buckley YM, Dexter N, Fortescue M, *et al.* (2015) A long-term experimental case study of the ecological effectiveness and cost effectiveness of invasive plant management in achieving conservation goals; Bitou Bush control in Booderee National Park in eastern Australia. *PLoS One* **10**, e0128482. doi:10.1371/journal.pone.0128482

Lindenmayer DB, Candy SG, Banks SC, Westgate M, Ikin K, Pierson J, *et al.* (2016a) Do temporal changes in vegetation structure predict changes in bird occurrence additional to time since fire? *Ecological Applications* **26**, 2267–2279. doi:10.1002/eap.1367

Lindenmayer DB, MacGregor C, Wood J, Westgate M, Ikin K, Foster C, *et al.* (2016b) Bombs, fire and biodiversity: vertebrate fauna occurrence in areas subject to military training. *Biological Conservation* **204**, 276–283. doi:10.1016/j.biocon.2016.10.030

Lindenmayer DB, Wood J, MacGregor C, Hobbs RJ, Catford JA (2017) Non-target impacts of weed control on birds, mammals, and reptiles. *Ecosphere* **8**, 1–19. doi:10.1002/ecs2.1804

Pyke G, Saillard R, Smith R (1995) Abundance of eastern bristlebirds in relation to habitat and fire history. *Emu* **95**, 106–110. doi:10.1071/MU9950106

14

Tasmania's forty-spotted pardalote: a woodland survivor

Sally L. Bryant

Summary

The problem

1. The forty-spotted pardalote is a highly specialised bird with a small total population size restricted to fragmented subpopulations occurring in a narrow habitat range in a few sites.
2. The species faces many threats and has exhibited a long pattern of extirpations and ongoing decline.
3. Monitoring, research and management effort has been discontinuous, largely because of episodic and intermittent funding.

Actions taken to manage the problem

1. Phases of survey and research effort have provided a reasonable evidence base about the pardalote's ecology, demography and threats.
2. Surveys have documented distribution and decline of subpopulations and identified important habitat.
3. Habitat of many important subpopulations has been protected through reservation on public and private land, policy changes and enhancement through plantings and nest boxes.
4. Much effort has been devoted to raising public awareness of the species' conservation needs.

Markers of success

1. Primary causes of decline identified and some now effectively managed.
2. Recovery plans developed, clearly defined and progress measured.
3. Enhanced community involvement and voluntary support.
4. New initiatives helping to mitigate threats.

Reasons for success

1. Favourable policy settings have facilitated permanent habitat protection on public and private land and helped develop widespread community support.

2. Occurrence of important subpopulations on islands has provided some inherent protection against some threats.
3. A reasonable evidence base about distribution, ecology and threats.
4. Ongoing conservation effort despite discontinuous program funding.

Introduction

The forty-spotted pardalote (*Pardalotus quadragintus*) formerly occurred across the eastern half of Tasmania, with small outlying populations on King Island and Flinders Island, stranded there with the rising waters of Bass Strait. The earliest accounts by John Gould and others in the 1840s suggested that the species was typically uncommon and naturally rare (Gould 1848) but it wasn't until detailed ecological and distributional information was collected over a century later that the extent of its rarity and dramatic decline in range became more apparent (Milledge 1980; Rounsevell and Woinarski 1983; Woinarski and Rounsevell 1983; Woinarski and Bulman 1985). By 1986, the forty-spotted pardalote had contracted to just six distinct subpopulations, all confined to Tasmania's east coast islands, headlands and peninsulas (Brown 1986). The last known record of the species from King Island was in 1887 (Green and McGarvie 1971), with this extirpation presumably through a dramatic loss of habitat and bushfires. Its population status on Flinders Island was poorly resolved, but precarious at 10–12 pairs (Brown 1986). It could no longer be found around Launceston, Bothwell or the Derwent Valley or anywhere in the north-east and was perilously close to extinction on Tinderbox, Coningham and Lime Bay in Tasmania's south-east (Milledge 1980; Brown 1986). The total population of the species numbered fewer than 3000 birds, mostly confined to Maria Island and Bruny Island, with those on Bruny being on private land at risk of clearing (Brown 1986; Fig. 14.1).

The forty-spotted pardalote (Fig. 14.2) has a highly specialised ecology, with a particular reliance on one tree species: white or manna gum (*Eucalyptus viminalis*) (Woinarski and Rounsevell 1983). This dependence means that, unlike the co-occurring more mobile

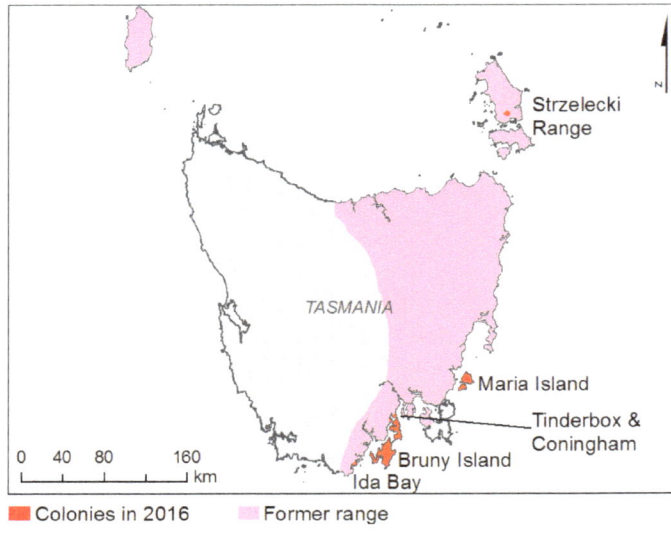

Fig. 14.1. Current and former range of the forty-spotted pardalote.

Fig. 14.2. Forty-spotted pardalote, 'Inala' Bruny Island (photo: Andrew Browne).

and generalist striated pardalote (*P. striatus*) and spotted pardalote (*P. punctatus*), the forty-spotted is philopatric and forms permanent breeding colonies around its white gum habitat. Although it feeds on invertebrates and lerp (an exudate produced mostly by psyllid insects), it is the sugary manna produced almost exclusively by white gum that is essential in its diet, especially during the breeding season when feeding chicks (Woinarski and Bulman 1985). Even though white gum remains a common tree species across eastern Tasmania, much of this forest and woodland is now fragmented and dominated by aggressive birds such as noisy miner (*Manorina melanocephala*) and laughing kookaburra (*Dacelo novaeguineae*), rendering it unsuitable to forty-spotted pardalotes (Rounsevell and Woinarski 1983; Bulman *et al.* 1986).

The national Recovery Program started in 1991 and, over a 25-year period, has concentrated on protecting and managing existing breeding colonies, and it is this clear direction that has contributed to the species survival in the wild today. Originally led by the Tasmanian Government, the program is now driven by the non-government sector, research institutions and the community in a model for successful partnerships. However, notwithstanding its success, the Recovery Program also contains sobering lessons. Over the course of this period, new threats have emerged that require more research effort and a shift in management priority.

Conservation management

In 1991, due to its small population and declining range, the forty-spotted pardalote was classified as Endangered and the first statutory national Recovery Plan was prepared

(Bryant 1991). This plan formalised the recovery process and it was thereafter managed collaboratively by the Tasmanian and Australian Governments through the Endangered Species Program under the federal *Endangered Species Protection Act 1992*.

The first conservation works focused on Bruny Island and were aimed at raising public awareness, growing and distributing 1500 white gum seedlings and identifying the highest priority colonies in need of protection (Bryant 1992). This program gained widespread community support and led to the largest breeding colony on private land (92 ha) being donated by the Dennes family (Bruny Island) and later proclaimed the Dennes Hill Nature Reserve. Other populations on state reserves, especially Maria Island National Park, were given greater prominence in management plans by the Tasmania Parks and Wildlife Service, and colonies on land subject to commercial forestry were managed by the Forest Practices Code. A private land swap on Walkers Gully and the reservation of Broughams Sugarloaf through the Regional Forestry Agreement secured the habitat of the Flinders Island population (Threatened Species Section 2012) and over time more than 60 perpetual covenants on private land have protected breeding colonies across the species' entire range, minimising its risk of extinction. Collectively, these initiatives led to a marked increase in the protection of known colonies from 55% in 1991 to 80% by 2016. These private land covenants were made possible through programs such as the Natural Heritage Trust, the Private Forest Reserves Program, the Protected Areas on Private Land Program and the Caring for Country Woodlands Birds for Biodiversity Program.

In addition to protecting habitat, many colonies have been enhanced through white gum plantings and installation of nesting boxes, especially on covenanted and Land for Wildlife properties. The public's awareness and involvement continues to grow through education, media and community events such as the biennial Bruny Island Bird Festival. As an added benefit, the habitat protected for the forty-spotted pardalote also aids the now Critically Endangered swift parrot (*Lathamus discolor*), which overlaps with the forty-spotted pardalote's breeding range on Bruny Island, Maria Island, Tinderbox and Tasman peninsulas (Saunders and Tzaros 2011).

Research

Research on the species has been intermittent. Several scientific papers and technical reports have been published and three Honours, one Masters and one Doctoral research thesis have been completed, with a second doctoral study now underway. Early research provided information on the pardalote's general ecology and life history, but less so on threats or reason for its decline (Woinarski and Rounsevell 1983; Woinarski and Bulman 1985). Much survey activity and some environmental modelling has helped clarify its distribution, size of subpopulations and extent of suitable habitat (Brown 1986; Brereton *et al.* 1997; Bryant 1997, 2010; Bryant and Webb 2012) and extension surveys have continued to search for new populations (http://www.hamishsaunders.com/), with a new colony being identified on the Tasmanian mainland at Ida Bay in 2015.

A more recent phase of autecological research has substantially extended ecological insights. This research has included examination of interspecific competition (Edworthy 2016), the species' ability to 'farm' manna (Case and Edworthy 2016), its genetic variation, the use of artificial nest boxes to augment habitat suitability and breeding productivity (Fig. 14.3), and the discovery of 'new' threats – notably the fly (*Passeromyia longicornis*) whose blood-sucking larvae parasitise nestlings (Edworthy and Kim 2015). This resurgence in research, in addition to that underway on other hollow nesting birds, is providing much-needed information to guide future management.

Fig. 14.3. Adult forty-spotted pardalotes feeding chicks in next box (photo: Andrew Browne, image ethically collected with the landholder's permission).

Recovery planning and policy

Two national recovery plans and two listing statements for the species have been jointly adopted under Tasmania's *Threatened Species Protection Act 1995* and the Commonwealth (federal) *Environment Protection and Biodiversity Act 1999* (Bryant 1991; Threatened Species Section 2006; Threatened Species Section 2012). These plans had a strong focus on habitat security and enhancement, raising awareness, and research and monitoring, and have remained in place well beyond their stated lifespans. Their implementation has paralleled government changes to tax legislation and rate rebating schemes, which have provided additional incentives to landowners to protect their land in the wake of improved land use control measures such as the *Land Use Planning and Approvals Act 1993* (LUPAA), local planning schemes (e.g. Kingborough Planning Scheme 2000) and Tasmanian land-use policies (e.g. Coastal Policy and Forest Practices Code). Unfortunately, some of the more recent changes mooted in a new draft state-wide planning scheme and the Tasmanian Government's cessation of covenanting private land have constrained options for these underlying protection mechanisms.

Monitoring

Monitoring has been irregular and problematic primarily because of a lack of funds for the specialist skills needed to detect the species in highly fragmented habitat dispersed across three islands and near-coastal mainland Tasmania. The bird's tiny size (9 cm) and habit of foraging high in the eucalypt canopy can make it challenging to find and distinguish from other pardalote species. Notwithstanding significant improvements in habitat security, an assessment conducted in 2010 found that in the 17 intervening years between survey periods, the total population had crashed by over 60% and the species now numbered fewer than 1500 birds (Bryant 2010). Although the causes of this decline were not entirely resolved, it coincided with a period of prolonged drought causing dieback of white gum and a noticeable increase in invasive bird disturbance in mainland colonies (Bryant 2010). Had regular monitoring been in place, this decline would have been detected sooner and remedial actions identified. Although a regular monitoring program is still not in place, extension surveys and nest box research undertaken since 2010 suggests that some subpopulations may have started to recover, whereas others are lost.

Several lessons have been learned. The first is that standardised repeatable monitoring is essential to track population trends and, if this can be aided by landholders, then a subset of colonies can be visited more regularly. Second, the sensitivity of white gum to drought means future restoration programs need to target areas of secure water availability to help withstand the effects of climate change. Finally, the proliferation in invasive and competitive species, including noisy miner (*Manorina melanocephala*), sugar glider (*Petaurus breviceps*), galah (*Eolophus roseicapilla*) and rainbow lorikeet (*Trichoglossus moluccanus*), means interventionist management and stronger biosecurity controls are needed, especially on islands.

Money

In 1990 a small grant from World Wide Fund for Nature initiated recovery works on Bruny Island but from then on nearly all funding has come from either the Australian Government through various endangered species programs or from natural resource management regional competitive bids. Between 1991 and 2010, the Tasmanian Government received ~A$85 000 in federal funding for recovery plan activities; however, larger sums were received by non-government organisations such as the Understorey Network, NRM South's Mountain to Marine Program, Threatened Species Network and BirdLife Australia for habitat restoration and protection. Several programs, such as the Private Forest Reserve Program and Protected Areas on Private Land Program, have facilitated the critical covenanting of private land and other bodies and agencies have given freely of their time for a wide range of policy and other support activities. A lack of committed funds for research, monitoring and continuous project management still impedes conservation of this species. In May 2015, the Australian National University successfully raised A$73 000 through a crowd funding program (https://pozible.com/project/193891) for nest box installation for three threatened hollow-dependent bird species including the forty-spotted pardalote, demonstrating there are new ways of supporting much-needed conservation research.

A summary of the key components of the Forty-spotted Pardalote Recovery Program from the 1980s to the present time (Fig. 14.4) shows that habitat security, habitat management and public awareness have been highly successful, whereas monitoring, scientific research and program funding have been insufficient to support timely conservation action.

The future

Aided by our efforts, Tasmania's forty-spotted pardalote has survived some major population declines; however, its future is still not assured. The survival of several colonies, such as those at Peter Murrell Reserve, Tinderbox and Coningham, remain precarious, and the Flinders Island population of ~20 birds is perilously close to extinction and needs urgent inter-

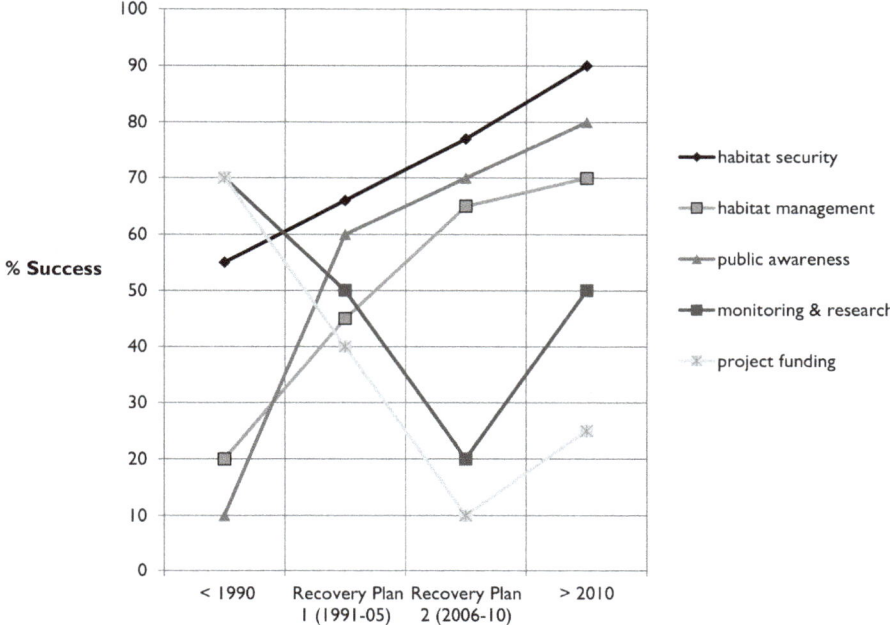

Fig. 14.4. Success of key components of the Forty-spotted Pardalote Recovery Program (estimates by Sally Bryant).

vention. Assisted colonisation could expand the number of safe havens and nest boxes could be deployed more widely to improve productivity and long-term resilience. We need to expand our efforts in combating the impact of invasive species, especially on offshore islands, and more research into all these focal areas is pivotal to help speed the recovery process.

Conclusion

Many of the past and current achievements of the Forty-Spotted Pardalote Recovery Program have been realised through national, state and local partnerships driven by committed people with a resounding willingness to save this unique bird. This program exemplifies the feedback loops that align government policy and regulation with community actions and expectation and the imperative that all need to be working together, constantly adopting and applying new learnings in an ever changing environment. Although people will always give freely of their time, without base funding it is difficult to maintain continuous project management and, without robust evidence, we are unlikely to uncover the solutions for this bird species' future.

Acknowledgements

This program has been supported by many government and non-government bodies, researchers, community groups, landholders and individuals. In particular, I acknowledge personnel from state and Commonwealth government agencies (especially those from threatened species units and parks and wildlife districts), local governments and regional natural resource management. Others include World Wide Fund for Nature, BirdLife

Australia, Recovery Team members, Forest Practices Authority, Bruny Island Environment Network, research scientists at the Australian National University and University of Tasmania, Tasmanian Land Conservancy, Threatened Species Network, Understorey Network, Tonia Cochran, Chris Tzaros, the Denne family, Friends of Peter Murrell Reserve, Friends of Maria Island, Hamish Saunders Memorial Trust Island Program, and the many committed landholders and volunteers across the species entire range. Thanks also to Peter Brown, Phil Bell and the editors for their very helpful comments and to Andrew Browne for use of his superb images.

References

Brereton R, Bryant SL, Rowell M (1997) 'Habitat modelling of the forty-spotted pardalote and recommendations for management'. Report to the Tasmanian Regional Forest Agreement Environment and Heritage Technical Committee, Hobart.

Brown PB (1986) 'The forty-spotted pardalote in Tasmania'. Technical report 86/4. National Parks and Wildlife Service, Hobart.

Bryant SL (1991) 'The Forty-Spotted Pardalote Recovery Plan: management phase'. Department of Parks, Wildlife and Heritage, Hobart.

Bryant SL (1992) 'Long term survival of the forty-spotted pardalote on Bruny Island'. World Wildlife Fund Final Report, project no. 157. Parks, Wildlife and Heritage, Hobart.

Bryant SL (1997) The status of colonies of the forty-spotted pardalote. *Tasmanian Bird Report* **26**, 45–50.

Bryant SL (2010) 'Conservation assessment of the endangered forty-spotted pardalote 2009–2010'. Report to Threatened Species Section, Department of Primary Industries Parks Water and Environment and Natural Resource Management South, Hobart.

Bryant SL, Webb M (2012) 'Status of the endangered forty-spotted pardalote *Pardalotus quadragintus* on Flinders Island, 2012'. Report to the Hamish Saunders Memorial Trust and Department of Primary Industries, Parks, Water and Environment, Hobart.

Bulman CM, Rounsevell DE, Woinarski JCZ (1986) 'RAOU Conservation Statement: the forty-spotted pardalote'. Royal Australasian Ornithological Union, Melbourne.

Case SB, Edworthy AB (2016) First report of 'mining' as a feeding behaviour among Australian manna-feeding birds. *The Ibis* **158**, 407–415. doi:10.1111/ibi.12350

Edworthy AB (2016) Competition and aggression for nest cavities between striated pardalotes and endangered forty-spotted pardalotes. *The Condor* **118**, 1–11. doi:10.1650/CONDOR-15-87.1

Edworthy AB, Kim A (2015) What's killing the endangered forty-spotted pardalote? *Tasmanian Geographic* April 2015, <https://www.tasmaniangeographic.com/whats-killing-pardalote>.

Gould JFRS (1848) *The Birds of Australia. Volume 2*. John Gould, London.

Green RH, McGarvie AM (1971) The birds of King Island with reference to other western Bass Strait islands and annotated lists of the vertebrate fauna. *Records of the Queen Victoria Museum* **40**, 1–43.

Milledge DR (1980) 'The distribution, status and ecology and evolutionary position of the forty-spotted pardalote'. Proceedings 78th annual conference, Hobart. Royal Australasian Ornithological Union, Melbourne.

Rounsevell DE, Woinarski JCZ (1983) Status and conservation of the forty-spotted pardalote, *Pardalotus quadragintus* (Aves: Pardalotidae). *Australian Wildlife Research* **10**, 343–349. doi:10.1071/WR9830343

Saunders DL, Tzaros CL (2011) 'National recovery plan for the swift parrot *Lathamus discolor*'. Birds Australia, Melbourne.

Threatened Species Section (2006) 'Fauna recovery plan: forty-spotted pardalote 2006–2010'. Department of Primary Industries and Water, Hobart.

Threatened Species Section (2012) 'Listing statement for *Pardalotus quadragintus* (forty-spotted pardalote)'. Department of Primary Industries, Parks, Water and Environment, Hobart.

Woinarski JCZ, Bulman C (1985) Ecology and breeding biology of the forty-spotted pardalote and other pardalotes on north Bruny Island. *Emu* **85**, 106–120. doi:10.1071/MU9850106

Woinarski JCZ, Rounsevell DE (1983) Comparative ecology of pardalotes, including the forty-spotted pardalote, *Pardalotus quadragintus* (Aves: Pardalotidae) in south-eastern Tasmania. *Australian Wildlife Research* **10**, 351–361. doi:10.1071/WR9830351

15

Broad-scale feral predator and herbivore control for yellow-footed rock-wallabies: improved resilience for plants and animals = Bounceback

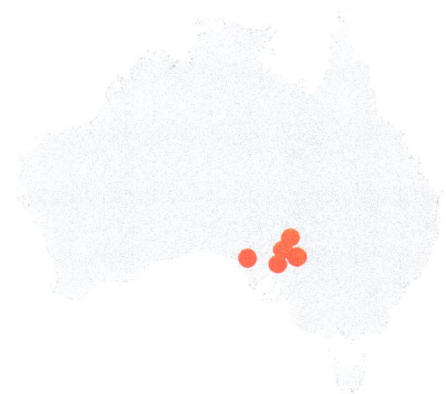

Robert Brandle, Trish Mooney and Nicki de Preu

Summary

The problem
1. Feral animals, particularly goats, rabbits, foxes and feral cats, have caused widespread habitat degradation and prevented ecological recovery on conservation reserves in South Australia's semi-arid ranges.
2. By the early 1990s, remnant populations of the yellow-footed rock-wallaby (*Petrogale xanthopus* ssp. *xanthopus*) were isolated and declining.

Actions taken to manage the problem
1. The Bounceback program was instigated to tackle key threatening processes and recover populations of the yellow-footed rock-wallaby.
2. Goat populations were reduced using musters followed by ground and aerial shooting, at least annually across multiple tenures and managers.
3. Large-scale rabbit warren ripping and blasting programs removed rabbits from target locations. This was followed by broad-scale reductions through the spread of rabbit haemorrhagic disease in the mid-1990s
4. Broad-scale biennial aerial baiting for foxes, alternating with ground baiting, has effectively removed resident fox populations across large areas.

Markers of success
1. Populations of yellow-footed rock-wallaby and other target species increasing.
2. The conservation status of the rock-wallaby has been downlisted.
3. Causes of decline were identified and some are now effectively managed.
4. Increasing community involvement and support has had collateral benefits of the program for neighbouring pastoral lands.

Reasons for success
1. Goat and fox control are likely to have contributed most to the remarkable recovery of the yellow-footed rock-wallaby across their range in South Australia.
2. The enthusiasm and commitment of South Australian Environment Department staff, land managers and volunteers has been a major driver of the success of the Bounceback program.

Introduction

Unpublished explorer records suggest that the yellow-footed rock-wallaby (*Petrogale xanthopus* ssp. *xanthopus*) was abundant before European influence on the Australian landscape. From the latter half of the 1800s up to 1919, however, hunting appears to have depleted populations (Copley 1983) along with habitat degradation due to high densities of livestock, especially around natural waterholes and springs. Predation by red foxes (*Vulpes vulpes*) and competition for resources by goats (*Capra hircus*) were implicated in further declines of the species after the fur trade was stopped (Lim *et al.* 1987).

Surveys and population studies conducted in the late 1970s and early 1980s identified a significant reduction in yellow-footed rock-wallaby numbers across their former range in South Australia, with many local populations extirpated and others at risk of loss (Copley 1983). This decline led to the species' being listed as Vulnerable under the South Australian *National Parks and Wildlife Act 1972* and as Vulnerable nationally under the *Environment Protection and Biodiversity Conservation Act 1999*, given similar declines elsewhere in its limited range beyond South Australia.

Conservation management

Research, biology, identification of key threats

The yellow-footed rock-wallaby is a habitat specialist with a penchant for sheer cliffs and steep-sided gorges (Lim *et al.* 1987). Deep crevices in these cliffs provide critical shelter from extreme summer heat and protection from ground-based predators such as foxes and dingoes (*Canis familiaris*). Research by Lethbridge and Sharp in South Australia (Lethbridge and Alexander 2008; Sharp *et al.* 2015) and on the black-footed rock-wallaby (*P. lateralis*) in Western Australia (Kinnear *et al.* 2010) has shown that controlling foxes makes a substantial difference to juvenile survival, and therefore to long-term colony viability. Anecdotal evidence supporting feral animal control being beneficial for the yellow-footed rock-wallaby came from a property in the Olary Ranges, where the efforts of one landholder in maintaining extremely low goat and fox numbers through ongoing shooting was probably an important driver in preventing the species disappearing from that region.

The impact of fox predation is likely to be the major factor preventing expansion of rock-wallaby colonies from refuge habitat with shelter sites that are inaccessible to predators. However, unmanaged goat populations can deplete rock-wallaby food resources in refuge habitats (Lethbridge and Alexander 2008), thus increasing the vulnerability of colonies to predation by forcing animals to forage further away from protection. Strong recovery of rock-wallabies in areas managed in the Bounceback program provides strong support for this, because rock-wallabies now occupy numerous rocky sites many kilometres away from areas with inaccessible shelter sites. This suggests that rock-wallabies have dispersed from core refugia and that formerly isolated populations are probably reconnecting across the landscape.

Bounceback Program

In the early 1990s, a small group of rangers and wildlife managers recognised the need for decisive intervention to protect the semi-arid ranges of South Australia from further species' losses and ongoing habitat degradation. This long-term conservation program, known as 'Bounceback', is based on the recognition that an integrated landscape-scale approach to threat management would provide broader benefit than a single species approach (Alexander and Naismith 2001). Initial focus was on reducing total grazing pressure from goats and rabbits, and predation pressure on rock-wallabies by foxes and cats in Ikara-Flinders Ranges National Park and one property in the Olary region. The success of the program in reducing grazing pressure was enhanced by the arrival of rabbit haemorrhagic disease in 1994 (Mutze et al. 2016). The decline in introduced herbivores also helped reduce the abundance of introduced predators, which were being targeted with dried meat 1080 baits and regular volunteer based shooting. The reduction in grazing pressure due to effective control of goat and rabbit population and ongoing management of kangaroos led to contour furrow and direct seeding being trialled to promote vegetation recovery on scalded areas of the park – the result of historic overgrazing by livestock.

By 1998, initial successes in reducing fox, goat and rabbit abundance led to South Australian Government and Australian Government support for an expanded program that incorporated additional parks and private landholdings in the northern Flinders Ranges. Within this framework, the parks were seen as core protection areas with a key management role played by dedicated park rangers. Neighbouring pastoral holdings and private sanctuaries were encouraged to be involved so that these core areas were buffered from re-invasion of pest species, as well as increasing the areas of managed habitat for recovering plants and animals (Alexander and Naismith 2001).

From 2000–2016, Bounceback has involved numerous landholders in coordinated pest management, including goat, fox and weed control. In 2016, management activities involved five South Australian Government-run parks, several of which are co-managed with the traditional Adnyamathanha owners, and 12 privately managed properties. Over 16 different organisations, including private companies, universities and volunteer organisations, have contributed to the success of Bounceback. In particular, volunteers from the Conservation and Wildlife Management Branch of the Sporting Shooters Association of Australia (SA) have been involved from Bounceback's inception and have made an enormous contribution to control of feral goats, foxes and feral cats in target areas (Fig. 15.1).

> '... I believe that the partnership has been so successful because the Conservation and Wildlife Management Branch developed and practised high standards of training and field operations, to ensure safety, animal welfare and effectiveness. Environment Department staff have supported the partnership with appropriate management and policies, such as operational sequencing of muster, ground shoot, and aerial shoot, to maximize benefits and effectiveness.' Kaz Herbst - Conservation and Wildlife Management Branch of the Sporting Shooters Association of Australia (South Australia; DENR 2012)

Planning and policy

The planning around Bounceback did not involve a recovery plan because the focus was broad-scale ecological restoration through threat management, with the yellow-footed rock-wallaby being the 'flagship species'. Although the initial aims were a significant

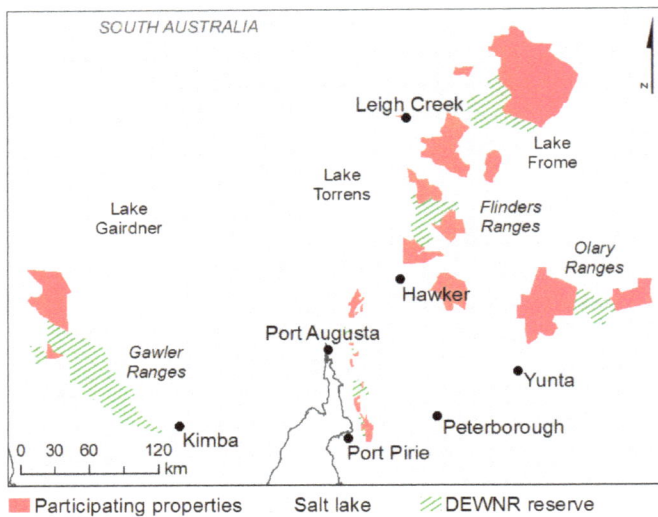

Fig. 15.1. Locations of Bounceback activities in South Australia.

reduction of goats, rabbits, foxes and cats, more comprehensive recovery indicators were subsequently developed. These included:

- increase in the diversity and density of perennial native grasslands, long-lived shrubs and trees
- increase in common fauna, such as kangaroos, emus, other birds, and reptiles, as well as threatened species such as the yellow-footed rock-wallaby
- successful reintroductions of one or more regionally extinct species
- reduction in populations of introduced plants and animals
- connected habitats resilient to drought and reduced rates of erosion.

Bounceback has had varying success in achieving these aims, and has also found it challenging to monitor changes effectively, especially at a broader scale. There have been some solid gains with reductions in introduced fauna populations across the range of the yellow-footed rock-wallaby. However, although the desired increase in condition of native grasslands and other habitats has been sustained in some locations, notably the Vulkathunha-Gammon Ranges National Park and the steep ranges in all areas, this has been harder to achieve in other areas where the early impacts of heavy stocking rates were severe. This is demonstrated through several herbivore exclusion experimental plots that demonstrate that high kangaroo numbers have hampered the recovery of native vegetation diversity, particularly so in particular perennial tussock and hummock grasslands.

Fauna reintroductions have met mixed fates. After only 7 years of the program (i.e. in 2005), the brush-tailed bettong (*Bettongia penicillata*) was reintroduced to two areas of Ikara-Flinders Ranges National Park, but this attempt failed with resource availability and foxes implicated (K. Bellchambers *pers. comm.*). Subsequently, fox control has become more effective and extensive, with aerial baiting commencing in 2007 and now covering more than 7000 km^2. This more-effective fox management, plus 14 more years of landscape recovery, provided confidence to trial translocations of two species with different ecological roles: a predator, western quoll or idnya (*Dasyurus geoffroii*); and a herbivore, common brushtail possum or virlda (*Trichosurus vulpecula*). Released during 2014 and

2015, populations of both species persist in the release environment. Monitoring has shown that feral cats may present a potential long-term threat to these reintroductions, particularly through droughts if either populations are reduced to a limited number of drought refuge areas (Pavey et al. 2017; Scheele et al. 2017). The current focus is therefore directed at researching and implementing effective cat management strategies to maximise early population expansion of reintroduced species.

The aim of increasing populations of existing common fauna has been met with measurable increases in larger terrestrial species such as the echidnas (*Tachyglossus aculeatusi*), sand goanna (*Varanus gouldii*), carpet python (*Morelia spilota*), blue tongue (*Tiliqua scincoides*) and sleepy lizard (*Trachydosurus rugosus*). However, increases in larger macropod species, such as the red (*Macropus* rufus) and western grey (*M. fuliginosus*) kangaroos and euro (*M. robustus*), have had some negative consequences. Their increased grazing pressure has resulted in some reversals for grasses that were recovering after stock removal and rabbit control, and the continuing high densities of macropods has the potential to impede vegetation recovery, increase erosion and promote weed invasion.

People, agencies, governance and accountability

Bounceback was born out of the enthusiasm of a coalition of park managers and Adelaide-based wildlife management staff to make real conservation improvement on recently acquired pastoral properties now known as Ikara-Flinders Ranges National Park and provide support for land managers' efforts in the Olary Ranges (DENR 2012). These instigators of the Bounceback Program and the gains that they could demonstrate were able to elicit enthusiasm and support from South Australian and Australian Government agencies, politicians, subsequent park staff and volunteer groups to expand the program across the ranges. The resourcing that this enthusiasm garnered enabled the program to employ a person to ensure that a coordinated approach was (and continues to be) maintained. A combination of ongoing state government financial support, with injections of extra funds through Australian Government environment/Landcare initiatives has also been crucial in taking the program beyond park boundaries and into privately managed land. A big plus for extending the program across the landscape has been the recognised benefits of fox, goat, rabbit and weed control to both conservation and agricultural production, such as successful lambing in flocks of sheep (*Ovis aries*).

As the Bounceback program has expanded and then been faced with budgetary constraints, some areas of focus, such as rabbit control, have suffered as decisions about landscape-scale effectiveness had to be made. Had rabbit haemorrhagic disease not been so effective in reducing the threat posed by rabbits, the program may have evolved quite differently, highlighting the need to have clear outcome-focused goals and the potential for a flexible response to the severity of those threats.

Decisions about management priorities are made in collaboration with on-the-ground managers who are also encouraged to attend annual workshops for information sharing and planning. The program has survived regular staff turnover through the dedication and commitment of all involved, including the assistance and goodwill of landholders who have been involved with the program since its inception. The traditional Adnyamathanha, Kokatha and Nadjuri owners have been excited and appreciative of the recovery of rock-wallabies and the return of locally extinct totem animals (idnya and virlda) made possible through Bounceback. Bounceback is now recognised and appreciated by the local communities where it works, the South Australian Government and non-government environmental organisations, and wildlife management researchers across Australia. This recognition comes from the involvement of community and volunteer groups, the adoption of effective management practices across land tenures at a landscape scale and the strong success story with yellow-footed rock-wallaby recovery.

Monitoring

Bounceback has invested in monitoring of both threatening processes (pests) and the response of native fauna and habitats to management intervention. Threat monitoring has involved car-based spotlight transects to monitor predator (foxes and cats; Fig. 15.2), and herbivores (kangaroos and rabbits), which are completed quarterly in Ikara-Flinders Ranges National Park, with less regular spotlight surveys in other locations. This work has included comparing data from unbaited control areas, and this comparison highlights the importance of rabbit haemorrhagic disease-related rabbit suppression (Fig. 15.3). Limitations with these data are that they focus on one geographic location and a limited group of habitat types within the Bounceback program that are not representative of other areas. Nevertheless, the story that these data tell is compelling and provides confidence in the effectiveness of the approach. Because aerial fox baiting provided more effective fox control from 2007, cat sightings have increased, and rabbits recovered, indicating that semi-dried meat baits are not effective for reducing cat populations in this region. The absence of foxes within the baited areas has been confirmed by camera trapping data since 2013, with these camera traps deployed primarily to monitor reintroduced species and predator activity. This camera trapping monitoring program is being expanded out across the Bounceback footprint to improve standardisation of methods and coverage across area and habitat types.

Fauna response monitoring has focused on the yellow-footed rock-wallaby, with almost annual helicopter surveys to 1989–2012 along fixed transects in the Flinders and Olary Ranges (Fig. 15.4) and mark–recapture studies in the Gawler Ranges, which estimated populations to have risen from 50 in 2005 to 350 by 2010 (DEWNR *pers. comm.*). Both methods have demonstrated population increases. However, reduced funding since

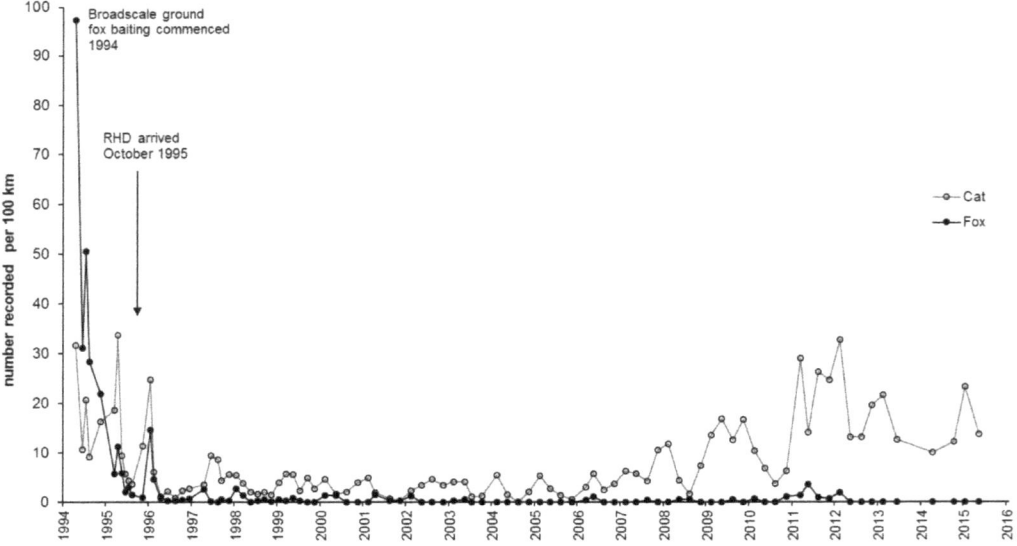

Fig. 15.2. Fox and feral cat response to 1080 fox baiting and rabbit haemorrhagic disease virus in Ikara-Flinders Ranges National Park.

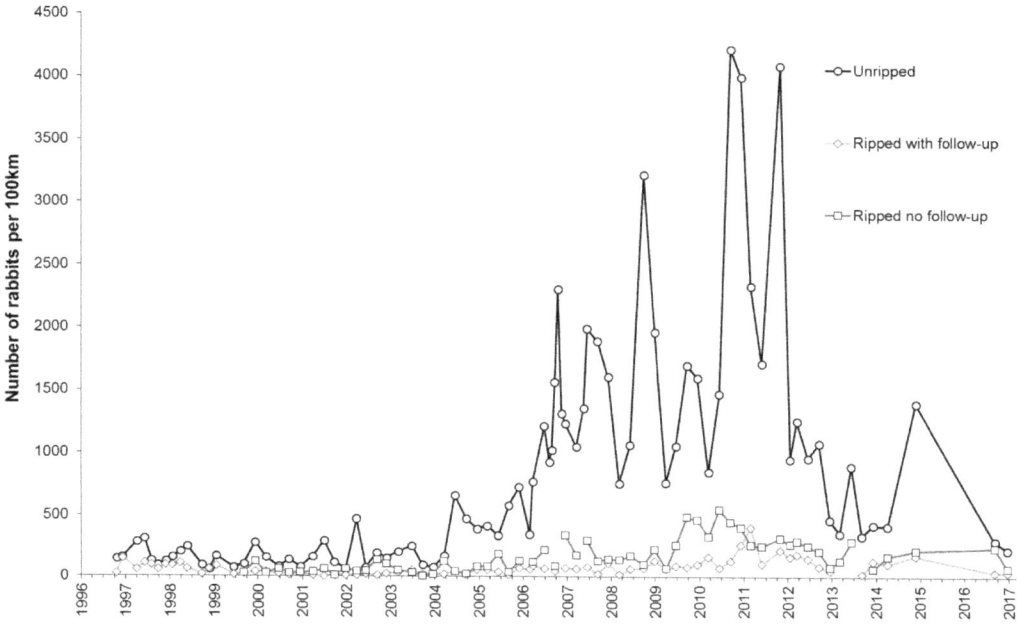

Fig. 15.3. Rabbit spotlight monitoring data in Ikara-Flinders Ranges National Park.

2012 has led to cessation and review of the required frequency of aerial and mark–recapture monitoring.

Although there is no reliable way to estimate populations of goats, the aerial shoot provides a good estimate of return for effort on which to monitor relative abundance within managed areas.

Since 1999, there has been some effort to monitor the impact of herbivores on vegetation. This work, based on fenced plots that exclude herbivores, has provided important information on the relative impacts of rabbits versus kangaroos and goats on perennial

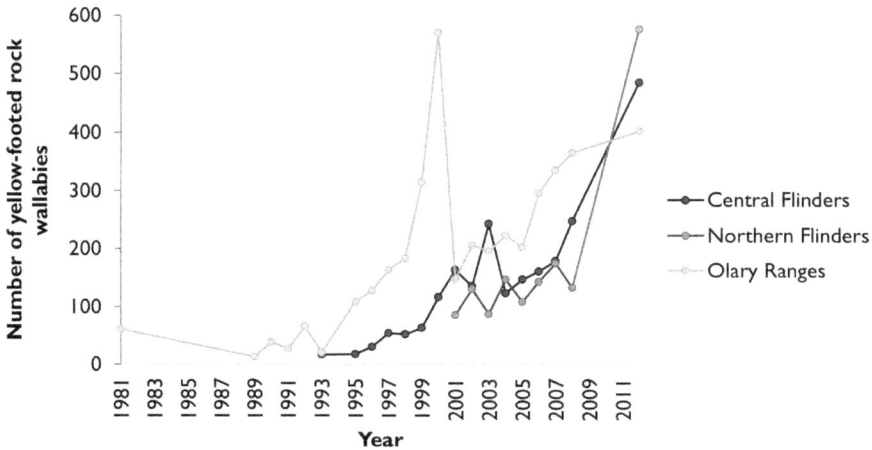

Fig. 15.4. Yellow-footed rock-wallaby response in three Bounceback regions.

grasses and shrubs and the need to also manage kangaroos when recovering degraded landscapes (DEWNR *pers. comm.*).

Photo points with associated vegetation monitoring sites have been established and revisited at regular intervals on all reserves in the Bounceback project area to track changes over time. Regeneration of long-lived palatable perennial native plant species such as bullock bush (*Alectryon oelifolius*), iga (native orange *Capparis mitchellii*) and various bluebush (*Maireana*) and saltbush (*Atriplex*) species is demonstrating positive signs of habitat recovery in many areas. Randomised land condition assessments using the South Australian Pastoral Assessment method have shown improvement across the four national parks participating in the Bounceback program (DEWNR *pers. comm.*).

Baseline investigations and research projects have provided opportunities to quantify changes in vegetation, bird communities and small mammal and reptile communities. Some of these baseline studies have been revisited and the results support the program's recovery aims of landscape recovery (DEWNR *pers. comm.*). However, these types of studies are best suited to decadal scale changes, and are not useful in guiding short-term (annual) management. For this reason, rapid herbivore impact assessment methods have been developed that can provide managers with feedback on the effectiveness of the previous year's herbivore control. Over time, these will build a picture of long-term trends for the recovery or decline of target plant groups and species. The proposed annual camera trapping will hopefully detect changes in long-term trends for species other than those being controlled. To date, camera trapping has shown that the echidna and sand goanna are significantly more likely to be recorded in Bounceback-managed areas.

Infrastructure

National Parks' historic infrastructure such as shearers' quarters and sheds have been an important part of the Bounceback program through providing storage for equipment, accommodation for volunteers contractors, and non-resident staff, and as venues for workshops and meetings.

Landscape-scale management and monitoring is greatly aided by aircraft, which provide relatively cheap and much more effective transport and delivery of baits and some management actions compared with on-ground delivery systems. The use of aircraft to monitor survival of reintroduced fauna also has been critical for determining cause of death of reintroduced individuals. For example, their use enabled us to determine that cats are the major threat to successful translocations of western quolls when foxes are absent.

Money

South Australian Government funding has ensured that the Bounceback program has a coordinator and ground-based staff support. Australian Government competitive grant programs including the Natural Heritage Trust, Caring for Our Country, Biodiversity Fund and the National Landcare Program have enabled Bounceback to provide incentives for private landholders to engage in coordinated feral animal and plant control programs, as well as contracting feral control specialists where required. South Australian Government ecologists have assisted with the development, collection and analysis of monitoring data, as well as encouraging university research projects aimed at answering specific questions relating to the program. There has been significant support from volunteer groups in removing feral predators and herbivores, collecting monitoring data, removing weeds and supporting all aspects of the program on private lands. The average annual value of the project in current dollars is likely to be in excess of A$750 000 per annum, corresponding to an investment of A$18 million since 1992.

The future

As a consequence of the Bounceback program, the conservation status of the yellow-footed rock-wallaby has improved. A recent regional review in South Australia rated the species as Rare but with an increasing trend (Regional Species Conservation Assessment Project – South Australian Arid Lands (Outback) Fauna, May 2013), and an IUCN red list evaluation during 2015 rated the species as Near Threatened (Copley *et al.* 2015).

Bounceback has demonstrated that we can recover a species by targeting the main threats to it at landscape scales. With ongoing commitment to manage goats to low levels at landscape scales and ongoing fox control through 1080 baiting, the future for rock-wallabies appears secure. The recent reintroduction of two mammal species is the result of the South Australian Government Environment Department partnering with the Foundation for Australia's Most Endangered Species (FAME): a not-for-profit fundraising body that has enabled the western quoll and the common brushtail possum to be reintroduced into Bounceback-managed lands. This has instigated a high level of community interest and investment into maintaining the hard fought gains of the last 24 years. Successful establishment of either of these species beyond 2017 will see plans for expanding these founder populations into other suitable areas.

The Bounceback Program is an important ongoing program for the regional environment arm of the South Australian Government and involving the community in natural resource management activities. Because threat management problems do not disappear, the ongoing maintenance of this program depends on its ability to maintain community support by remaining relevant and demonstrating effective conservation outcomes to funding bodies. Although core management activities can maintain rock-wallabies and (potentially reintroduced species with some extra investment in cat management), the broader aims of the program will rely on reducing total grazing pressure across formerly degraded landscapes. This is likely to require added investment in management and innovative policy development.

Conclusion

The main drivers for the success of Bounceback in assisting the recovery of the yellow-footed rock-wallaby include: thinking broad scale; dedicated staff and volunteers; partnering with the community to deliver outcomes; ongoing commitment to maintaining (and when possible expanding the program); learning from mistakes; adopting innovation; and sharing success.

Acknowledgements

Staff and other long-term contributors include Alan Harbour, Andy Sharp, Arthur Coulthard, Bob Starkey, Brenton Arnold, Cathy Zwick, Chris Havelberg, Chris Holden, Christine Arnold, Craig Nixon, Damien Pearce, Danny Doyle, Darren Crawford, Darren McKenzie, Darren Wilson, David Peacock, Dick Hoare, Dylan Koerner, Frank Bernhardt, Graham Miller, Greg Mutze, Greg Patrick, Ian Falkenberg, Jae Ellis, Jason VanWeenen,

John McDonald, Justin Jay, Kaz Herbst, Keith Bellchambers, Kristian Coulthard, Lorraine Edmunds, Lucy Dodd, Matt Kennewell, Mark Lethbridge, Michael Freak, Michael Trebilcock, Nicki de Preu, Peter Alexander, Peter Baker, Peter Copley, Peter Watkins, Scott Jennings, Sian Johnson, Steven Kowalick, Stuart Beinke, Tania McKenzie, Tim Strangways, Tom McIntosh, Trevor Naismith, Wade Johnson.

Past and present participating pastoral and conservation properties include Alpana Station, Angepena Station, Arkaba Station, Arkaroola Wilderness Sanctuary, Aroona Sanctuary, Buckaringa Sanctuary, Boolcoomatta Reserve, Commodore Station, Bunkers Conservation Reserve, Edeowie Station, Glenlyle Station, Gum Creek Station, Hiltaba Reserve, Holowiliena Station, Holowiliena South Station, Merna Mora Station, Motpena Station, Nantawarrina Indigenous Protected Area, Narrina Station, Pinda Springs, Plumbago/Mt Victor Stations, Rasheed Run Wilpena, Pinda Springs, Rawnsley Park Station, Warraweena Sanctuary, Wertaloona Station, Willow Springs Station, Witchelina Reserve, Wooltana Station, Yadlamalka Station, Yankaninna Station.

Other partnering organisations include the Adnyamathanha community, Arkaroola Wilderness Sanctuary, Bushwalking and 4WD clubs, Australian Wildlife Conservancy, Bush Heritage Australia, Conservation & Wildlife Management Branch of the Sporting Shooters Association of Australia (SA), Friends of Parks groups, Landscape Partnerships, Nature Foundation SA, University researchers, Yellow-footed Rock-wallaby Preservation Association, Zoos SA, Foundation for Australia's Most Endangered Species (FAME).

References

Alexander P, Naismith T (2001) Ecological Recovery in the Flinders Ranges based on integrated predator and competitor control strategies. In *A Biological Survey of the Flinders Ranges, South Australia 1997–1999*. (Ed. R Brandle). pp. 35–37. National Parks and Wildlife SA, Department for Environment and Heritage, Adelaide.

DENR (2012) 'Celebrating 20 years of Bounceback'. Department of Environment and Natural Resources, Adelaide, <http://www.naturalresources.sa.gov.au/aridlands/plants-and-animals/native-plants-and-animals/bounceback>.

Copley P (1983) Studies on the yellow-footed rock-wallaby, *Petrogale xanthopus* Gray (Marsupialia: Macropodidae) I. Distribution in South Australia. *Wildlife Research* **10**, 47–61. doi:10.1071/WR9830047

Copley P, Ellis M, van Weenen J (2015) IUCN web page: *Petrogale xanthopus* (Ring-tailed Rock Wallaby, Yellow-footed Rock Wallaby). IUCN, Gland, Switzerland, <http://www.iucnredlist.org/details/16750/0)>.

Kinnear JE, Krebs CJ, Pentland C, Orell P, Holme C, Karvinen R (2010) Predator-baiting experiments for the conservation of rock-wallabies in Western Australia: a 25-year review with recent advances. *Wildlife Research* **37**, 57–67. doi:10.1071/WR09046

Lethbridge MR, Alexander PJ (2008) Comparing population growth rates using weighted bootstrapping: guiding the conservation management of *Petrogale xanthopus xanthopus* (yellow-footed rock-wallaby). *Biological Conservation* **141**, 1185–1195. doi:10.1016/j.biocon.2007.09.026

Lim TL, Robinson AC, Copley PB, Gordon G, Canty PD, Reimer D (1987) The conservation and management of the yellow-footed rock-wallaby *Petrogale xanthopus* Gray, 1854. Special Publication No. 4. National Parks and Wildlife Service and Department of Environment and Planning, Adelaide.

Mutze G, Cooke B, Jennings S (2016) Density-dependent grazing impacts of introduced European rabbits and sympatric kangaroos on Australian native pastures. *Biological Invasions* **18**, 2365–2376. doi:10.1007/s10530-016-1168-4

Pavey CR, Addison J, Brandle R, Dickman CR, McDonald PJ, Moseby KE, *et al.* (2017) The role of refuges in the persistence of Australian dryland mammals. *Biological Reviews of the Cambridge Philosophical Society* **92**, 647–664. doi:10.1111/brv.12247

Scheele BC, Foster CN, Banks SC, Lindenmayer DL (2017) Niche contractions in declining species: mechanisms and consequences. *Trends in Ecology & Evolution* **32**, 346–355. doi:10.1016/j.tree.2017.02.013.

Sharp A, Norton M, Havelberg C, Cliff W, Marks A (2015) Population recovery of the yellow-footed rock-wallaby following fox control in New South Wales and South Australia. *Wildlife Research* **41**, 560–570. doi:10.1071/WR14151

16

Recovering the mountain pygmy-possum at Mt Blue Cow and Mt Buller

Linda Broome, Dean Heinze and Mellesa Schroder

Summary

The problem
1. The mountain pygmy-possum is a highly specialised marsupial with a small total population now restricted to fragmented subpopulations occurring in a narrow range of habitats in a few mainland alpine and subalpine sites.
2. Discovered as a living animal in 1966, many of the known populations have subsequently declined in the face of a complex range of threats of local, regional and global origins.

Actions taken to manage the problem
1. Long-lasting and detailed survey and research effort have been undertaken to map populations and habitats, and understand ecology.
2. Monitoring of the population size and genetic heterogeneity has been sustained for decades.
3. Management partnerships with one of the major stakeholders, ski-resort managers, have helped raise profile and reduce some acute threats.
4. Innovative and targeted management responses, including gene pool mixing and specially designed culverts to maintain dispersal.

Markers of success
1. The previous declining trend was first stabilised then reversed so that the population has returned to previous levels.
2. The causes of decline were identified and most are now being effectively managed.
3. A recovery plan has been developed and is being implemented by a well-managed and enduring recovery team.
4. The community has been engaged in the project and is providing ongoing support.

Reasons for success
1. The species has an intriguing ecology and a history with strong public appeal.
2. A robust evidence base about ecology and threats has been developed from intensive and long-lasting research programs by dedicated research champions.
3. Effective monitoring has been linked to strategic and practical conservation management developed with buy-in by stakeholders and managers who have been prepared to adopt innovative, but strongly evidenced-based, actions.
4. Long-term support and resourcing through a mix of government and non-government funds.

Introduction

The mountain pygmy-possum (*Burramys parvus*) is a small terrestrial marsupial endemic to the high country of the Australian mainland (Fig. 16.1). It was described from fossil remains found at Wombeyan Caves, near Mittagong, New South Wales in 1895, found as a living animal in a ski lodge at Mt Hotham Victoria in 1966 and located in the surrounding habitat in 1971. It was found in the southern Kosciuszko region of New South Wales in 1970 and on the highest peak of the Australian mainland, Mt Kosciuszko (2228 m) in 1972 (Calaby *et al.* 1971; Mansergh and Broome 1994). A population was discovered on Mt Buller, Victoria in 1995 (Heinze and Williams 1998) and another in northern Kosciuszko National Park as recently as 2010 (Schulz *et al.* 2012). The lower altitudinal limit of currently known populations is around 1300 m in the montane zone at Mt Buller and 1200 m in northern areas of Kosciuszko National Park. This distribution roughly corresponds with the lower limit of the winter snowline, around 1200–1400 m (Slatyer *et al.* 1985; Broome *et al.* 2012 and references therein). Fossil and sub-fossil remains recovered from caves in Victoria and New South Wales suggest that at the height of the last Pleistocene glacial period (~20 000 years BC), and perhaps more recently, the mountain pygmy-possum had a much wider distribution in south-eastern Australia. Since that time, its range appears to have contracted with a gradually warming climate and receding snowline (Caughley 1986; Heinze *et al.* 2004).

Due to the small number of records and lack of knowledge of its distribution, the mountain pygmy-possum was first listed under Schedule 12 Part 2 (Vulnerable and Rare) under the New South Wales *National Parks and Wildlife Act 1974*. Following further

Fig. 16.1. The mountain pygmy-possum (*Burramys parvus*) eating seed of mountain plum-pine (*Podocarpus lawrencei*) (photo: Linda Broome).

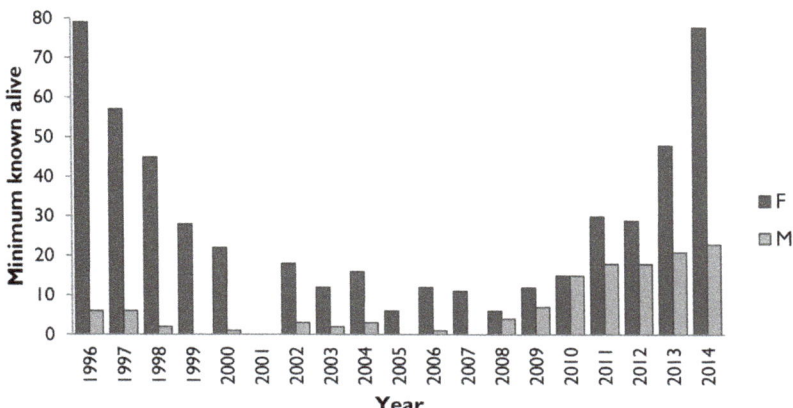

Fig. 16.2. Abundance of mountain pygmy-possums (minimum known alive or MKA) recorded during the annual monitoring program from the Federation and Wombat areas at the Mt Buller, 1996–2014. Note: the figures exclude any individuals that were introduced (translocated) to Mt Buller.

surveys of its distribution and abundance and assessment of ongoing threats, it was transferred to Schedule 1 (Endangered Species) of the *Threatened Species Conservation Act 1995*. It is listed as Threatened under the Victorian *Flora and Fauna Guarantee Act 1988* (with Critically Endangered categorisation under the Victorian advisory list), and Endangered nationally under the Commonwealth (federal) *Environment Protection and Biodiversity Conservation Act 1999*.

From 1997–2009, four monitored populations in the Kosciuszko region declined on average by 43%, with the largest decline of 77% at Mt Blue Cow. Declines on Mt Hotham and the Bogong High Plains were slight (~3%). However, at Mt Buller, the population declined from around 350 adults in 1996 to a low of around 40 adults in 2005–2008 (88% decline). Due to these declines and ongoing threats, a revision of the IUCN status of mountain pygmy-possum was made in 2009 to 'Critically Endangered' (Broome *et al.* 2012).

The situation has improved since 2009, with recovery of the populations at Mt Blue Cow and Mt Buller by 2016 (Figs 16.2, 16.3). Several other populations in New South Wales have increased since 2010, but populations in the Mt Hotham resort are currently low and the population at Falls Creek has not recovered following fires in 2003 and 2006 (L Broome and D Heinze unpublished).

The total population size of the species is currently estimated to be ~2750 adults (2080 females and 670 males), with ~950 in Kosciuszko National Park (650 females, 300 males), 1600 in the Mt Bogong to Mt Higginbotham region, and 200 at Mt Buller (Heinze *et al.* 2004; Broome *et al.* 2013; L Broome and D Heinze unpublished). Populations of the three regions are highly genetically distinct (Mitrovski *et al.* 2007).

Conservation management
Victoria
Mapping of distribution, research on ecology and conservation requirements, and long-term monitoring of mountain pygmy-possums was commenced by Ian Mansergh and colleagues in 1982. The largest population was in the ski resort of Mt Hotham; populations also occurred at Falls Creek Ski Resort and a scattering of sites across the Bogong High Plains, located between Mt Hotham and Mt Bogong (Fig. 16.4). Key threats were identified

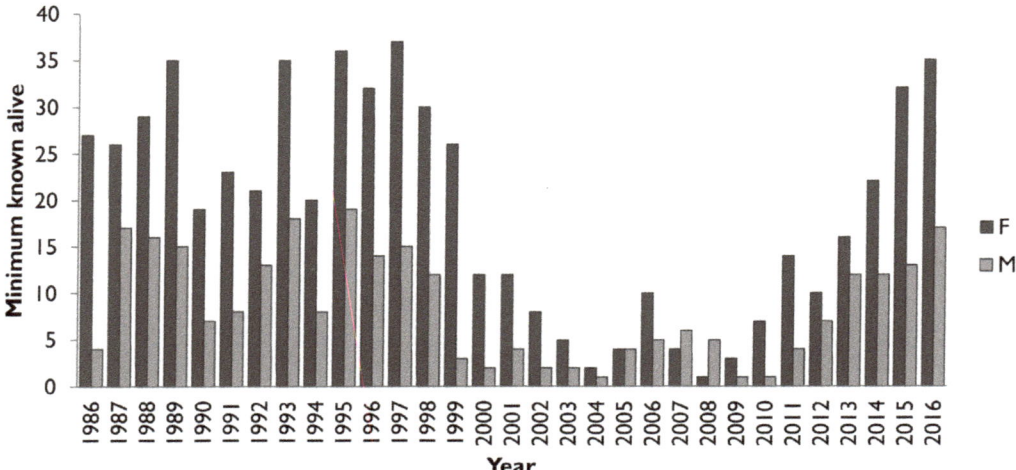

Fig. 16.3. Abundance of mountain pygmy-possums (minimum known alive or MKA) recorded during the annual monitoring program at Mt Blue Cow 1986–2016. Total habitat area ~5 ha.

as loss and fragmentation of habitat (boulder fields with associated shrubby heath) due to construction of roads, buildings and other structures and 'grooming' of ski slopes, in which rocks and vegetation are removed to provide a smooth skiing surface. A major access road at Mt Hotham had interrupted downhill dispersal movements of males and juveniles from the female breeding sites above the road. By constructing an artificial boulder field linked to a rock-filled culvert under the road, normal dispersal movements and social structure were resumed (Mansergh and Scotts 1989). Christened the 'tunnel of love' it received widespread media attention and became the prototype for 'possum crossing' design throughout the Australian Alps.

Ski resort development had already disturbed and isolated much of the habitat when the mountain pygmy-possum population was first found at Mt Buller in 1995 and Dean

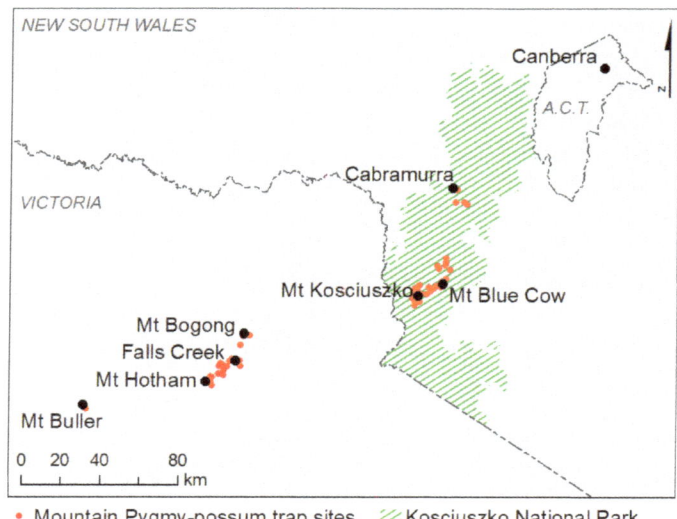

Fig. 16.4. Distribution of the Mountain Pygmy-possum showing locations referred to in the text.

Heinze commenced a research and monitoring program in 1996. The population declined from 1996 (Fig. 16.2) and, when only three females were trapped at the Federation ski area in 2005, the Mt Buller Alpine Resort developed the Mt Buller Mountain Pygmy-Possum Recovery Plan. The planning process brought together conservation scientists, land managers, ski lift operators and statutory authorities to provide input into a comprehensive plan, which was soon implemented. Key actions included the construction of 'possum crossings' between key areas of habitat that had been fragmented due to ski field developments, ongoing planting of thousands of food-producing plants in disturbed habitat and adjacent areas, as well as an integrated predator-control program targeting cats, foxes and wild dogs. The program includes 12 summer bait stations that are checked and refreshed weekly from November to May, as well as six winter bait stations spread across the Resort. In addition, specialised shooting is undertaken for 1 week, as well as trapping for 1–2 weeks. From 2005 to 2017, over 155 feral cats were removed by trapping and shooting.

Between 2005 and 2010, modest demographic recovery was observed in the population, although genetic variation declined to less than half what it was in 1996 (Mitrovski *et al.* 2008). To remedy this decline, six males were translocated to the population from Mt Hotham (a separate regional population or evolutionary significant unit) each year in 2010 and 2011. The translocation of the six males in 2011 led to a 25% transfer of central genes in the 2012 juvenile cohort, slightly above the 20% gene flow level judged ideal for a successful genetic rescue (Weeks *et al.* 2012). This is the first time that such a wild translocation has been undertaken in Australia for the genetic rescue of a population. The combination of tackling threatening processes on the ground with 'gene pool mixing' or 'genetic rescue' and long-term monitoring has been a key element to the recovery of the Mt Buller mountain pygmy-possum population. The population is now comparable to what was observed when it was first discovered in 1996.

New South Wales – Kosciuszko National Park

Preliminary surveys of abundance and distribution of mountain pygmy-possums in New South Wales were undertaken from 1980 to 1984, with an estimate of 500 adults in 8 km^2 of rock and heath habitat, all in Kosciuszko National Park (Caughley 1986). In 1985, approval was granted to develop ski resort facilities at Mt Blue Cow, with an Environmental Impact Statement concluding that the area of habitat mapped (8 ha) represented only 1% of the total population. However, due to ongoing concerns, an ecologist, Linda Broome, was engaged in 1986 to study the mountain pygmy-possums and suggest ways to ameliorate potential adverse effects. The concern was justified, because subsequent more detailed habitat mapping and continued surveys of distribution and abundance from 1986 to 2012 revealed that the largest colonies were located at the high elevations and southerly aspects favoured by ski resorts. Rather than supporting 1% of the population, the Mt Blue Cow area supports around 20% of the total population in southern Kosciuszko National Park, with 18% in the Charlotte Pass Ski Resort (Broome *et al.* 2013).

Radio-tracking of possums, studies on hibernation, and assessment of damage to the slow-growing mountain plum-pine (*Podocarpus lawrencei*), which is an important part of the habitat, resulted in installation of several 'possum crossings' across roads and ski runs, restriction of snow-grooming and exclusion of important habitat from snow sports activities (Broome 2001; NPWS 2002). There were no sustained declines in annual population numbers until 1999 (for males) and 2000 (for females) (Fig. 16.3). Feral cats had been seen occasionally during many nights of radio-tracking from 1986 to 1989, but had not appeared to be abundant. However, following discovery of a litter of kittens and evidence of cat predation on birds and skinks in 1999, and with continued low possum numbers in 2001, cat control was commenced in 2002, with cage traps set during winter throughout the resort. Thirty cats were trapped during the first winter, 20 in 2003 and three to seven every subsequent winter. Bushfires in February 2003 burnt around 80% of the vegetation cover at Mt Blue Cow. Most shrub species re-sprouted over the following years but all fire-affected mountain plum-pine plants were killed (some of these burnt plants were sectioned and were aged at up to 500 years old). Due to the extremely dry conditions and severity of burning of the humus layer, very few mountain plum-pines established from seeds, so over 1000 cuttings were taken from the remaining plants, propagated *ex situ* and planted. These plants are slowly growing but it will be many years before they provide food and cover.

With only one female possum trapped in 2008, additional introduced predator control during the summer months was initiated at Mt Blue Cow in December 2010, using soft-jaw leg-hold traps. This has also been coupled with greater awareness of the ski resort community to report feral cat sightings as part of the 'See a cat catch a cat' program. Once a cat is reported, a cage trap is deployed to the locality and the lodge monitors the cage until a cat is trapped. These measures immediately increased the numbers of cats trapped from five in 2010 to 20 in 2011. Between two and 10 foxes have also been trapped annually since 2010. The mountain pygmy-possum population increased steadily from 2010 until, by 2016, the numbers at Mt Blue Cow had recovered to pre-decline numbers (Fig. 16.3).

Identification of key threats

To attribute any single cause to population declines and recovery of threatened (or any) animal species is often not possible, because threats can be multiple, can interact and can vary on a local site level. Statistically, there is seldom enough data on the level of each threat to enable robust modelling. The decline and recovery at Mt Blue Cow and Mt Buller coincided with the Millennium Drought of 1997–2009 (CSIRO 2010) and to some extent this was also reflected in other Kosciuszko National Park sites, but not as strongly at sites around Mt Hotham and the Bogong High Plains (D. Heinze unpublished). A recent PhD study (Bates 2016), which included low elevation sites in northern Kosciuszko National Park, showed that the strongest factor limiting the occupation of boulder fields by mountain pygmy-possums was a requirement for permanent water in, or close to, the habitat. Dry conditions may have contributed to the population declines and recovery on these mountain peaks, where boulder field streams are reliant on snowmelt, local runoff and summer rainfall. The declines may have been further exacerbated by relatively low numbers of migratory bogong moths (*Agrotis infusa*), which are a key part of the diet, observed in some of the drought years at Mt Blue Cow and Mt Buller during the 2000s (L. Broome and D. Heinze unpublished). Bogong moths aestivate during summer in the boulder fields occupied by possums, and their numbers are related to plant growth conditions (soil moisture, temperature and rainfall) in their winter breeding grounds on the inland plains (R. Gibson, L. Broome, M. Hutchinson, unpublished.). Low moth numbers could also lead to an interaction with predation if possums were required to spend more time searching for alternative food sources outside the shelter of boulder fields.

It is possible that cat numbers have increased with successive years of low snow cover since the early to mid-1990s. Winter mortality of cats is possibly much higher in years of deep, prolonged cover, which is occurring with decreasing frequency with the rise in global temperatures. Loss of shrub cover caused by the February 2003 bushfire and slow regeneration of the habitat at Mt Blue Cow may have limited response to the removal of cats from 2002 to 2009. However, mountain pygmy-possum numbers have increased since 2010 when we started the intensive cat-control program and there have been increases in mountain pygmy-possum numbers at two other cat-controlled sites well above what they were in the years before the Millennium Drought. Therefore, we are confident the recovery has been aided by the removal of feral cats and that predation from cats is a key threat to mountain pygmy-possum populations. The inclusion of additional 'possum crossings' by the resort operators in recent years has possibly also contributed to the recovery at Mt Blue Cow.

At Mt Buller, it appears that the mountain pygmy-possum population may already have been in decline when it was discovered in 1996 and this was most likely due to extensive ski field developments in the area during the 1980s and 1990s (Mitrovski *et al.* 2008). Genetic analysis showed that very few breeding males were successful and the very highly skewed sex ratios and low population numbers resulted in inbreeding, with significant loss in genetic diversity. This also resulted in lower numbers of pouch young and lower winter survival rates of individuals (Weeks *et al.* 2012). Males have increased and sex ratios have moderated since recovery actions were undertaken (Fig. 16.2). The relative contributions of increased genetic fitness and reduced predation rates on the wide-ranging males due to reconnecting habitat with predator-safe movement corridors and predator control, is yet to be determined.

Planning and policy

Management of the Mt Blue Cow mountain pygmy-possum habitat was initially achieved through a ski slope plan developed as part of the lease agreement. However, the vision of the first resort owners was of green ski slopes devoid of boulders and vegetation, with the possum regarded as a hindrance to development. The first Recovery Team, consisting of representatives from resort management, New South Wales National Parks and Wildlife Service and independent scientists was established after the enactment of the NSW *Threatened Species Conservation Act 1995*, which required recovery plans to be developed for all listed species. Annual meetings were held in which mountain pygmy-possum population trends, development proposals and management actions were discussed (often heatedly) and the first approved Recovery Plan was published in 2002 (NPWS 2002). The plan, based on detailed science, mapping of habitat and monitoring, was a valuable summary of the status of the possum and its interaction with resort management activities. It provided a detailed statement of ongoing management and recovery actions and responsibilities. The total estimated cost of recovery from 2002 to 2007, including in-kind staff time was approximately A$1.2 million. The meetings and the Recovery Plan served an important role in establishing communication and understanding between the resort operators, their environmental consultant, National Parks and Wildlife Service staff and scientists. However, the production and approval process of the Recovery Plan was very time consuming, and having served its function for the first 5 years, it was replaced by a Priority Action Statement in 2007. The recovery actions are now incorporated into the NSW Office of Environment and Heritage Saving our Species Program. Ongoing information sharing and consultation is facilitated by direct communication between National Parks and Wildlife Service resorts monitoring and assessments staff, Threatened Species Officers responsible for monitoring and managing the mountain pygmy-possum, the resort operators and their ecologists. The current resort operators have embraced the conservation of the mountain pygmy-possum and assist with recovery actions. Community support and awareness is quite strong, with skiers galvanised by a long-term volunteer, Karen Watson, who scold the occasional transgressor into the mountain pygmy-possum exclusion zone from the chairlifts. The public also report cat sightings in winter, enabling National Parks and Wildlife Service staff to target cats with cage traps.

In Victoria, a management strategy and guidelines for the conservation of the mountain pygmy-possum was developed and published by the Victorian Department of

Conservation, Forests and Lands in 1989. An Action Statement (No. 2) under the *Flora and Fauna Guarantee Act 1988* was published in 1991 and was revised in 2016. A large part of the possum population in Victoria occurs within three alpine resorts (Mt Buller, Mt Hotham and Falls Creek), which are areas that are primarily managed for recreation. These resorts also have obligations to manage their natural assets. Mt Buller has a dedicated Recovery Plan for the mountain pygmy-possum: it includes short-term (5 years) and long-term objectives (30 years) and these are assessed and revised as necessary every 5 years. Mt Hotham is currently preparing a recovery plan for all threatened fauna species that occur within their resort. Each resort engages an ecologist to conduct annual population monitoring. The Victorian Recovery Team includes representatives from each of the four relevant land management authorities (Mt Buller, Falls Creek and Mt Hotham Alpine Resorts, and Parks Victoria), two wildlife biologists with expertise on the species, a conservation geneticist and a representative from Zoos Victoria. The team is chaired by a representative from the Victorian Government (DELWP).

A National Recovery Plan prepared under the Commonwealth *Environment Protection and Biodiversity Conservation Act 1999* was published in 2016 (DELWP 2016). The production of the document and process of endorsement through agencies of two states and the Commonwealth was a long process: it took 5 years from when the plan was first drafted until it was endorsed. However, it is an important document for a species such as the mountain pygmy-possum, which is subject to multiple threats, occurs across state boundaries and has a high public profile. The National Recovery Plan summarises the status, recovery objectives, actions, responsibilities and estimated cost of recovery ($8 160 000) for the next 5 years. The mountain pygmy-possum is one of the 20 priority mammal species in the Australian Government's Threatened Species Strategy targeted for action to have an improved trajectory by 2020 (Australian Government 2015).

The monitoring and recovery programs at Mt Blue Cow and Mt Buller have been helped considerably by the long-term involvement of key players who have been involved in research, monitoring and drafting of recovery plans from the beginning.

Funding for the monitoring program at Mt Blue Cow was secured as part of the ski resort lease agreement in the form of an annual environmental levy. Additional funds for cat trapping were secured through the National Parks and Wildlife Service 'Find it and Fix it Program' and Centenary Funds (winter surcharge for park use) and in 2015–2016 a grant from the Commonwealth Threatened Species Commissioner under the National Landcare Program. Ongoing funding for feral animal control and monitoring actions for the next 5 years will be provided by the NSW Office of Environment and Heritage Saving our Species Program. Monitoring at Mt Blue Cow and the other Kosciuszko National Park sites has relied on the assistance of a team of up to 20 dedicated volunteers each year who have contributed over 7000 volunteer days helping with trap checking, recording of data and washing of over 16 600 traps. Many of the volunteers return each year, which is invaluable because they know the procedures and sites and can help train new volunteers.

The future

The mountain pygmy-possum is the only Australian mammal restricted to alpine and subalpine environments. Research in the 1980s and 1990s indicated it was a cold climate specialist, dependent on snow cover and therefore under severe threat from climate change. Bioclimatic models suggested its habitat would literally disappear off the top of the mountain with a 1°C rise in mean annual temperature (see citations in Broome *et al.* 2012). At the time, breeding females were not known to occur below 1400 m elevation and were mostly confined to elevations above 1600 m. The discoveries in 2010–2012 of breeding females in northern Kosciuszko National Park at elevations as low as 1200 m, where snow cover is thin and short in duration, together with the fine-scale habitat modelling of boulder field temperatures that show they can provide cool refugia even when surface temperatures are high (Shi *et al.* 2015), gives us some hope that the threat of climate change is not quite as dire as originally thought. However, the low elevation populations in northern Kosciuszko National Park are genetically restricted (Weeks 2014), so may require genetic supplementation at some point. Mountain pygmy-possums require a cool, moist environment with plentiful water (Broome *et al.* 2012; Bates 2016) and cannot cope physiologically with temperatures above 28°C (Fleming 1985). Bogong moths, which support high densities of possums, are susceptible to rising temperatures and increased frequency of droughts. However, the diet of the possum is quite broad (Gibson 2007; Hawke 2015) and it is possible it may survive in low densities even without bogong moths. Nevertheless, the predicted decrease in snow cover with rising temperatures is clearly cause for great concern. Hennessy *et al.* (2003) predicted that the total alpine area with at least 100 days of snow cover duration decreases by more than 60% with a 1°C rise in temperature above 1990 levels and is non-existent with a 3°C rise. Preliminary modelling of the survival of Kosciuszko National Park mountain pygmy-possums on the southern monitoring sites indicates that survival rates of females are reduced by up to 25% when duration of snow cover is less than 100 days (Broome *et al.* 2012). In addition to providing insulation at high elevations and perhaps decreasing survival of feral predators, snow also provides a source of water as it melts in spring and early summer. The predicted loss of snow cover with climate change, along with increasing frequency of drought, fires and high temperatures does not provide an optimistic outlook for the mountain pygmy-possum.

In 2009, the future looked so bleak that a funding drive was commenced through the Foundation for National Parks and Wildlife and Australian Geographic to raise funds for a captive breeding program with a difference: attempt to breed and adapt mountain pygmy-possums to warmer temperatures. This is still an objective in the National Recovery Plan and it may be possible to transfer populations to moist lowland forest environments in future if necessary. However, the greatest impediment is likely to be predation from cats. The still high genetic diversity within the southern Kosciuszko National Park and Bogong High Plains populations suggests that restriction to boulder field habitats may be an evolutionary recent event, perhaps precipitated by the arrival of feral cats and foxes. If we could effectively control these predators across the landscape, we may see the mountain pygmy-possum naturally expand its range to less rocky, heath habitats and even to lower elevations.

Conclusion

Since its discovery as a living animal in 1966, the mountain pygmy-possum has been the focus of a large amount of survey effort, research and public engagement, particularly because of its occurrence within ski resorts at Mt Hotham, Falls Creek and Mt Buller in Victoria, and at Mt Blue Cow and Charlotte Pass in New South Wales. Its high profile and apparent dependence on snow cover and its charismatic behaviour and endearing nature has led it to becoming an icon of the Australian Alps. Recovery of the populations at Mt Buller and Mt Blue Cow, which declined in the late 1990s, has been assisted by evidence-based management, cooperation with ski resort operators and long-term support and resourcing through a mix of government and non-government funds. The future survival of the mountain-pygmy-possum will be reliant on continued management of habitat and control of feral predators, but ultimately is likely to depend on the extent to which climate change can be limited.

Acknowledgements

Many people and organisations have assisted with the monitoring, research and management of the mountain pygmy-possum populations at Mt Buller and Mt Blue Cow. Over 600 volunteers have assisted with long-term monitoring, with nine helping for 10 years and two for longer (Neil McElhinney (21 years) and Karen Watson (17 years). Student research projects by Karen Watson, Rebecca Gibson, Hayley Bates, Haijing Shi, Tahneal Hawke have contributed immensely to understanding the ecology of the possum. Management actions could not have occurred without the assistance of ski resort and agency staff: Andrew Kennedy, Tanya Bishop, Mark Feeney (Perisher Blue Ski Resort), Rolf Klicker, Didj Hopkins (Charlotte Pass Snow Resort), Louise Perrin and Tom Kelly (Mt Buller Alpine Resort) Ben Derrick (Falls Creek Alpine Resort), Georgina Boardman and Tom Pelly (Mt Hotham Alpine Resort), Dave Woods, Tim Greville, Andrew Miller, Virginia Logan (National Parks and Wildlife Service, Jindabyne), Charlie Pascoe (Parks Victoria), Glen Johnson and Jerry Alexander (DELWP), Andrew Weeks (CESAR), Ian Mansergh, Alexandra Olejniczak, Rudi Pleschutschnig (Kingfisher Taxidermy). Cat trappers: Ian Eggleton, Bill Morris Andrew Miners and Dottie the cat detection dog. Outside agency funding sources have included: Foundation for National Parks and Wildlife, Snowy Hydro, Department of Environment and Energy, and Transgrid.

References

Australian Government (2015) Threatened Species Strategy. Australian Government, Canberra,<http://www.environment.gov.au/biodiversity/threatened/publications/threatened-species-strategy>.

Bates H (2016) Assessing environmental correlates of populations of the endangered mountain pygmy-possum (*Burramys parvus*) in Kosciuszko National Park, New South Wales. PhD thesis. The University of New South Wales, Sydney, Australia.

Broome LS (2001) Seasonal activity, home range and habitat use of the mountain pygmy-possum *Burramys parvus* (Marsupialia: Burramyidae) at Mt Blue Cow, Kosciuszko National Park. *Austral Ecology* **26**, 275–292. doi:10.1046/j.1442-9993.2001.01114.x

Broome L, Archer M, Bates H, Shi H, Geiser F, McAllan B, *et al.* (2012) A brief review of the life history of, and threats to, *Burramys parvus* with a pre-history based proposal for ensuring that it has a future. In *Wildlife and Climate Change: Towards Robust Conservation Strategies for Australian Fauna* (Eds D Lunney, P Hutchings) pp. 114–126. Royal Zoological Society of NSW, Mosman, New South Wales.

Broome L, Ford F, Dawson M, Green K, Little D, McElhinney N (2013) Re-assessment of mountain pygmy-possum *Burramys parvus* population size and distribution of habitat in Kosciuszko National Park. *Australian Zoologist* **36**, 381–403. doi:10.7882/AZ.2013.009.

Calaby JH, Dimpel M, McTaggart-Cowan I (1971) The mountain pygmy possum, *Burramys parvus* Broom, in Kosciusko National Park. *Australian Wildlife Research* **13**, 507–517.

Caughley J (1986) Distribution and abundance of the mountain pygmy-possum, *B. parvus* Broom, in Kosciusko National Park. *Australian Wildlife Research* **13**, 507–517. doi:10.1071/WR9860507

CSIRO (2010) 'Climate variability and change in South-Eastern Australia: a synthesis of findings from Phase 1 of the South Eastern Australian Climate Initiative (SEACI)'. CSIRO, Canberra.

DELWP (2016) 'National Recovery Plan for the mountain pygmy-possum *Burramys parvus*'. Australian Government, Department of Environment, Land, Water and Planning, Canberra.

Fleming MR (1985) The thermal physiology of the mountain pygmy-possum *Burramys parvus* (Marsupialia: Burramyidae). *Australian Mammalogy* **8**, 79–90.

Gibson R (2007) The role of diet in driving population dynamics of the mountain pygmy-possum, *Burramys parvus*, in Kosciuszko National Park, NSW. BSc (Hons) thesis. Division of Botany and Zoology, The Australian National University, Canberra.

Hawke T (2015) Dietary analysis of an uncharacteristic population of the mountain Pygmy-possum (*Burramys parvus*) in Kosciuszko National Park, New South Wales, Australia. BSc (Hons) thesis. The University of New South Wales, Sydney, Australia.

Heinze D, Williams L (1998) The discovery of the mountain pygmy-possum *Burramys parvus* on Mount Buller, Victoria. *Victorian Naturalist* **115**, 132–134.

Heinze D, Broome L, Mansergh I (2004) A review of the ecology and conservation of the mountain pygmy-possum *Burramys parvus*. In *The Biology of Australian Possums and Gliders*. (Eds RL Goldingay and SM Jackson) pp. 254–267. Surrey Beatty & Sons, Chipping Norton, New South Wales.

Hennessy K, Whetton P, Smith I, Bathols J, Hutchinson M, Sharples J (2003) 'The impact of climate change on snow conditions in mainland Australia'. CSIRO Atmospheric Research, Aspendale, Victoria.

Mansergh I, Broome L (1994) *The Mountain Pygmy-possum of the Australian Alps*. New South Wales University Press, Sydney.

Mansergh I, Scotts D (1989) Habitat continuity and social organisation of the mountain pygmy-possum restored by tunnel. *The Journal of Wildlife Management* **53**, 701–707. doi:10.2307/3809200

Mitrovski P, Heinze D, Broome L, Hoffmann AA, Weeks AR (2007) High levels of variation despite genetic fragmentation in populations of the endangered mountain pygmy possum *Burramys parvus* in alpine Australia. *Molecular Ecology* **16**, 75–87. doi:10.1111/j.1365-294X.2006.03125.x

Mitrovski P, Hoffman AA, Heinze DA, Weeks AR (2008) Rapid loss of genetic variation in an endangered possum. *Biology Letters* **4**, 134–138. doi:10.1098/rsbl.2007.0454

NPWS (2002) 'Approved Recovery Plan for the mountain pygmy-possum, *Burramys parvus*'. New South Wales National Parks and Wildlife Service, Sydney.

Schulz M, Wilks G, Broome L (2012) An uncharacteristic new population of the mountain pygmy-possum *Burramys parvus* in New South Wales. *Australian Zoologist* **36**, 22–28. doi:10.7882/azoo.36.1.e2048x7427x54461

Shi H, Paull D, Broome L, Bates H (2015) Microhabitat use by mountain pygmy-possum (*Burramys parvus*): implications for the conservation of small mammals in alpine environments. *Austral Ecology* **40**, 528–536. doi:10.1111/aec.12220

Slatyer RO, Cochrane PM, Galloway RW (1985) Duration and extent of snow cover in the Snowy Mountains and a comparison with Switzerland. *Search* **15**, 327–331.

Weeks A (2014) 'Population genetic structure of *Burramys parvus* in the Kosciuszko National Park'. Report to the Office of Environment and Heritage, Queanbeyan, New South Wales.

Weeks A, Kelly T, Griffiths J, Heinze D, Mansergh I (2012) 'Genetic rescue of the Mt Buller mountain pygmy-possum population'. Report to Department of Sustainability and Environment, Melbourne.

5

17

Wild orchids: saving three Endangered orchid species in southern New South Wales

Helen P. Waudby, Matt Cameron, Geoff Robertson, Rhiannon Caynes and Noushka Reiter

Summary

The problem

1. Three Endangered orchid species that occur in southern New South Wales have small disjunct populations, with estimates ranging from 12 to 1000 individual plants in each.
2. Small population size, grazing by exotic herbivores, inappropriate land management and competition from weeds threatened remaining populations.
3. The species' ecological requirements, including their association with mycorrhizal fungi and pollinators, were largely unknown.

Actions taken to manage the problem

1. On-ground management has included changes to grazing regimes via fencing, weed and fire management, and thinning of regenerating white cypress pine (*Callitris glaucophylla*).
2. Land managers and the community have been engaged via ongoing liaison around orchid habitat requirements and appropriate land management activities.
3. Techniques were developed to propagate orchids for future translocations and to augment existing populations. Investigation of the insect pollinators of each species has commenced.

Markers of success

1. Propagation techniques have been established for all three species, and several hundred orchids have been reintroduced to existing sites.

2. Community support is strong. The Country Women's Association of one small town has erected a statue to celebrate their endemic donkey orchid, the Oaklands diuris.

Reasons for success

1. Ongoing funding and in-kind support has been provided by collaborators and investors over the years, which has allowed continuity of on-ground work, such as weed control and population augmentation.
2. Land managers and local communities have been strongly engaged in the conservation program thanks to ongoing efforts by agency staff.
3. Collaboration between research and management has been strong with research being clearly directed toward meeting a management outcome (techniques to support orchid propagation).

Introduction

Globally, orchids are the largest and best known of the world's flowering plant families (Reiter *et al.* 2016), and among the most beautiful and mysterious of Australian flora. Unfortunately, many species are imperilled, with orchids constituting 17% of all Australia's threatened plants (Backhouse 2007). Over 300 orchid taxa are listed as threatened under the Commonwealth (federal) *Environment Protection and Biodiversity Conservation Act 1999*. Three Endangered orchids of southern New South Wales (listed under the *Biodiversity Conservation Act 2016 (NSW)*) include the sandhill spider orchid (*Caladenia arenaria*), crimson spider orchid (*Caladenia concolor*) and Oaklands diuris (*Diuris callitrophila*). All three species have small disjunct populations (Fig. 17.1). The isolation and small size of these populations have likely occurred through land clearing, weed competition, and grazing by livestock and other exotic animals, such as rabbits (NSW NPWS 2003; NSW DEC 2004). These species are now primarily threatened by the small number

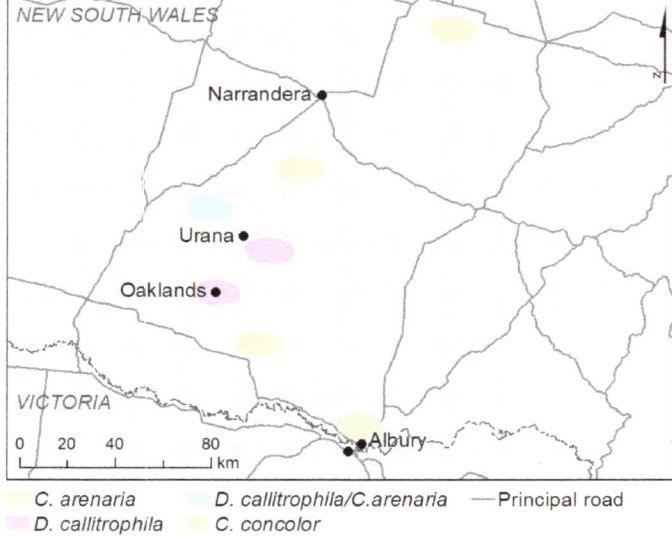

Fig. 17.1. Location of major towns and roads in proximity to orchid populations.

Fig. 17.2. (A) The sandhill spider orchid (*Caladenia arenaria*) has five pale yellow sepals and petals with red-brown tips. (B) The Oaklands diuris (*Diuris callitrophila*) typically supports up to nine whitish, mauve or purple blotched flowers per stem, each with two long lime-green hanging sepals. (C) The crimson spider orchid (*Caladenia concolor*) is a dark red or maroon spider orchid with five spreading petals and sepals (photos: Matt Cameron).

and size of populations, which increases the risk of extinction from threatening and stochastic processes (Swarts and Dixon 2009; Swarts *et al.* 2009).

The sandhill spider orchid (Fig. 17.2A) has five yellow-shaded sepals and petals that emit a soft floral bouquet. Over summer and early autumn, plants retreat underground, persisting as a tuber, and emerging in early winter as a single hairy leaf. Flowers typically emerge in the first week of September. Despite extensive surveys, the sandhill spider orchid is known only from the Riverina region of New South Wales. The total population is estimated at fewer than 2000 plants (NSW DEC 2004).

The Oaklands diuris (Fig. 17.2B; also sometimes known as the 'Oaklands doubletail') belongs to a group of orchids referred to as 'donkey orchids', owing to the prominent petals that resemble donkey ears. The species flowers in spring, producing an inflorescence of up to 50 cm in height, with white, mauve and purple flowers (Jones 2006). Seed pods are present in November, maturing and then dehiscing by mid-December (M. Cameron unpublished). Plants are then dormant for the rest of summer, retreating to an underground tuber (Jones 2006). Flowering can vary dramatically from year to year, depending on winter and spring rainfall. The Oaklands diuris grows at three sites, all in southern New South Wales. These populations are small, ranging from 50 to 800 individuals. The species is no longer found at two other sites near Urana, where it occurred in the 1990s (G. Robertson unpublished). The best known population is located in a reserve on the edge of the small rural town of Oaklands (the basis of the species' common name), where the local Country Women's Association commissioned a sculpture to highlight the importance of the species to the local community.

The dark red or maroon crimson spider orchid (Fig. 17.2C) is said to produce a scent reminiscent of a hot motor or mandarin. In our experience, the scent resembles a sweet, citrus-based cleaning spray. The crimson spider orchid is dormant in summer and for most of autumn. The species produces one sparsely haired leaf in late autumn or winter and flowers in spring. Fewer than 260 individuals are believed to exist in the wild, in fewer than five populations, in Victoria and New South Wales (NSW NPWS 2003). The only known New South Wales population is scattered through granite ridges in grassy boxy woodland at Nail Can Hill Crown Reserve, on the edge of the town of Albury (Fitzgerald 1882; Coates 2003). Historically, crimson spider orchids were probably common in the area (Coates 2003) with reports that in the 1940s Albury residents would win prizes at the local show by entering displays festooned by armfuls of crimson spider orchids (Scannell 2012). Now, the Albury population is believed to consist of fewer than 50 individuals (M. Cameron unpublished).

Conservation management

In New South Wales, recovery programs commenced for the crimson (NSW NPWS 2003) and sandhill spider orchids (NSW DEC 2004) in the late 1990s, and for the Oaklands diuris in 2001 (G. Robertson unpublished). Recovery efforts and foci have adapted over time in response to fluctuating staff and financial resources, but strong engagement with land managers, local government and other state government agencies by the Office of Environment and Heritage has ensured that core activities (i.e. maintenance of fencing, herbivore and weed management, appropriate land-use practices and monitoring) have been maintained over the years.

Initially, recovery efforts for the sandhill spider orchid focused on undertaking surveys to locate additional populations. Protection of existing populations from grazing and forestry activities through fencing and liaison with Forestry Corporation of NSW was also central to recovery efforts. Oaklands diuris recovery actions have focused on working with public and private land managers to improve management of orchid sites, including fencing of sites to exclude herbivores and thinning of white cypress pine (*Callitris glaucophylla*) regeneration, which may degrade orchid habitat if too dense (M. Cameron unpublished). Recovery efforts for the crimson spider orchid have included extensive survey work to locate new populations (none have been found to date), genetic analyses (Broadhurst *et al.* 2008), and close liaison with public land managers and other stakeholders around appropriate fire and trail management in the Albury Ranges.

Since 2012, project partners have implemented additional conservation management activities for the species, with work including orchid propagation to boost existing populations and establish new sites. Consequently, researchers at the Australian Botanic Garden Mount Annan (for the Oaklands diuris) and the Royal Botanic Gardens Victoria (for the spider orchids) have developed symbiotic propagation techniques. In 2016, the Royal Botanic Gardens Victoria began studying the pollinators of each species, to help inform the choice of reintroduction sites (N. Reiter unpublished).

Research, biology, identification of key threats

At the time of their listing, all three species were known only from relatively small populations (estimates ranged from 12 to 1000, depending on the species) occupying limited geographic areas. Surveys indicated that habitat at these sites was, in most cases, in reasonable condition, with native plant species dominant in the understorey. However, a substantial range contraction has likely occurred for all three species over the past century. For example, the sandhill spider orchid once occupied sandhills among cypress pine on the Edwards, Columbo, Yanco and Murrumbidgee Rivers (Riverina region), but extant populations now only survive in the eastern part of its historical range (NSW DEC 2004). Historical causes of declines of the surviving populations are probably related to grazing management and exotic animals, such as rabbits (*Oryctolagus cuniculus*). However, one of the primary issues now for these species, especially for the crimson spider orchid, is small population size, which may cause inbreeding depression leading to low seed set and germination (Broadhurst *et al.* 2008). Consequently, augmenting the populations with propagated individuals has become a vital recovery strategy (Swarts and Dixon 2009; Reiter *et al.* 2016).

Orchids sometimes have bizarre and complex sex lives (Rasmussen 1995), which can include unique one-on-one relationships with their pollinators (Peakall and Beattie 1996; Phillips *et al.* 2009). To identify pollinators, many hours must be spent in the field, observing visitations by insects (and other animals) to flowering orchids. Orchid seed is microscopic, being many times smaller than a grain of salt, and will not germinate without the presence of a mycorrhizal fungal partner (Rasmussen 1995). The fungus forms a symbiotic (or 'interdependent') relationship with the orchid by providing germinating seed with extra nutrients and water via hyphae (long finger-like tendrils that the fungus uses to penetrate the cells of the seed) (Warcup 1981). It is a fascinating relationship, but one that presents problems for orchid conservationists who must be able to locate and isolate the appropriate fungus to grow more orchids (Rasmussen 1995). The presence of both the mycorrhizal fungi and the pollinators is important for seed set to occur and for seed to germinate (Reiter *et al.* 2016).

Previously, none of these orchid species had been propagated with their mycorrhizal fungi, and their pollinators were completely unknown (Fig. 17.3). Consequently, undertaking the research to successfully isolate and culture the mycorrhizal partner and to develop the techniques to successfully grow the orchids has been critical to their recovery. The Australian Botanic Garden Mount Annan had early success propagating Oaklands diuris, producing around 200 plants for trialling reintroduction methods. Subsequently, the Royal Botanic Gardens Victoria successfully propagated the two spider orchids. As of 2016, the Royal Botanic Gardens Victoria has propagated around 400 crimson spider orchids, 600 sandhill spider orchids and 100 Oaklands diuris for reintroduction and/or assessment of pollinators (N. Reiter unpublished).

Planning and policy

In New South Wales, the crimson spider orchid is the subject of a draft State Recovery Plan (NSW NPWS 2003) and a National Recovery Plan (Coates *et al.* 2002). A State Recovery Plan is in place for the sandhill spider orchid (NSW DEC 2004) as well as national conservation advice (Threatened Species Scientific Committee 2015). A draft Recovery Plan was prepared for the Oaklands diuris, but was never formalised because of a move away from

Fig. 17.3. A wasp pollinator visiting a sandhill spider orchid (*Caladenia arenaria*) (photo: Noushka Reiter).

formal recovery plans in New South Wales to the more streamlined Priority Action Statements, which were introduced in 2007 (NSW OEH 2013). In New South Wales, threats and key recovery actions have been identified for all three species under the state's Saving our Species Program, which was formed following a statutory review of Priority Action Statements in 2010 (NSW OEH 2013).

People, agencies, governance, and accountability

Over the last 18 years, the Office of Environment and Heritage has spent considerable time engaging with the land managers and land owners of sites where these orchids occur. Good relationships exist between the Office of Environment and Heritage and land managers on private and public land, which has resulted in changes to grazing regimes, fencing, trial applications of fire, and thinning of white cypress pine. At a site supporting Oaklands diuris and sandhill spider orchids, stock are removed when these orchids are above ground. Intensive grazing at other times assists in maintaining habitat quality, and ensures the landholder is able to make productive use of the area. This strategic grazing regime has resulted in an increase in abundance and distribution of both species across the management site.

Managing

In 2015, the orchid recovery work received funding from the New South Wales Environmental Trust Saving our Species Partnership Grants Program for the Wild Orchids Project. This project aims to:

1. Elucidate the pollination biology of the species.
2. Propagate each species to supplement existing wild populations.
3. Reintroduce each species to suitable sites to increase the number of wild populations.
4. Continue to manage threats to existing populations through fencing, working with land managers, and managing weeds.
5. Improve community understanding and appreciation of all three species.

Importantly, the continuation of orchid recovery work is reliant on people (including community groups and individuals) being committed to the recovery work beyond the duration of funding cycles. Current funding finishes in 2021, but partners have pledged in-kind contributions to support recovery work until 2025.

Monitoring

One of the challenges of monitoring orchids is that they do not emerge or flower every year (Kindlmann and Balounová 2001; Tremblay *et al.* 2009), particularly if climatic conditions preceding flowering have been suboptimal. However, an absence of emergent plants does not mean that orchid populations have declined and dormancy should be considered when evaluating monitoring data (Tremblay *et al.* 2009), because viable tubers may be underground awaiting better conditions. Monitoring must be repeatable and long term to account for this variability and ensure that observed patterns are actually related to management activities. Surveillance monitoring over the years indicates that, in spite of improved habitat management after the commencement of the Recovery Program, at least 10-fold declines occurred in emergent populations of the sandhill spider orchid and Oaklands diuris during the 'Millennium Drought' (from 2000 to 2010). The crimson spider orchid also declined through this period, although not as much (G. Robertson unpublished). One of the benefits of the funding provided for the Wild Orchids Project by the NSW Environmental Trust (2015–2021) is that it will support the development of long-term and rigorous monitoring programs, which will be used to assess population trends over time.

Money

Over nearly 20 years, the orchid recovery project has been supported by staff from the Office of Environment and Heritage, numerous partner agencies and dedicated community members. Estimating the value of these in-kind contributions is exceedingly difficult to do accurately. The dedication of these natural resource managers kept these species 'on the agenda' and meant that the orchid conservation programs have received virtually continuous funding for recovery work since 2012. A 3-year project (2012–2014) was managed by the Office of Environment and Heritage and funded by the former Murray Catchment Management Authority via the Australian Government's Caring for our Country Program, to the value of around A$49 000. The orchids were funded for a further 2 years (2015–2016) by the Saving our Species Program to a value of around A$25 000. These initial grants meant that managers were in a good position to apply for larger grants. In 2015, the conservation programs for the three orchid species attracted nearly another A$600 000 via the NSW Environmental Trust's Saving our Species Partnerships Grants Program (the Wild Orchids Project). The program supports 10-year projects that are funded by the Trust for the first 6 years, and by partners for the final four. The Wild Orchids Project provided an opportunity to consolidate the efforts of partners and calculate the actual costs of orchid recovery. The seven partners (all Wild Orchids Consortium members) have committed over A$800 000 in in-kind contributions over 10 years to the conservation of these orchids.

The future

The latest iteration of the orchid recovery work is the Wild Orchids Project. This project is supporting propagation and pollinator studies. Most reintroductions are planned to take place in 2019. Over the duration of the project, at least seven reintroductions are planned within the species' current range. Support for the project is relatively secure with in-kind commitments from consortium members until 2024/2025.

To date outcomes from the Wild Orchids project include:

- successful isolation of mycorrhizal fungi and propagation of each species for re-introduction and *ex-situ* collections
- initiation of pollination studies of all three species
- the establishment of a more rigorous monitoring program
- considerable community education and engagement.

Several questions about the role of hybridisation with other orchid species need to be resolved for certain populations. Other challenges include ensuring that funders and the community understand the nature of orchid populations and orchid monitoring with low flowering years not necessarily meaning that populations are in decline.

Conclusion

The project's successes to date can largely be attributed to three main factors:

- ongoing resourcing of the project by the Office of Environment and Heritage and associated collaborators and funding bodies
- collaborating with people with appropriate expertise, which was critical to establishing propagation techniques for all three species
- engaging with the people with the key responsibilities for management of land that supported key orchid populations.

People love orchids, and will go to great lengths for them. In 2012, the Oaklands Country Women's Association commissioned a sculpture to celebrate the locally occurring Oaklands diuris. The 'Orchid in a Teacup' sculpture was completed in 2013 and continues to be a point of interest for visitors to Oaklands. As natural resource managers dealing with environmental problems every day, we need to remember that positive messages can be found among the gloom and that these are the stories to which community responds. Ultimately, we hope that the work being done for these species will result in such improved prospects for survival that they may one day be removed from threatened species schedules altogether.

Acknowledgements

We thank Karen Sommerville (Australian Botanic Garden Mt Annan), Shona Cowley (NSW Department of Primary Industries – Lands), Steve Campbell (Forestry Corporation of NSW), Dave Pearce (NSW National Parks and Wildlife Service), Jo Lynch and Martin Driver (Australian Network for Plant Conservation), Anthea Packer and Parklands Albury Wodonga, Paul Scannell (Albury Botanic Gardens), Albury City, Urana Shire (Federation Council), landholders, and the OEH Ecosystems and Threatened Species team. Thanks to staff of the former Murray Catchment Management Authority and the current Murray Local Land Services, including Alison Skinner, Elisa Tack, Tracy Michael, Cassie Douglas, David Costello, and Patricia Bowen. We thank Marc Freestone (Royal Botanic Gardens Victoria) for pollinator baiting assistance, Glen Johnson (Department for Environment, Land, Water and Planning) for help with Victorian crimson spider orchids, and volunteers from the Australasian Native Orchid Society Victoria for work in the Royal Botanic Gardens Victoria laboratory and shadehouse. Thanks to Graham Brown and Rod Peakall for pollinator study assistance. We are grateful to the Oaklands community, including the Oaklands Country Women's Association and Sarah Moloney. The orchid project has been funded by the Australian Government, the NSW Government's Saving our Species Program, and the NSW Environmental Trust Saving our Species Partnership Grants Program.

References

Backhouse GN (2007) Are our orchids safe down under? A national assessment of threatened orchids in Australia. *Lankesteriana* 7, 28–43.

Broadhurst LM, Scannell PK, Johnson GA (2008) Generating genetic relatedness maps to improve the management of two rare orchid species. *Australian Journal of Botany* 56, 232–240. doi:10.1071/BT07101

Coates F (2003) Action Statement 143 – crimson spider-orchid, *Caladenia concolor*. Victorian Department of Sustainability and Environment, Melbourne, <http://www.depi.vic.gov.au/environment-and-wildlife/threatened-species-and-communities/flora-and-fauna-guarantee-act-1988/action-statements>.

Coates F, Jeanes J, Pritchard A (2002) 'Recovery Plan for twenty-five threatened orchids of Victoria, South Australia and New South Wales, 2003–2007'. Victorian Department of Sustainability and Environment, Melbourne.

Fitzgerald RD (1882) *Australian Orchids*. Thomas Richards, Government Printer, Sydney.

Jones DL (2006) *A Complete Guide to Native Orchids of Australia, Including the Island Territories*. Reed New Holland, Sydney.

Kindlmann P, Balounová Z (2001) Irregular flowering patterns in terrestrial orchids: theories vs empirical data. *Web Ecology* 2, 75–82. doi:10.5194/we-2-75-2001

NSW DEC (2004) '*Caladenia arenaria* Fitz. recovery plan'. New South Wales Department of Environment and Conservation, Sydney.

NSW NPWS (2003) 'Draft recovery plan for the crimson spider orchid (*Caladenia concolor*), including populations at Bethungra and Burrinjuck to be described as two new species'. New South Wales National Parks and Wildlife Service, Sydney.

NSW OEH (2013) 'Saving our Species technical report'. New South Wales Office of Environment and Heritage, Sydney.

Peakall R, Beattie AJ (1996) Ecological and genetic consequences of pollination by sexual deception in the orchid *Caladenia tentactulata*. *Evolution* 50, 2207–2220. doi:10.1111/j.1558-5646.1996.tb03611.x

Phillips RD, Faast R, Bower CC, Brown GR, Peakall R (2009) Implications of pollination by food and sexual deception for pollinator specificity, fruit set, population genetics and conservation of *Caladenia* (Orchidaceae). *Australian Journal of Botany* 57, 287–306. doi:10.1071/BT08154

Rasmussen HN (1995) *Terrestrial Orchids: From Seed to Mycotrophic Plant.* Cambridge University Press, New York.

Reiter N, Whitfield J, Pollard G, Bedggood W, Argall M, Dixon K, Davis B, *et al.* (2016) Orchid re-introductions: an evaluation of success and ecological considerations using key comparative studies from Australia. *Plant Ecology* **217**, 81–95. doi:10.1007/s11258-015-0561-x

Scannell P (2012) The threatened crimson spider orchid, Albury NSW. *Australasian Plant Conservation* **20**, 10–11.

Swarts ND, Dixon KW (2009) Terrestrial orchid conservation in the age of extinction. *Annals of Botany* **104**, 543–556. doi:10.1093/aob/mcp025

Swarts ND, Sinclair EA, Krauss SL, Dixon KW (2009) Genetic diversity in fragmented populations of the critically endangered spider orchid *Caladenia huegelii*: implications for conservation. *Conservation Genetics* **10**, 1199–1208. doi:10.1007/s10592-008-9651-9

Threatened Species Scientific Committee (2015) *Approved Conservation Advice for* Caladenia arenaria *(sand-hill spider-orchid)*. Threatened Species Scientific Committee, <http://www.environment.gov.au/biodiversity/threatened/species/pubs/9275-conservation-advice-0110 2015.pdf>.

Tremblay RL, Perez M-E, Larcombe M, Brown A, Quarmby J, Bickerton D, *et al.* (2009) Dormancy in *Caladenia*: a Bayesian approach to evaluating latency. *Australian Journal of Botany* **57**, 340–350. doi:10.1071/BT08163

Warcup J (1981) The mycorrhizal relationships of Australian orchids. *New Phytologist* **87**, 371–381. doi:10.1111/j.1469-8137.1981.tb03208.x

18

Population enhancement plantings help save the Tumut grevillea

John Briggs and Dave Hunter

Summary

The problem
1. The Tumut grevillea is confined to nine natural subpopulations distributed along a 6 km stretch of the Goobarragandra River, in southern New South Wales. Seven of the subpopulations occur on private land and in 2012 these contained 80% of the total population of 924 plants. Within the last 10 years, one small isolated colony of eight plants has been discovered on private land near Gundagai, also in southern New South Wales. Implementation of conservation actions for this species is thus highly reliant on the successful engagement of landholders.
2. Land clearing, domestic stock grazing and weed proliferation have heavily impacted on this species in the past. Blackberry infestations and stock grazing continue to threaten most of the remaining populations.

Actions to manage the problem
1. Detailed surveys were first conducted in 1993 to ascertain the full distribution of the species, obtain baseline population size information and assess threats. Since then, a total of five comprehensive censuses have been conducted to monitor population size and structure trends.
2. Relevant landholders were informed of the presence of the species on their land and of its threatened status.
3. Consent of landholders was obtained to allow protective actions such as fencing and weed control to be undertaken on their land.
4. Propagation of the species has been relatively easy and this has facilitated enhancement plantings that have been undertaken to bolster existing subpopulations and establish new populations.

Markers of success
1. The earlier declining trend was first stabilised and is now increasing.

2. The causes of decline were identified and are now being effectively managed.
3. A Recovery Plan was developed and implemented by a Recovery Team that is enduring and well managed.
4. Community involvement and support has been fostered and this interest continues.

Reasons for success
1. It was essential to make personal contact and maintain good relationships with all landholders with the species on their property, as well as with some others with potential habitat on their land.
2. The conservation program could not have been undertaken without the cooperation of most landholders in either allowing actions to be undertaken on their land to manage the main threats or to allow enhancement plantings.
3. There has been a high survival rate of enhancement plantings that have been undertaken on five properties adjoining the Goobarragandra River. Natural recruitment from two of these plantings has been so successful that 80% of the current population of 1096 plants is comprised of plantings and the progeny of plantings.
4. The enthusiasm and dedication of the members of the early Recovery Teams and later the New South Wales Office of Environment and Heritage staff leading the conservation program has also been important in ensuring management actions have been implemented and outcomes monitored.

Introduction

The Tumut grevillea (*Grevillea wilkinsonii*) is a very distinctive species occurring in a long-settled part of New South Wales (NSW). Despite this, it was discovered only as recently as 1982 by a local naturalist (Makinson 1993). It was not formally described until 1993 (Makinson 1993). At that time it was only known from a single small population on a road verge and an adjoining Travelling Stock Reserve on the Goobarragandra River, ~20 km east of Tumut on the NSW South-west Slopes. Because of its rarity, it was immediately recognised as threatened and was first formally listed as Endangered in Briggs and Leigh (1988) and again in 1996 (Briggs and Leigh 1996). It was subsequently listed as Endangered under the NSW *Threatened Species Conservation Act 1995* and under the *Environment Protection and Biodiversity Conservation Act 1999* due to its restricted geographic distribution and small population size, with ongoing threats. Despite a loss of 50% of the natural population and 75% of enhancement plantings undertaken since 2000 due to severe floods in September 2010 and March 2012, the conservation program is still considered successful in that the total population has increased from 644 individuals in 1998 to over 1096 plants in 2015 (J. Briggs unpublished). Over 80% of the existing population comprises plantings and the progeny of plantings and two of the enhanced populations continue to expand. It is anticipated that two additional planted populations will also begin to expand through natural recruitment within the next few years. Indeed, in January 2017, the first 11 naturally recruited seedlings were observed at one of these more recently planted sites.

Conservation management

The first efforts to protect the Tumut grevillea occurred in 1991 when members of the Australian National Botanic Gardens, the Society for Growing Australian Native Plants, the

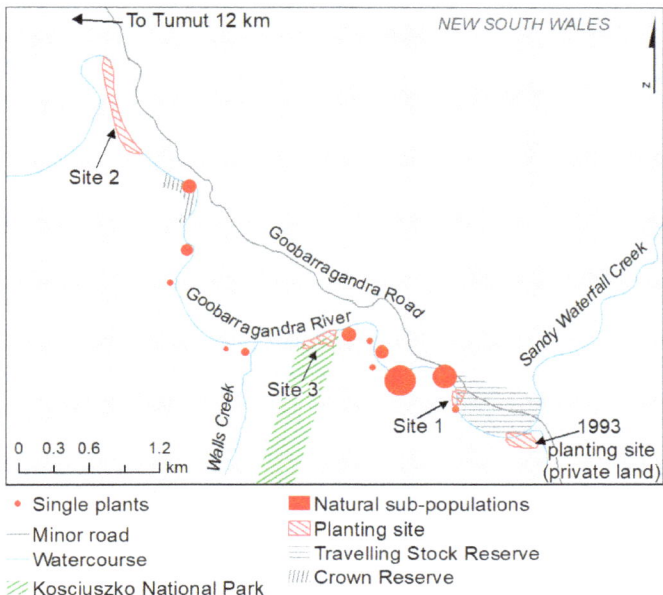

Fig. 18.1. The distribution of the main natural populations of Tumut grevillea and of the enhancement plantings that are contributing to its successful conservation.

local community and the NSW National Parks and Wildlife Service successfully negotiated with Tumut Shire Council to prevent proposed roadworks from destroying part of the population. Following those negotiations, the population adjoining the road easement (Fig. 18.1) was fenced to protect it from damage by road works and from domestic stock grazing. In 1992, a formal Recovery Team was established to coordinate conservation actions for the species. The first plantings of this species into the wild date back to 1993 when staff from the Australian National Botanic Gardens undertook a small trial planting on a Travelling Stock Reserve (which also contained part of the natural population) and on adjoining private land (Butler and Makinson 1993). In 1995, additional plantings were undertaken on 12 properties in the Goobarragandra River valley. Part of the 1993 planting has done particularly well – expanding from eight plants to 350 adult and subadult plants plus at least another 100 seedlings by 2012 (Taws 2013). In 2000, as part of the Recovery Program for the species, the NSW Office of Environment and Heritage (OEH) commenced an enhancement planting project at Site 1 (a private land site) where 13 natural plants survived. Over the next several years, 50 new individuals were successfully established at Site 1 and by 2005 the first natural recruitment from these plantings was observed.

Following this success, in 2008 planting commenced within the natural range of the species at two other sites: Site 2 on private land (Fig. 18.2) and Site 3 on public land. Post-2000, plantings at all sites (a total of 307 plants) were progressing very well until a major flood in 2010, followed by a record flood event in March 2012. These flood events washed away 80% of the plantings at Sites 2 and 3, as well most of the riparian zone plantings at Site 1. The floods also destroyed ~50% of the original natural population, which is now reduced to ~200 plants. Fortunately at Site 1, over half the plantings extended up a low ridge adjoining the river and these escaped the flood. Despite the floods, there are now over 100 plants (planted and naturally recruited) established there – a major increase on the 13 original plants at the site. There are also plans to undertake an enhancement planting at the

Fig. 18.2. (A) The handsome flowers of the Tumut grevillea, a species only discovered in 1982. (B) OEH staff re-planting Tumut grevillea above the flood zone at Site 2 in 2013. (C) Plantings are on both private land, as here at Site 2, and in public reserves.

Gundagai site to bolster the extremely low population of eight plants and where no natural recruit has been observed in recent years.

In addition to the enhancement planting action, other actions have been successfully implemented at several sites to control major threats. These have included fencing to exclude domestic stock grazing, spraying of blackberry to prevent smothering of existing plants and to increase area of habitat available for colonisation by the Tumut grevillea. Seeds were collected for propagation of new plants for enhancing plantings and for long-term storage in case of an unforeseen major loss of the wild population.

Research, biology, identification of key threats

The major threats to the Tumut grevillea of domestic stock grazing and habitat invasion by blackberry are obvious and are relatively easy to manage, providing landholder agreement can be obtained to allow actions to be implemented. Fortunately, most of landholders have been highly supportive of the conservation program. One landholder, whose property contains a relatively large stretch of known habitat for the Tumut grevillea, has not been willing to cooperate in protecting the species, and the threat of grazing there has remained unmanaged. As a consequence of the grazing and the 2010 and 2012 floods, the population on that property has declined from 230 plants in 1998 to 88 in 2012.

Although many grevilleas are known to be fire sensitive, with frequent fire posing a major threat, fire does not seem to be a frequent event in the habitat of the Tumut grevillea. No Tumut grevillea population has been burnt since the Recovery Program began and its susceptibility to fire is thus not known.

Major flood events in 2010 and 2012 caused a drastic reduction in both the natural and planted populations and scoured the river bank, greatly reducing the amount of riparian habitat. These floods have led to a revised planting strategy which is now targeting suitable habitat above the 2012 flood level.

The primary objective of the conservation program for the Tumut grevillea is to maintain or expand existing populations and to establish new populations that will ultimately be self-sustaining or expand through natural recruitment. Recruitment within planted populations has been most successful where the plantings have been on sites dominated by other native vegetation and with some tree and shrub cover that has resulted in a less dense groundcover that has allowed seedling establishment (limited recruitment has occurred where there is a dense grass groundcover). A significant challenge to a future expansion of the conservation program is that very few additional sites remain within the species' known natural range that retain substantial native vegetation and that would thus be immediately suitable as future re-establishment sites. There is, however, potential for future trials to combine replanting sites with both the Tumut grevillea and other native vegetation to create more suitable conditions for the species to recruit naturally.

The Tumut grevillea has been relatively easy to propagate from cuttings and to establish plants *in situ*. More recently, plants have been propagated from seed, which has produced more vigorous and robust planting stock. Rather than using cuttings, the program is now primarily using seeds, which has significantly improved the survival rates and the genetic diversity within the plantings.

Two genetic studies (Gleeson 1994; A. Westman *pers. comm.*) have been undertaken on the Tumut grevillea. Both found a very low level of genetic diversity in remaining populations of the species. Sourcing seed from as many subpopulations as possible is therefore likely to be beneficial in retaining as much of the remaining genetic diversity as possible within translocated populations.

Planning and policy

The first Recovery Team was established in 1992 and this team developed and published the first formal Recovery Plan in 1993 (Butler and Makinson 1993). This plan was then revised in 1995 (Butler 1995). The focus of the Recovery Plan was on propagation and enhancement planting. In the late 1990s, the Recovery Team was reconstituted and produced an updated NSW and national Recovery Plan (NSW NPWS 2001). This plan included many additional actions aimed at achieving protection and appropriate management of all the known sites as well as maintaining the option for further enhancement plantings. This plan included six clear specific objectives and numerous detailed actions to achieve those objectives. Each action was accompanied by specific performance criteria. Following changes to the NSW TSC Act which introduced Priority Action Statements to replace recovery plans, the objectives and actions within the recovery plan were reviewed and incorporated into the Priority Action Statements. The Priority Action Statements have since been re-badged into what is now known as the NSW Saving our Species Program. Almost all the actions in the 2001 Recovery Plan have either been completed or are continuing to be implemented.

People, agencies, governance, and accountability

The first recovery team was established in 1992 and was coordinated by staff from the Australian National Botanic Gardens and the NSW National Parks and Wildlife Service and operated until 1995. The recovery team was re-activated in 1995 by the NSW OEH and continued until 2007 after which the NSW Recovery Planning process was replaced by the

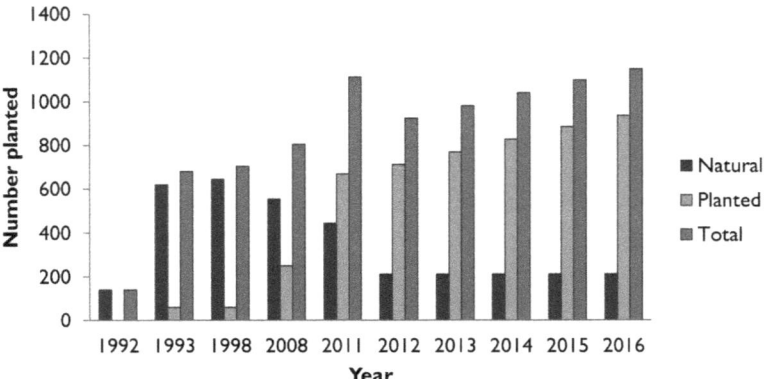

Fig. 18.3. Numbers of Tumut grevillea growing naturally (last counted 2012) and of planted origin. Numbers of natural plants has probably changed little in the last 5 years because little recruitment has been observed in the riparian zone following the 2012 flood. In contrast, numerous seedlings in a range of size classes are evident at the site of largest population derived from planting (also not counted since 2012).

Priority Action Statement scheme. The stability of leadership provided by these government-led recovery teams and subsequently the Saving our Species Program has almost certainly contributed to the success of the conservation program. Recovery Teams comprised both conservation agency staff and staff from other relevant government departments, landholders and other interested members of the local community. The key players from the first Recovery Team were included in the membership of the subsequent Recovery Team. The team generally met annually and all members contributed to the development of a Recovery Plan. Actions were not included within the plan without the agreement of affected landowners or those nominated to implement particular actions. Whilst broad community support for the recovery of the Tumut grevillea was sought, direct involvement of the community was limited because most of the actions were focused on private land and the privacy of landholders needed to be respected. Development of the current conservation project for the Tumut grevillea under the NSW Saving our Species Program involved the input of an expert panel comprised of some representatives from the previous Recovery Team. The cost-effectiveness of this project has been assessed using criteria applied to all NSW Threatened species and the project has then been ranked as a high priority for funding for the next five years under the Saving our Species Program.

Monitoring

Actions completed before finalisation of the Recovery Plan were listed in the Plan (NSW NPWS 2001). Actions completed and their costs in each financial year since the Saving our Species Program commenced for this species in 2013 are recorded in the Saving our Species database. A summary of implementation and population trend is made public in the form of an annual report card. It is intended that an entire population count will be undertaken every 5 years and a report produced. The population count includes assigning individuals to one of three height class categories (<0.2 m, 0.2–1 m and >1 m). Five complete censuses have been undertaken since 1994 (Taws 2013). In the intervening period, annual counts were made by the NSW OEH of surviving plants at the translocation sites commenced in 2008 (Sites 2 and 3). Surveillance type monitoring is undertaken annually at all sites to detect unforeseen threats that require control. Figure 18.3 shows the overall popu-

Table 18.1. Major management and natural events influencing recovery progress for the Tumut grevillea.

Year	Action
1982	Species discovered
1993	First complete survey increased known range and population size. Sixty seedlings planted into TSR and private land site by SGAP.
2001–2006	NSW OEH undertook enhancement plantings on private land site – 50 mature plants established. First recruitment from plantings observed in 2005.
2008	NSW OEH commenced translocation plantings at two new sites.
2010 and 2012	Major flood events in September 2010 and March 2012 resulted in the loss of half the natural population and 80% of plantings undertaken since 2008.
2013–2017	NSW OEH continues translocation plantings at the two new sites, but targeted to above 2012 flood levels.

lation trend since 1998, including breakdown of the number of plants that are natural and those that are planted or derived from plantings. Major events that have influenced the population trend are listed in Table 18.1.

Infrastructure

The conservation program for the Tumut grevillea could not have been undertaken without the cooperation of most landholders in either allowing actions to be undertaken on their land to control the main threats or to allow enhancement plantings.

Money

The initial recovery work for the Tumut grevillea was supported through the in-kind contributions of staff from the Australian National Botanic Gardens and members of the Society for Growing Australian Plants, which also covered the cost of producing the first tubestock for the 1993 trial planting. Between 1993 and 1995, funding from the Commonwealth's (federal) Endangered Species Program was obtained to support the development of an expanded Recovery Plan and to commence implementation of the approved actions in that plan. Between 2003 and 2013, there were no significant resources available to support implementation of major actions such as extensive weed control and fencing. During this time, NSW OEH provided in-kind support in providing staff to implement low cost actions and maintain contact with landholders. The voluntary assistance of one of the members of the Recovery Team in propagating plants at no cost was invaluable in allowing the planting program to continue until 2006. Funding from the NSW OEH Saving our Species Program since 2013 has enabled more costly actions such as additional fencing, the spraying of larger infestations of blackberry and more extensive plantings to be undertaken.

The future

The Saving our Species Program was launched by the NSW OEH in 2013. Under this Program, a conservation project was developed for the Tumut grevillea by an expert panel and government funding to implement the program for the grevillea commenced in the same year. More recently, the NSW Government announced a funding package for the Saving our Species Program for the next 5 years, which has assured ongoing funding of A$15 000 per year for the Tumut Grevillea Program until June 2022. Each financial year, an action plan and funding requirements are developed to implement the conservation project that has a stated corporate goal of maintaining at least one viable population in the wild in 100 years' time. The NSW OEH produces annual report cards on what has been achieved and these are publicly available. Although funding for the conservation program is not guaranteed beyond June 2022, by then most of the major and most costly works such fencing, control of major weeds and the establishment of new populations should have been completed. Future ongoing maintenance could be expected to be relatively low in cost.

At present, the implementation of the conservation program on private land is entirely reliant on the goodwill of the landholders and there are no formal agreements in place to ensure the work can continue. Most landholders are happy to cooperate with the conservation program, but are reluctant to formalise arrangements. To implement the conservation program the NSW OEH has taken a risk management approach in investing on private land sites where a current or future owner may chose not to continue to support the Program. So far, after 20 years, no landholder has ceased to cooperate. The development of suitably worded formal agreements that are acceptable and non-threatening to landholders is a task to be considered in the future.

Conclusion

The fact that landholders within the range of the Tumut grevillea are largely very cooperative has been critical to the success of the conservation program. Continuity of leadership since 1995 from within the Threatened Species team of the NSW OEH and the securing of funding between 1999 and 2003, and again since 2013, for implementing actions has without doubt greatly contributed to the ongoing success of the conservation program for the Tumut grevillea. Voluntary contributions from some members of the Recovery Teams enabled some actions such as weeding, plant propagation and planting to be undertaken in the early stages of the program. The commitment of many of those involved in the Recovery Program since it commenced in 1992 also has been very important to its success. Being a large attractive, easily propagated plant located near a large country town where there is community interest in conservation projects also has been an advantage for the Program. Past successes have encouraged expansion of the conservation work to include establishing new sites, both in the Goobarragandra valley and in areas nearby.

Acknowledgements

The successful implementation of recovery actions for this species will be largely dependent on the ongoing cooperation of those landholders who have populations of the Tumut grevillea growing on their land. The cooperation of these landowners to date, including allowing field surveys on their properties, has made an invaluable contribution to the recovery process. The support of other neighbouring landowners in allowing populations to be established on their land is also making a major contribution. For privacy reasons, landholder names are not being publicised.

The numerous members of the Recovery Team that operated until 2007 are thanked for their enthusiastic contributions to the development of the Recovery Plan and in implementing many of the actions. The financial support of the Australian Government through its Endangered Species Program in the late 1990s and more recent financial support from the NSW OEH through its Saving our Species Program has been vital in funding the implementation of many of the management actions that are assisting the Tumut grevillea to recover.

References

Briggs JD, Leigh JH (1988) *Rare or Threatened Australian Plants: 1988 Revised Edition*. Special Publication No. 14. Australian National Parks and Wildlife Service, Canberra.

Briggs JD, Leigh JH (1996) *Rare or Threatened Australian Plants: 1995 Revised Edition*. CSIRO Publishing, Melbourne.

Butler G (1995) 'Species Recovery Plan for *Grevillea wilkinsonii*: previously *Grevillea* sp. nov. (Tumut) (Tumut grevillea)'. 2nd edn. Australian National Botanic Gardens, Canberra.

Butler G, Makinson R (1993) 'Species Recovery Plan for *Grevillea wilkinsonii* (Tumut grevillea)'. 1st edn. Australian National Botanic Gardens, Canberra.

Gleeson T (1994) Patterns of genetic variation in *Grevillea wilkinsonii*, *G. acanthifolia* and *G. ramosissima*. BSc (Hons) thesis, Australian National University, Canberra, Australia.

Makinson RO (1993) *Grevillea wilkinsonii* (Proteaceae) a new species from southern NSW. *Telopea* **5**, 351–358. doi:10.7751/telopea19934978

NSW NPWS (2001) *Grevillea wilkinsonii* (a shrub) Recovery Plan. New South Wales National Parks and Wildlife Service, Sydney.

Taws N (2013) 'Survey of the Tumut Grevillea, *Grevillea wilkinsonii* after Record Flood Heights in 2012'. Unpublished report by Greening Australia prepared for the NSW Office of Environment and Heritage, Department of Premier and Cabinet, Sydney, <http://www.environment.nsw.gov.au/resources/nature/surveys/tumut-grevillia-grevillea-wilkinsonii-consultants-survey-report-record-flood-2012.pdf>.

19

The spiny daisy: the disappearance and re-emergence of a unique Australian shrub

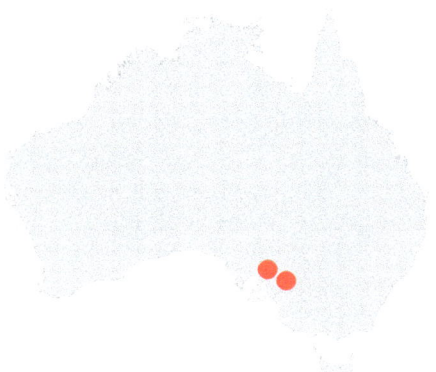

Doug Bickerton, Erica Rees, Tim Field, Amelia Hurren and Christophe Tourenq

Summary

The problem
1. The spiny daisy is confined to six degraded roadside populations.
2. All colonies are clonal, constraining options for maintaining genetic diversity.
3. The species' ecological requirements were largely unknown.

Actions taken to manage the problem
1. Weed control and site maintenance has been undertaken at the six sites with natural populations.
2. Plants have been translocated to more secure sites with fewer or more readily manageable threats.
3. Creation of a grass-root community driven project that included private, non-government organisation and government partners.

Markers of success
1. The once small population is increasing both at natural sites and translocated sites.
2. Most causes of decline have been identified and are being effectively managed.
3. A Recovery Plan has been developed and is implemented under the direction of a well-managed and committed Recovery Team. The community is heavily involved and providing substantial support.

Reasons for success
1. Community has been engaged, with interest piqued by the connection to the historically significant Burke and Wills expedition of 1860 and the long apparent disappearance of the species.

2. Commitment by government staff who were determined not to see the species lost again.
3. A strong partnership between the state environment agency, non-government organisations and private enterprise and volunteer support.
4. Dedication to maintaining the wild populations and the propagation of plants in nurseries.

Introduction

The spiny everlasting or spiny daisy (*Acanthocladium dockeri*) is a low spreading shrub, usually no more than 0.5 m in height, whitish-grey in colour, with spines on the outer branchlets, fine hairs on the oval leaves and branches, and yellow composite flowers (Fig. 19.1). The shrub is currently known only from eastern South Australia (Fig. 19.2), and is the only species within the genus *Acanthocladium* (SHSA 2017). The species is Critically Endangered under the *Environmental Protection and Biodiversity Conservation Act 1999* (DoE 2016) and Endangered under South Australia's *National Parks and Wildlife Act 1972*.

The spiny daisy was first collected in 1860 by Hermann Beckler, botanist on the ill-fated Burke and Wills expedition, from sandhills near the explorers' Darling River depot camp in

Fig. 19.1. (A) The original voucher specimen of spiny daisy (*Acanthocladium dockeri*) collected in 1860 by Hermann Beckler. (B) Detail of inflorescence and foliage. (C) Spiny daisy (*Acanthocladium dockeri*) growing on a road verge (photos: A, Reproduced with permission from the National Herbarium of Victoria (MEL), Royal Botanic Gardens Melbourne; B, C, Christophe Tourenq).

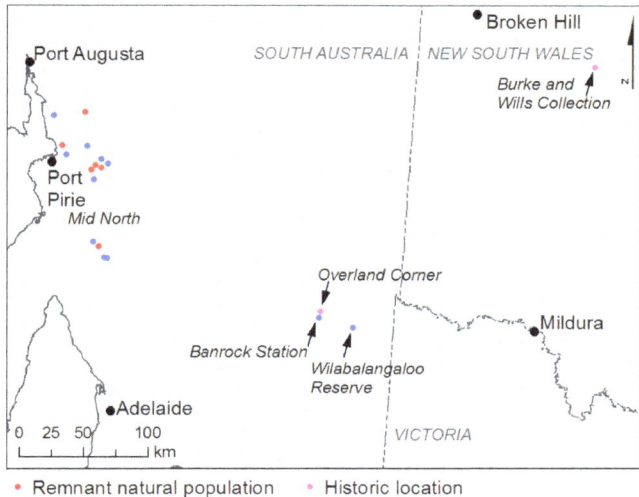

Fig. 19.2. The current and historical distribution of spiny daisy (*Acanthocladium dockeri*), including remnant populations and translocation sites.

western New South Wales (Burbidge 1958; Davies 1992) (Fig. 19.2). The species was not recorded again until 1910, found at Overland Corner in South Australia's Riverland (Davies 1992) (Fig. 19.2). Despite concerted searches in the 1990s (Davies 1992), the spiny daisy was not seen and was thought to be extinct. But in a remarkable twist, the shrub was re-discovered by chance in 1999, on a roadside in South Australia's Mid-North, by farmer and amateur botanist Paul Slattery (Clarke *et al.* 2013). By 2012, another five populations had been discovered in the Mid-North region (Fig. 19.2), all growing on roadsides in highly modified and vulnerable habitat. These six locations are the only known wild populations of the species (Clarke *et al.* 2013), and each occupies less than 0.2 ha (Fig. 19.3).

The spiny daisy is a 'thorny' species to recover, not just because of its spines. This species has some particular idiosyncrasies that pose conservation challenges. Its reproduction in the wild appears to be solely vegetative, and each of the six remnant populations are clonal; hence overall genetic diversity is low (Adams 2013). Seed production and seed viability are also low, both in the wild and in nurseries; germination trials have not been successful, and cross-pollination has not been achieved (Jusaitis 2008). Management of the remnant populations on narrow and degraded roadsides is challenging (Rees 2012a), and it is also difficult to ascertain the preferred habitat for the species (Clarke *et al.* 2013). The roadside populations are all adjacent to cropping paddocks and are subject to snail attack and herbicide overspray (Clarke *et al.* 2013). The spiny daisy is also unremarkable in appearance and easily overlooked, making searching for new populations difficult. Despite these challenges, the Recovery Team has made some positive steps towards improving the conservation status of this species and engaging the community in its recovery.

Conservation management

The Spiny Daisy Recovery Team has been operating since 1999 and, despite some changes in membership and governance, continues to meet regularly to implement the Recovery Plan (Clarke *et al.* 2013). Since 1999, Recovery Team project officers have worked with local councils, the Department of Planning, Transport and Infrastructure (DPTI) and

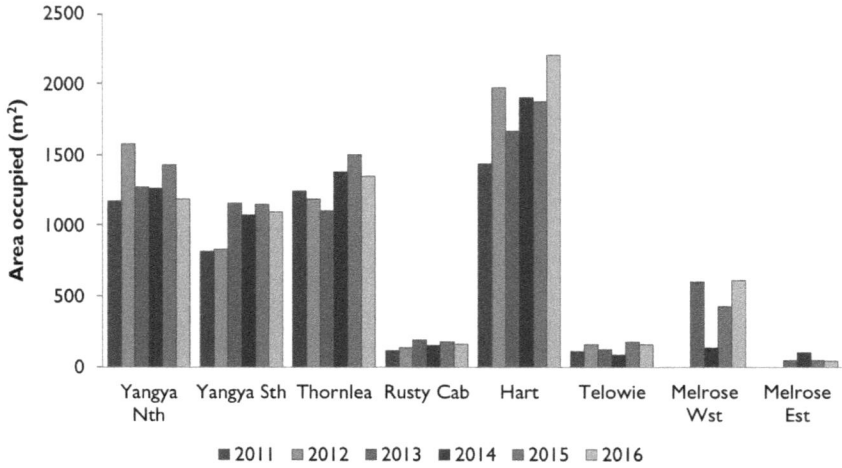

Fig. 19.3. The approximate area of the six remnant spiny daisy populations since Trees For Life has been responsible for their maintenance. Note that the Yangya and Melrose populations are found on both sides of the road. Spiny daisy extends into the adjacent paddocks at Yangya, Thornlea and Melrose.

landowners adjacent to remnant populations to raise awareness of the status and vulnerability of the species, and to form agreements on suitable management practices. A variety of signage has been used including Road Marker System signs, markers to delineate the grading edge on unsealed roads and special spiny daisy signs. On-site discussions and toolbox meetings have been held with Council, DPTI staff and farmers. As a result of this engagement, local landholders and council staff have expressed a sense of personal responsibility and pride that the six remnant populations occur on roadsides in their area. Regular, concerted and ongoing weed and snail control is gradually paying off. At some sites, there has been a reduction in weeds, with the spiny daisy spreading beyond the signage that marked its initial extent (E. Rees unpublished).

Given that natural populations are few, small and tenuous, establishing new populations of spiny daisy has been a key goal of the Recovery Team. The initial focus of translocations was to boost extant populations or to establish new, more secure sites in the Mid-North. The Department of Environment, Water and Natural Resources (DEWNR) undertook research to explore propagation techniques (Jusaitis 2008) and a community nursery was established at Blyth, mainly under the guidance of local naturalist Dave Potter. Since 1999, 11 conservation plantings at new sites and three augmentation plantings at remnant sites have been established in the Mid-North (Fig. 19.2). Planting success has been variable, with some producing numerous ramets (suckers), some progressing slowly after a few years and four failing. Friable soil and moisture seem to be the main factors that determine ramet production, which is the main indicator of reproductive success. Another seven plantings were made for educational purposes, including at Botanic Gardens in Adelaide, Melbourne and Canberra.

Perhaps the most remarkable success has been the establishment of new populations of spiny daisy in South Australia's Riverland, ~200 km from the nearest known remnant population. In 2013, The Recovery Team investigated the potential to re-establish the species near Overland Corner (Fig. 19.2), where it had been recorded 100 years ago. DEWNR ecologists searched the region for potential translocation sites, and met with Banrock Station's two conservation ecologists. The Banrock Station complex (Fig. 19.2) is situated on the River Murray, and comprises vineyards, a cellar door facility and 1375 ha of internationally recognised (Ramsar) wetland, which is actively managed for conservation and restoration (Tourenq *et al.* 2016). Banrock Station is located just across the river from Overland Corner. Given the favourable environmental credentials of landowner Accolade Wines, Banrock Station was selected by the Recovery Team as a spiny daisy translocation site. For Accolade Wines, the spiny daisy's connection to the famous Burke and Wills story and the fascinating detective-style drama surrounding the species' disappearance and reappearance motivated their decision to partner with government and non-government organisations in the Recovery Program.

A memorandum of understanding was officially signed between DEWNR and Accolade Wines in 2013. Cuttings were collected from the six remnant populations and propagated in Banrock Station's native plant nursery. In July 2014, a formal planting ceremony was held at Banrock Station, with dignitaries, partner representatives and local community members.

Three sites on Banrock Station, 250–400 m^2 in area, were targeted for translocation. In 2014, the soil was tilled, fences and/or tree guards were installed, and 396 plants were planted. Cuttings were sourced from all six Mid-North remnant sites, and all clones but one were successfully propagated. Drip irrigation was provided during the first season and weeds were controlled. The plants are now well established and native species from the surrounding bush are colonising the sites. All have been a great success, in part due to methodical site selection but also thanks to the added attention that the on-site conservation ecologists have been able to provide.

Following the closure of the community native plant nursery in the Mid-North, Banrock Station's nursery took over the propagation of spiny daisy cuttings, as sites for new populations are investigated and prepared. In 2015, cuttings were planted at National Trust South Australia's Wilabalangaloo property in the Riverland (Fig. 19.2), thanks to some enthusiastic staff and volunteers. Further translocations are being planned in the Riverland.

Research, biology and identification of key threats

All six remnant spiny daisy populations are located at highly insecure roadside sites. At some sites, the daisy extends into the neighbouring croplands, where tillage and other cropping practices seem not to deter the emergence of numerous ramets. Roadsides are a particularly difficult location for a threatened plant, being exposed to regular threats from road management activities. Furthermore, the management of roadside populations poses significant work safety issues for staff, because Work Zone Traffic Management procedures must be followed, which often makes volunteer participation impractical. The Recovery Team's strategy has been to manage and monitor remnant sites, while also establishing translocated populations in more secure locations (Clarke *et al.* 2013).

Another key threat to the persistence of the species is the lack of genetic diversity, because each population is clonal. To date, four of the six clones have been established at single-clone translocation sites, and mixed-clone translocations also have been established to trial natural cross-pollination. However, in-field and nursery cross-pollination trials have not yielded successful recruitment (Jusaitis 2008; Potter and Linke *pers. comm.*).

As almost nothing is known of the spiny daisy's historical distribution, further *in-situ* and *ex-situ* studies of its habitat requirements are needed. Several of the translocations have trialled different soil types to gauge the adaptability of the species. Initial indications are that the spiny daisy might be able to survive in heavier soils than previously expected.

People, agencies, governance and accountability

The recovery project has undergone many transformations since its inception, and a flexible approach to management has meant that a range of non-government organisations has partnered with the South Australian Government to ensure continued recovery effort. Initiated by community-led Threatened Plant Action Group following the 1999 re-discovery, and funded by the National Heritage Trust, the project's primary focus was to identify the threats to the species, understand its biology and compile the first Site Action Plans for management of remnant populations (Robertson 2002). DEWNR later took on the project and contracted Greening Australia to maintain sites.

From the beginning, the Recovery Team has included key experts and community members, who have been the drivers in maintaining the species' public profile. Of particular note are Kevin and Lorna Jaeschke who have been strong community advocates for ecosystem restoration generally, and in particular the Spiny Daisy Recovery Program. The Jaeschkes live near one of the remnant populations, and have been instrumental in maintaining a secure translocated population at the Hart Field Day site, which receives over 600 visitors annually, as well as establishing two populations on their own property (Clarke *et al*. 2013).

Since 2011, the not-for-profit organisation Trees For Life has been contracted to maintain all threatened flora sites in South Australia's Northern and Yorke NRM Region, and in 2015 it took on the leadership of the Recovery Team. Funding is provided by the National Landcare Programme, through the NRM Board, and DEWNR provides ongoing assistance with planning and mapping.

The partnership with Banrock Station has provided a substantial boost to the Recovery Program, raising the profile of the species, engaging with the wider community and inspiring community groups in the Riverland to participate in the program. As a result, spiny daisy populations have been established on conservation reserves with secure tenure and fewer threats, well-resourced staff and access to volunteers for assistance. One of Banrock Station's translocation sites is located within walking distance of the cellar door, which welcomes around 60 000 visitors per year. This has provided a unique opportunity to raise the public profile of the species and highlight the conservation efforts of partners. Similarly, the Wilabalangaloo Reserve translocation sites are easily accessed by the public, and the inclusion of this site has enabled volunteers to participate in the conservation and management of spiny daisy.

Project officers have helped generate considerable interest in the project through local media stories. The unique history of the species' rediscovery has been promoted at regional fairs, agricultural field days and even wine shows, with plant displays, distribution of fact sheets and invitations for community participation.

Monitoring

The Spiny Daisy Recovery Team meets twice a year to report on their progress against the Recovery Plan objectives. This regular reminder of the team's objectives helps keep the project on track and allows for evaluation of actions and modification of approach as needed. Conservation management priorities are allocated via the Recovery Plan and site-specific translocation plans.

Each autumn, monitoring is conducted on remnant and translocation sites to report on trends in area of occupancy, growth rate, suckering and any change in threatening processes. As the species is clonal, the counting of individual plants at remnant sites is irrelevant and mostly not feasible. Instead the focus is on monitoring the changes in area of occupancy (i.e.

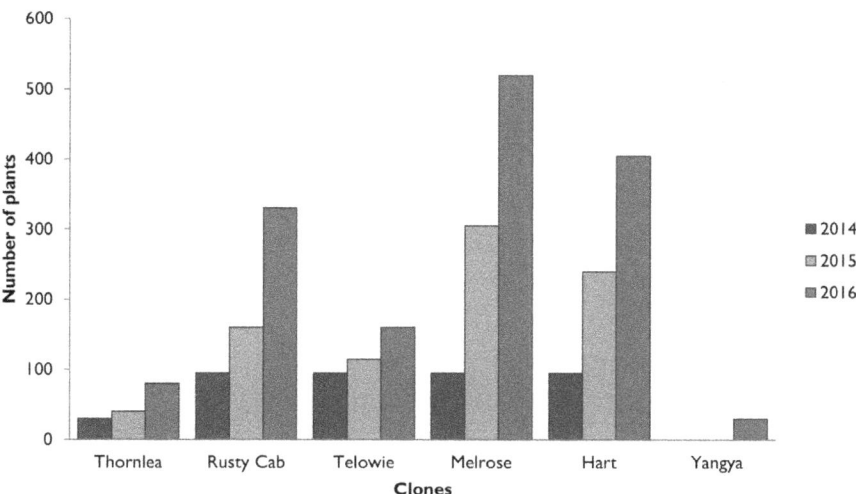

Fig. 19.4. The number of ramets (suckers) produced at Banrock Station's translocation sites in 2015 and 2016, compared with the number of juveniles planted in 2014. Names of clones denote the origin of the translocated stock.

length and width) of populations and the density within that area (Clarke and Haase 2012). These data indicate that the size of remnant populations is being maintained or is gradually increasing (Fig. 19.3), and weed infestations are being controlled and reduced.

Monitoring at Banrock Station shows that supplying water to plants for the first 6 months played an important role in establishment. As summer abated, the watering cycle was reduced and ultimately turned off when suckering was observed. In 2015, a year after planting, the survival rate was 100% and 472 ramets had been produced. Two years after the translocation, the survival rate of the original plants was 90% and another 618 ramets had been produced across all three sites (Fig. 19.4). The new translocation site at Wilabalangaloo also recorded a 93% survival rate 1 year after the translocation in 2015 (Fig. 19.5).

After 2 years, the 2014 plants at Banrock Station have more than doubled in height and width (T. Field unpublished), with the more loamy site displaying more growth than the others. There has also been a notable difference in translocation success between clones. In 2016, Banrock Station staff planted a further 76 plants to supplement the two least successful clones in the original plantings. The re-establishment of spiny daisy populations near the site of its first South Australian record is heartening, and has significantly expanded the distribution of the species.

Money

Commonwealth (federal) grants such as Natural Heritage Trust, Caring for our Country and National Landcare Program have provided the majority of the funding needed for the project. South Australian Government in-kind support has been substantial, both from the NRM regions and DEWNR's Threatened Species Unit, ensuring the continuity of the project between funding shortages. Non-government grants have been provided by Foundation for Australia's Most Endangered, Wildlife Conservation Fund and Threatened Species Network, and the recent inclusion of funding from the Banrock Station Environmental Trust to the project has provided a cost-effective injection of funds to assist with the Riverland translocation sites. Finally, at least six different community groups have selflessly

Fig. 19.5. The future: a new translocation at National Trust South Australia's Wilabalangaloo property in the Riverland in 2015 with Berri–Barmera Landcare volunteers (photo: Christophe Tourenq).

volunteered their time over the years to assist with weeding, planting and maintenance of sites (Fig. 19.5).

The future

With state government budgetary constraints and uncertainty about the future of Australian Government funding, it can be difficult to maintain continuity of expertise and guidance for recovery programs. However, the Spiny Daisy Recovery Team has produced an array of technical documents over the course of the project that provide a legacy for future members, authorities and the community. These include the National Recovery Plan (Clarke *et al.* 2013), monitoring manuals (Clarke and Haase 2012), site action plans (Robertson 2002; Rees 2012a), research (Jusaitis 2008; Adams 2013) and translocation guides (e.g. Clarke 2006; Rees 2012b). A 'Species Studbook' (a register that documents the source and history of individuals within a translocated population) is kept by the Trees for Life project officer, and is available to the team as needed. At Banrock Station, the translocation sites are now included in the Ramsar site management operations.

The spiny daisy translocations have been established at secure sites in the Mid-North and the Riverland, significantly improving the future prospects for the species. Of primary importance is the degree to which land managers of translocation sites have accepted the responsibility for caring for the new populations. The main challenge that remains is the limited genetic diversity. How the species will adapt to climate change is uncertain, and it is possible that adaptation will be affected by this lack of genetic diversity (Clarke *et al.* 2013).

Translocations to secure sites will continue to be a priority for the Recovery Team, and there appears to be a strong adoption of the Spiny Daisy Project by the Riverland community, which the Recovery Team must capitalise on as opportunities for secure sites in the Mid-North are diminishing.

Following the growing interest generated for media and sponsors and arrival of new partners (Banrock Station, Landcare groups and National Trust South Australia), it is hoped that further research can be undertaken to explore the species' ecology. The trials on heavy clay soils at Banrock Station are a prelude to *in-situ* and *ex-situ* (laboratory) experiments under different soil, salinity and climatic environments. The factors that led to the current highly fragmented and precarious persistence of natural populations are poorly resolved. Resolution of the pattern and cause of this species' historical range contraction may be best reached with cross disciplinary investigations beyond conservation biology into social sciences (history).

Conclusion

The spiny daisy has a unique link to the well-known, but tragic, story of the Burke and Wills expedition, with an important place in Australia's post-settlement history. This has been used to great advantage by the project partners to promote the recovery of the spiny daisy. In times when the conservation of biodiversity makes the headlines mostly in the obituary section of the media, the 'feel-good' aspect of the story, the ongoing passion and dedication of the Recovery Team members and the constructive partnership between all stakeholders have been vital for the success of the project. Being seen as a project 'carrying hope' helped mobilise the resources of a private company and to trigger a successful and high-profile public relations campaign. Through this community involvement, the careful management of the few, small and insecure natural populations, and successful translocations to new sites, the conservation outlook for the spiny daisy has been substantially enhanced.

Acknowledgements

The authors wish to acknowledge and/or thank the following people for their substantial contributions to the success of this project: Threatened Plant Action Group (especially the late Yvonne Steed); Mark Adams (Evolutionary Biology Unit, SA Museum); Annie Bond, Anne Brown, Amber Clarke, Uta Grehn, Bernie Haase, Anthony Pieck, David Potter, Paul Slattery, Wendy Stubbs and Jean Turner (ex-Recovery Team); Cheryl Hill (Foundation for Australia's Most Endangered); Pete Copley, Meg Robertson, Andy Sharp (DEWNR); Manfred Jusaitis and Ellen Ryan-Colton (ex-DEWNR); Matt Humphries, Kevin and Lorna Jaeschke, Aimee Linke (Recovery Team members); Suzanne Blake, Anita Poddar and Alison Searle (Accolade Wines); Banrock Station Environmental Trust; National Trust South Australia; Leanne Rathbone, Helga Kieskamp, Kim Lohmann, Jimi Curtis, Robin Parker, Richard Raams, Linc Gore and Sandi Hall (Berri–Barmera Landcare); Frank Udovicic, Josephine Milne, and Wayne Gebert (National Herbarium of Victoria); and Sav Sarro and Mark Rudiger (Big River Toyota, Berri).

References

Adams M (2013) 'An assessment of the extent of clonal reproduction in the Melrose population of spiny daisy (*Acanthocladium dockeri*)'. Department of Environment, Water and Natural Resources, Northern and Yorke Region, Clare, South Australia.

Burbidge NT (1958) A monographic study of *Helichrysum* subgenus *Ozothamnus* (Compositae) and of two related genera formerly included therein. *Australian Journal of Botany* **6**, 229–284. doi:10.1071/BT9580229

Clarke A (2006) 'Translocation proposal for *Acanthocladium dockeri* 2006–2007: Translocation 3 Caltowie-Stone Hut Road Reserve (all four genotypes)'. Department for Environment and Heritage, Northern and Yorke Region, Clare, South Australia.

Clarke A, Haase B (2012) '*Acanthocladium dockeri* (spiny daisy) monitoring manual'. Department of Environment, Water and Natural Resources, Northern and Yorke Region, Clare, South Australia.

Clarke A, Robertson MA, Pieck A (2013) 'Recovery Plan for *Acanthocladium dockeri* (spiny daisy)'. Department of Environment, Water and Natural Resources, Clare, South Australia.

Davies RJ-P (1992) *Threatened plant species of the Murray Mallee, Mount Lofty Ranges and Kangaroo Island Regions of South Australia*. Conservation Council of South Australia, Adelaide.

DoE (2016) *Acanthocladium dockeri*. Species profile and threats database. Department of the Environment, Canberra, <http://www.environment.gov.au/sprat>.

Jusaitis M (2008) Flowering and seed production in the endangered spiny daisy, *Acanthocladium dockeri*. *Australasian Plant Conservation* **17**, 14–15.

Rees E (2012a) 'Revised site action plans for the Thornlea, Yangya, Rusty Cab, Hart, Telowie and Melrose populations of the nationally endangered spiny daisy (*Acanthocladium dockeri*)'. Trees for Life, Adelaide.

Rees E (2012b) 'Ongoing site action plans for the translocated populations of the nationally endangered spiny daisy (*Acanthocladium dockeri*)'. Trees for Life, Adelaide.

Robertson MA (2002) 'Spiny daisy *Acanthocladium dockeri* recovery. Site action plans 2002'. Department for Environment and Heritage/Northern Areas Council, Adelaide.

SHSA (2017) Family: Compositae, *Acanthocladium*. Electronic Flora of South Australia genus fact sheet. State Herbarium of South Australia, Department of Environment, Water and Natural Resources, Adelaide, <http://www.flora.sa.gov.au/cgi-bin/speciesfacts_display.cgi?form=speciesfacts&name=Acanthocladium>.

Tourenq C, Field T, Searle A (2016) Restoration of Banrock Station Ramsar wetlands, South Australia: over 20 years of successful involvement by a private agribusiness. In *Restoring Life on Earth: Private-sector Experiences in Land Reclamation and Ecosystem Recovery*. (Eds MJ Mulongoy and J Fry) pp. 25–36. Technical Series No. 88. Secretariat of the Convention on Biological Diversity, Montreal, Canada.

20

The path to recovery for the 'extinct' Lord Howe Island phasmid

Hank Bower, Nicholas Carlile, Rohan Cleave, Chris Haselden, Dean Hiscox and Lisa O'Neill

Summary

The problem
1. The Lord Howe Island phasmid was considered extinct from 1930 to 1960 after introduced black rats eliminated it from Lord Howe Island.
2. The remaining population discovered on Balls Pyramid in 2001 was considered to be small and highly vulnerable.

Actions taken to manage the problem
1. Surveys of Balls Pyramid were undertaken to search for the Lord Howe Island phasmid.
2. Captive populations were established in various zoos in Australia and overseas.
3. Weeds were controlled on Balls Pyramid.
4. Palatability trials of a range of plants on Lord Howe Island were conducted.

Markers of success
1. Extinction was probably averted and the population is now increasing.
2. A Recovery Plan was developed and is being implemented by a well-managed Recovery Team.
3. Captive populations are increasing.
4. The community has become involved and is providing ongoing support.

Reasons for success
1. Outstanding communication programs that have galvanised support for conservation efforts of the species.
2. The captive populations are well managed and are flourishing.
3. The threat from invasive species on Balls Pyramid, including control of weeds, has been well managed.

Fig. 20.1. A male Lord Howe Island phasmid feeding on a melaleuca during the 2005 survey (photo: Dean Hiscox).

Introduction

Rarely is there an example of a species recovery that starts with a listing of extinction. Although we long for such an outcome for iconic large fauna, this seems increasingly unlikely as time passes. However, traits of some invertebrate fauna, such as small home range, mean that it is possible to overlook key refuges (Crew 2012) including physically isolated areas (Goodall *et al.* 2009).

The Lord Howe Island phasmid (*Dryococelus australis*) is a shiny black, heavy bodied, flightless stick insect, with females measuring up to 150 mm and males reaching 120 mm in length (Honan *et al.* 2007; Fig. 20.1). Hatchling nymphs are pale to mid green, becoming darker as they moult, until black at the final instar stage (Honan *et al.* 2007). Maturity is reached by about 7 months (Honan 2007).

Rediscovery

The Lord Howe Island phasmid (also known as the land lobster) is listed as Critically Endangered on the IUCN Red List and is considered to be the world's rarest insect (Priddel *et al.* 2001, 2003). It was formerly common and widespread on Lord Howe Island: an isolated oceanic island with land area of ~16 km^2. Early reports described it living in the Island's palm forests and other unique vegetation communities. However, it was extirpated on Lord Howe Island by the 1930s following the accidental introduction of black rats (*Rattus rattus*) in 1918. In the 1960s, climbing parties found remains of dead phasmids on Balls Pyramid, a 27 ha 550 m rock stack 23 km south-east of Lord Howe Island. The first live specimens on Balls Pyramid were found during a targeted nocturnal survey in February 2001 (Priddel *et al.* 2003; Fig. 20.2). A small population was found on a single rock ledge with a freshwater seep supporting bushes of Lord Howe Island melaleuca (*Melaleuca*

Fig. 20.2. Surveying Balls Pyramid is a major exercise in skill and logistics (photo: Nicholas Carlile).

howeana). Follow-up surveys in 2002, 2003 and 2005 estimated the Lord Howe Island phasmid population to be ~40 individuals, although logistic constraints mean that a thorough survey of other ledges with water seeps and melaleuca bushes has never been conducted. In 2012, during a nocturnal survey for Lord Howe Island gecko (*Christinus guentheri*) and Lord Howe Island skink (*Oligosoma lichenigera*) on Balls Pyramid, several adults were observed 25 m away from the original melaleuca refuge bush (R. Bray *pers. comm.*).

Soon after the rediscovery of live animals, the Lord Howe Island phasmid was listed as Critically Endangered under the Commonwealth (federal) *Environment Protection and Biodiversity Conservation Act* 1999 and the New South Wales *Threatened Species Conservation Act* 1995. The New South Wales National Parks and Wildlife Service prepared a draft Recovery Plan, with the main recovery actions being to: (1) undertake surveys to estimate the size of the wild population; and (2) establish a captive population to prevent its extinction and to limit *in situ* research because of the extreme fragility of the habitat, difficulty of access and risks to both phasmid populations and researchers. Recreational climbing of Balls Pyramid is prohibited under the Lord Howe Island Permanent Park Preserve Plan of Management 2010.

Conservation management
Captive breeding

To safeguard against extinction, two adult pairs of the Lord Howe Island phasmid were taken from Balls Pyramid in February 2003 to establish two captive populations (Melbourne Zoo and Insektus – a Sydney education and consultancy company). Within a month, the condition of both females had deteriorated to the extent that the Sydney female died and the Melbourne female was revived only after specialist veterinary intervention. Critically, 30 eggs had been laid by these females. These and a few hatchlings from the 248 eggs laid following the recovery of the Melbourne female (Carlile et al. 2009) have become the source population for the ongoing captive breeding population of the species.

The four insects taken from Balls Pyramid in 2003 have been the sole genetic base for the entire captive population (Honan et al. 2007). Low genetic diversity in captive breeding programs can lead to inbreeding depression and compromised health and reproductive success. Despite this, and after several early challenges that led to ongoing refinement of husbandry protocols (Honan et al. 2007), a population was established at Melbourne Zoo and it is now in its 13th generation (Fig. 20.3). Over 14 000 nymphs have successfully hatched and the zoo has also developed a husbandry manual to share the lessons learned in their captive management program (Honan et al. 2007). Captive management of the Lord Howe Island phasmid is complex, but the program is viewed as being a major success both in the wider conservation community and the zoo fraternity (Cleave et al. 2014).

In 2008, a second captive population was established in an enclosure at the Lord Howe Island Board plant nursery. Additional island populations also have been established at the Lord Howe Island Central School (in 2013) and the Lord Howe Island Museum (in 2014).

Fig. 20.3. Captive facilities at Melbourne Zoo have been critical for establishing the species in captivity (photo: Rohan Cleave).

The adult population in the Lord Howe Island Board captive facility has reached 120 mature individuals at times (Lord Howe Island Board *pers. comm.*), providing replacement animals for the additional display sites on the island. In 2015 and 2016, eggs were sent from Melbourne Zoo to international zoos, including San Diego Zoo in the USA, Bristol Zoo in the UK, and Toronto Zoo in Canada. This helped secure captive populations and highlight the remarkable story and plight of this species. Bristol Zoo and San Diego Zoo have successfully hatched the eggs that were sent to them and offspring have reached adulthood, with females in these overseas colonies laying eggs for the first time.

As one of 21 'fighting extinction' species targeted by Zoos Victoria, the captive management program for the Lord Howe Island phasmid at Melbourne Zoo aims to increase community awareness of the plight of this invertebrate and encourage support for its conservation. There is also an Australasian Species Management Program for the Lord Howe Island phasmid to manage captive breeding for all facilities where they are held (see http://www.arazpa.org.au/ASMP_About.htm).

The captive breeding program has two primary goals: (1) to ensure that the species does not go extinct – a genuine risk given the precarious situation of the population on Balls Pyramid; (2) to serve as a source population for reintroduction to Lord Howe Island if the introduced black rat can be eradicated there. Planning for a rodent eradication program is well underway (Lord Howe Island Board 2009) and the required funding is largely secured. However, the eradication program faces many challenges and it may be some years before reintroduction to the wild at Lord Howe Island becomes possible. Both of these key objectives rely on having a robust population of healthy captive insects.

Palatability trials of Lord Howe Island plants

Captive populations of the Lord Howe Island phasmid at Melbourne Zoo have now been successfully reared on a variety of host plants including Lord Howe Island melaleuca and tree lucerne (*Chamaecytisus prolifer*), with other species also trialled such as Moreton Bay fig (*Ficus macrophylla*) and blackberry (*Rubus fruticosa*) (Honan 2007). Captive populations on Lord Howe Island have been fed mainly Lord Howe Island melaleuca and Lord Howe Island banyan (*Ficus macrophylla* subsp. *columnaris*).

A native plant palatability study found the adults of the Lord Howe Island phasmid in captivity preferred Lord Howe Island Banyan and nymphs preferred Lord Howe Island melaleuca, with several other common plant species on Lord Howe Island also eaten at differing rates (McGrath 2014). Numerous other plants on the island were identified as suitable for the Lord Howe Island phasmid to consume at different developmental stages. These plant species are widely distributed across a range of habitats and plant communities from lowlands to high elevations (McGrath 2014). This provides hope that, should the species be successfully reintroduced to Lord Howe Island following the eradication of rodents, food should not be a limiting factor for the species.

To promote optimum nutritional fitness in captivity, Melbourne Zoo recently investigated the effect of different host plants on morphology, fecundity and longevity. Analysis of host plant leaf nutritional values and comparison with the contents of stick insect frass indicated that animals are adapted to low nutrient food (Zoos Victoria *pers. comm.*).

Targeted weed control of known habitat on Balls Pyramid

Between 2003 and 2005, it was observed that the main habitat and fodder bush of the small population on Balls Pyramid was being invaded by an exotic creeper, five-leaved morning glory (*Ipomoea cairica*). The creeper is unpalatable for the Lord Howe Island phasmid

Fig. 20.4. Habitat on Balls Pyramid showing the central melaleuca bushes being smothered by the invasive weed, five-leaved morning glory from the right of the bushes. Left unattended, the main habitat of the Lord Howe Island phasmid can quickly become engulfed. (photo: R. Brouwer).

(McGrath 2014). Since 2002, targeted weed control has reduced the vine cover by 95% and prevented the crucial wild refuge from being degraded (Fig. 20.4). Fortunately the upper reaches of Balls Pyramid appear to be naturally free of the creeper.

Weed eradication and habitat recovery on the main island of Lord Howe is critical before a Lord Howe Island phasmid reintroduction. Much of the Lord Howe Island melaleuca community suffers from invasion by introduced ground asparagus (*Asparagus aethiopicus*). Eradication of this invasive scrambler is a goal of the Island 'War on Weeds' program, with over 20 ha of coastal melaleuca forest already treated successfully (see http://www.lhib.nsw.gov.au/environment/environmental-programs/weed-eradication-program).

Media involvement and community engagement

The dramatic and unexpected conservation success of the Lord Howe Island phasmid has been well reported. Several publications and productions tell this remarkable tale. Animator Jilli Rose and Bespoke Animation in 2013 produced a 20 minute animation film 'Sticky', which was nominated for the 'Best short animation documentary' at national and international film festivals (see https://vimeo.com/76647062). Rick Wilkinson's book *The Return of the Phasmid: Australia's Rarest Insect Fights Back from the Brink of Extinction* (Wilkinson 2014) documents the history of the species, from its original description, assumed extinction, later discovery of dead specimens, rediscovery of live insects and the struggle to keep the captive colony productive in the early years of management. Rohan Cleave and illustrator Coral Tulloch produced an award-winning children's book, *Phasmid: Saving the Lord Howe Island Stick Insect* (Cleave and Tulloch 2015; Fig. 20.5). A video documenting the struggles of a hatchling Lord Howe Island phasmid has been viewed more than 28 million times on Zoos Victoria's Vimeo webpage (https://vimeo.com/14413689). Even fiction has drawn on the drama, with an adult murder mystery book climaxing on Balls Pyramid including the phasmid in the pivotal part of the plot (Maitland 2008). A British children's television series 'Octonauts' by Brown Bag Films recently featured the Lord Howe Island phasmid and its rock stack refuge (see http://iview.abc.net.au/programs/octonauts/ZX9194A007S00#). These books and productions have made the conservation success story accessible to millions of people worldwide.

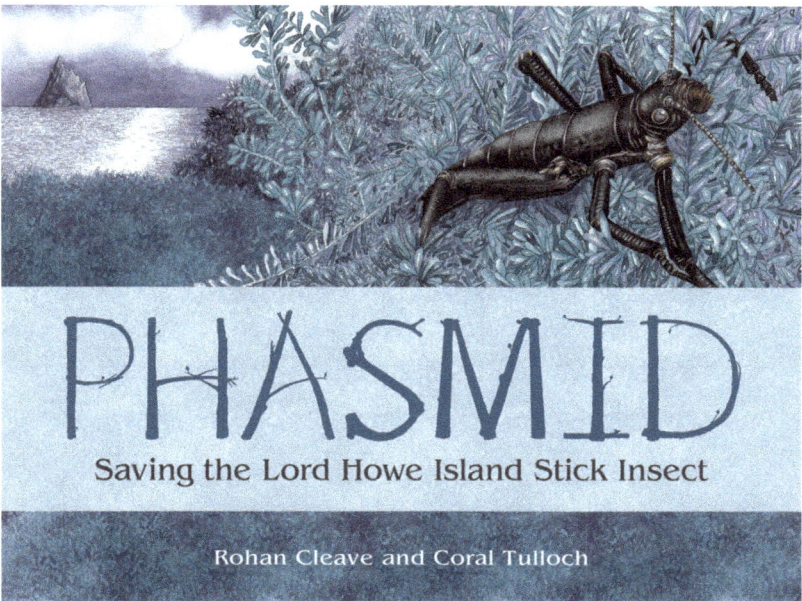

Fig. 20.5. Front cover illustration of the children's phasmid book by Rohan Cleave and Coral Tulloch.

At every stage of the recovery of the species, individual government departments or institutions have been responsible for funding aspects of the program. Although initial actions aimed to rescue the species from a single vulnerable refuge, efforts are now focused on the final goal necessary for survival of the species in the wild: its reintroduction to suitable native habitat on Lord Howe Island.

Funding for the captive program has come primarily from Melbourne Zoo, who committed over A$2.5 million in cash and staff time to the program supported by smaller amounts from the New South Wales Government and a private company (Insecktus).

The future
Rodent eradication
The introduction of rats onto Lord Howe Island led to the global extinction of many endemic species and localised extinction of others. Rodents are still having significant detrimental impacts on threatened species and World Heritage areas on Lord Howe Island. Following rediscovery of the Lord Howe Island phasmid, and with increased government recognition of the widespread impacts of rodents (e.g. Auld *et al*. 2010), a Draft Lord Howe Island Rodent Eradication Plan (Lord Howe Island Board 2009) was developed. The eradication of rodents will benefit many species directly and may also present an opportunity to simultaneously eradicate other introduced species such as the masked owl (*Tyto novaehollandiae castanops* x *novaehollandiae*).

In 2012, the Lord Howe Island Board received significant funding to implement the Rodent Eradication Program from the New South Wales Government Environment Trust and the Australia Government Caring for Our Country Program (Lord Howe Island Board 2016). It is hoped that once all environmental impacts are assessed, the baiting program may begin as early as June 2018. Following eradication, a 2-year period of ongoing monitoring for surviving rodents is required before the program can be considered successful (Russell *et al.* 2016). Until then, the captive population of Lord Howe Island phasmid will need to be carefully managed, with multiple breeding locations, to underpin reintroduction.

Targeted surveys to collect more genetic material

Since the establishment of the captive population(s) of the Lord Howe Island phasmid, a concern has been the limited number of founding individuals. Collection of more individuals from Balls Pyramid could broaden the limited genetic diversity of the captive population and potentially improve the health and quality of insects for reintroduction. Several attempts to land on Balls Pyramid from 2009 to 2014 for such a collection were thwarted by poor sea conditions but a single adult female was captured during an Australian Museum expedition in March 2017, one of just 16 animals found during a search of the whole island over 7 days. Laboratory studies will now quantify improvements in genetic diversity of the captive populations.

Risks of success: reintroduction of ecological equivalents to restore natural balance

Lord Howe Island has suffered major upheavals in its species composition and ecological processes. There may be some risks that reintroduction of phasmids to Lord Howe Island may be too successful, because the natural predators that may have formerly controlled its population are now missing. Before the modern ecological upheavals, the Lord Howe Island boobook owl (*Ninox novaeseelandiae albaria*) was a major predator of the Lord Howe Island phasmid adults on Lord Howe Island. However, the Lord Howe Island boobook owl went extinct on Lord Howe Island in the 1950s (Hutton 1991), probably due to nest hollow competition from the larger introduced masked owl (Priddel and Carlile 2010). Eradication of the masked owl is planned as part of the rodent eradication, but an ecological equivalent to the extinct Lord Howe Island boobook owl may be required to constrain the population size and ecological impact of a reintroduced Lord Howe Island phasmid population. Such an equivalent species is available on Norfolk Island where a hybrid population of the Norfolk Island boobook owl (*Ninox novaeseelandiae undulata*) remains. This species has reached maximum capacity on Norfolk Island due to the loss of natural habitat (Garnett *et al.* 2011). Offspring of the remaining population on Norfolk Island could be used for a reintroduction to Lord Howe Island (Olsen *pers. comm.*).

Conclusion

This dramatic story started with a species assumed to be extinct. This was followed by its unexpected discovery on an unlikely refuge. Galvanised by the rediscovery of a colony of the Lord Howe Island phasmid, the Recovery Team collected several animals for captive breeding and the establishment of an *ex-situ* population to safeguard the species survival. The program was almost doomed from the start, with the pair sent to Sydney soon dying and the female at Melbourne Zoo needing specialist treatment by Patrick Honan and the Zoo's veterinary staff to nurse it back to health.

A drama of good (island restoration experts) and evil (rodents) is in the advanced planning phase and will hopefully play out shortly as an island-wide rodent eradication. There is then potential for a final chapter, with the much anticipated reintroduction to Lord Howe Island, almost 100 years after it was banished.

The unlikely hero in this tale is an invertebrate, not a charismatic vertebrate. The 'bit players' in this drama are dedicated explorers, scientists, zoo specialists and ordinary people whose interest and care keeps this species in the public eye and carries us towards a 'fairy tale' ending. For society, this story also provides some redemption in repairing a little of the damage wrought by humans on this planet.

Acknowledgements

We acknowledge David Roots, the climber whose images first alerted the world to the potential ongoing existence of the Lord Howe Island phasmid. David Priddel led the rediscovery team, stewarded it in the early years, post-rediscovery, and negotiated the difficult process of obtaining permits for the collection of an 'extinct' species and removing live specimens to the Australian mainland via strict air transport protocols. We particularly acknowledge the efforts of Patrick Honan (now with Museum Victoria), as well as Stephen Fellenberg (Insektus), who established initial protocols for management of captive populations, with Honan's dedication in coaxing the last remaining captive female through an unknown illness. Norman Dowsett, Robert Anderson, Zoe Marston, Sarah Silcocks, Kate Pearce and Jessie Sinclair of Melbourne Zoo's Invertebrate Department have, over the years, dedicated their husbandry skills to saving this species. We also thank the Lord Howe Island Board field team for husbandry of the species on Lord Howe and the weed team's management of the *in-situ* population on Balls Pyramid. Thanks also to Michael Magrath in developing this chapter.

References

Auld TD, Hutton I, Ooi MKJ, Denham AJ (2010) Disruption of recruitment in two endemic palms on Lord Howe Island by invasive rats. *Biological Invasions* **12**, 3351–3361. doi:10.1007/s10530-010-9728-5

Carlile N, Priddel D, Honan P (2009) The recovery programme for the Lord Howe Island phasmid (*Dryococelus australis*) following its rediscovery. *Ecological Management & Restoration* **10**, 124–128. doi:10.1111/j.1442-8903.2009.00450.x

Cleave R, Tulloch C (2015) *Phasmid: Saving the Lord Howe Island Stick Insect*. CSIRO Publishing, Melbourne.

Cleave R, Pearce K, Courtney P, van Weerd H (2013) A conservation program for the Lord Howe Island stick insect (*Dryococelus australis*) at Melbourne Zoo. *International Zoo News* **60**, 244–252.

Cleave R, Birkett J, Pearce K, Dowsett N, Anderson R, Silcocks S, *et al.* (2014) Connecting people with invertebrates, observations of the Lord Howe Island Stick Insect: *Dryococelus australis*. *Thylacinus* **38**, 2–4.

Crew B (2012) The elusive stick giant. In *Zombie Tits, Astronaut Fish and Other Weird Animals*. pp. 217–222. NewSouth Publishing, Sydney.

Garnett ST, Olsen PD, Butchart SHM, Hoffmann AA (2011) Did hybridization save the Norfolk Island boobook owl *Ninox novaeseelandiae undulata*? *Oryx* **45**, 500–504. doi:10.1017/S0030605311000871

Goodall J, Maynard T, Hudson G (2009) Lord Howe Islands phasmid or stick insect. In *Hope for Animals and Their World*. pp. 299–305. Grand Central Publishing, New York, USA.

Honan P (2007) The Lord Howe Island stick insect: an example of the benefits of captive management. *Victorian Naturalist* **124**, 258–261.

Honan P, Cleave R, Dowsett N, Anderson R, Marston Z (2007) Husbandry manual for the Lord Howe Island stick insect. Unpublished manual produced by Zoos Victoria, Melbourne.

Hutton I (1991) *Birds of Lord Howe Island: Past and Present*. I. Hutton, Coffs Harbour.

Lord Howe Island Board (2009) 'Draft Lord Howe Island Rodent Eradication Plan'. Lord Howe Island Board, New South Wales.

Lord Howe Island Board (2016) 'Lord Howe Island Rodent Eradication Project Draft Public Environment Report'. EPBC 2016/7703, Lord Howe Island Board, New South Wales.

Maitland B (2008) *Bright Air*. Allen and Unwin, Sydney.

McGrath S (2014) 'Food plant preferences and their life-history effects in the Lord Howe Island stick insect *(Dryococelus australis)*.' MSc thesis, University of Melbourne, Parkville, Victoria, Australia.

Priddel D, Carlile N (2010) Return of the Lord Howe Island phasmid to Lord Howe Island, Australia. In *Global Re-introduction Perspectives: Additional Case-studies from around the Globe*. (Ed. PS Soorae) pp. 17–20. IUCN/SSC Re-introduction Specialist Group, Adu Dhabi, United Arab Emirates.

Priddel D, Carlile N, Humphrey M, Fellenberg S, Hiscox D (2001) Draft Recovery Plan for the Lord Howe Island stick-insect *(Dryococelus australis* (Montrouzier)). New South Wales National Parks and Wildlife Service, Sydney.

Priddel D, Carlile N, Humphrey M, Fellenberg S, Hiscox D (2003) Rediscovery of the 'extinct' Lord Howe Island stick-insect *(Dryococelus australis* (Montrouzier)) (Phasmatodea) and recommendations for its conservation. *Biodiversity and Conservation* **12**, 1391–1403. doi:10.1023/A:1023625710011

Russell J, Binnie H, Oh J, Anderson D, Samaniego-Herrera A (2016) 'Optimising eradication confirmation with rapid eradication assessment'. Report to the Department of Conservation, Wellington, New Zealand.

Wilkinson R (2014) *The Return of the Phasmid: Australia's Rarest Insect Fights Back from the Brink of Extinction*. Christopher Beck Books, Brisbane.

21

Against the flow: the remarkable recovery of the trout cod in the Murray–Darling Basin

Jarod P. Lyon, Mark Lintermans and John D. Koehn

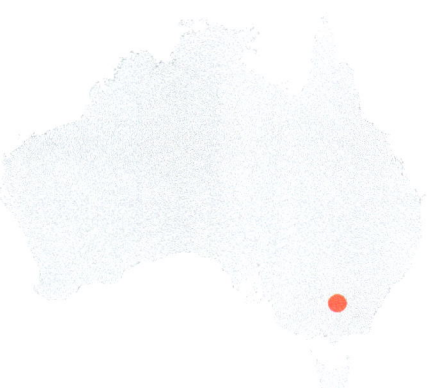

Summary

The problem
1. The trout cod, which exists in southern tributaries of the Murray–Darling Basin was decimated by river regulation, competition with introduced species and human take.
2. The cod is highly aggressive so is readily caught by anglers.

Actions taken to manage the problem
1. Hatchery-produced fish were used to establish new populations.
2. Regulations on angling including closures were introduced.
3. Angler education, particularly how to identify it from the closely related Murray cod.
4. Habitat was rehabilitated, especially by re-instating structural woody habitats.

Markers of success
1. The previous declining trend has now been reversed and large parts of its former range re-occupied.
2. A Recovery Plan has been developed and is being implemented under the guidance of a skilled and committed Recovery Team.
3. The angler community has become involved and is providing ongoing support.

Reasons for success
1. The accumulation of ecological knowledge and its use to influence recovery.
2. The development and use of a population model to improve outcomes from reintroductions.
3. The re-establishment of populations.
4. A coordinated long-term approach to recovery.

Fig. 21.1. Trout cod are large river fish that suffered from over-fishing and habitat loss but have been able to recover as a result of regulation and reintroductions (photo: Gunther Schmida).

Introduction

Internationally, freshwater fish populations have declined due to habitat loss, altered flow and temperature patterns, in-stream sedimentation, introduced fishes, population fragmentation due to in-stream barriers and over fishing (Kondolf *et al.* 2008; Koehn and Lintermans 2012). In an Australian context, a relevant case study is the Australian trout cod (*Maccullochella macquariensis*), which was once considered widespread in the southern tributaries of the Murray–Darling Basin (Cadwallader and Gooley 1984), but which is now listed nationally as Endangered under the Commonwealth (federal) *Environment Protection and Biodiversity Conservation Act 1999* and Endangered by the IUCN Red List (www.iucnredlist.org) (Berra 1974; Harris and Rowland 1996).

Trout cod are a long-lived (>20 years), large-bodied species, with a maximum size of 16 kg and 850 mm total length (Harris and Rowland 1996; Fig. 21.1). They occupy a range of habitats, but are strongly associated with large woody instream habitats (Nicol *et al.* 2007; Koehn and Nicol 2014). River regulation, the introduction of fish species such as European carp (*Cyprinus carpio*) and redfin (*Perca fluviatilis*), over-fishing and habitat destruction all played key roles in the decline of the species (Koehn *et al.* 2013). A wide range of management actions was undertaken to halt, and indeed reverse this decline, and recovery of the species has been observed (Fig. 21.2). Key among these actions was implementing of fishing closures and regulations, and a well-managed and science-based re-introduction program (Lyon *et al.* 2012). As such, the population has now recovered from only two sites with low abundances in the early 1990s, to its current status with established, breeding and expanding populations across significant areas of its historical range.

Conservation management

Management of river systems has changed markedly over the past 50 years in Australia. Indeed, many hundreds of millions of dollars are spent annually to restore natural function in aquatic environments through delivery of interventions such as environmental flows, riparian replanting and water quality improvements (Lintermans 2013). Particular life history characteristics of trout cod that contributed significantly to the species decline included a strong association with instream woody habitat (Nicol *et al.* 2007; Koehn and Nicol 2014), their relatively large size and their aggressive, predatory nature, making them

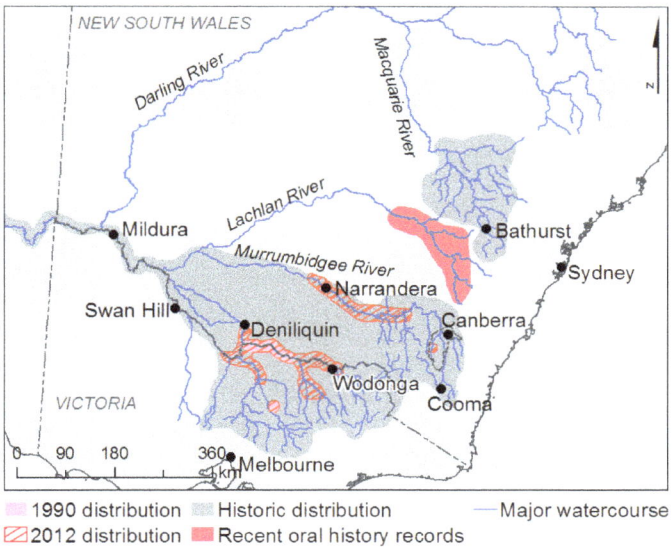

Fig. 21.2. Changes in the distribution of the trout cod over time demonstrating its large former range, the nadir in the 1990 and more recent recovery.

easily captured by anglers. Research revealed that removal of such instream woody habitat, and increasing angling pressure, were both major causal factors in the decline of the species. By the late 1980s, monitoring showed that the species was limited in range to a small area of the Murray River downstream from Lake Mulwala (where they were in low numbers) and a small, translocated population located over a 10 km reach of Seven Creeks, in north-east Victoria.

Early conservation actions to ameliorate against the threat from over-fishing, included total fishing closures at both remnant locations – one of which (Seven Creeks) continues today. Another key initiative was an education program to enable anglers to distinguish trout cod from the closely related and morphologically similar Murray cod (*Maccullochella peeli*), which often occurred in the same habitats, so as accidental take was minimised. In addition, a population modelling program was undertaken, based on the best scientific knowledge at the time, to guide reintroduction attempts through a targeted restocking program (see Fig. 21.2 and Lyon *et al.* 2012).

Increased understanding of the biology and ecology of trout cod were influential in the development and prioritisation of recovery actions: improved knowledge of biology and breeding improved hatchery outputs (Ingram and Rimmer 1992; Koehn *et al.* 2013); breeding in the wild (Koehn and Harrington 2006); and the use of radio-telemetry to define habitat use (Nicol *et al.* 2007; Koehn and Nicol 2014) and movement patterns (Koehn and Harrington 2006; Ebner *et al.* 2007, 2009). As a result of this work (see Fig. 21.2), the drastic decline of the species was halted, and a remarkable recovery has begun. The trout cod has now established healthy, breeding populations in several key waterways where it was absent only a couple of decades before (see Lyon *et al.* 2012; Koehn *et al.* 2013). These populations are now self-sustaining, genetically robust, and expanding naturally into previous habitats. Ongoing fisheries enforcement and education, along with provision of environmental flows, re-snagging programs, and improvements in river health through state and federal initiatives and angling licence fee grants have all contributed to the ongoing recovery (Koehn *et al.* 2013).

Planning, people, governance

A range of state and federal initiatives have contributed to the recovery of the species, including funds provided through state government (i.e. the Victorian Waterway Management Strategy) and federally (i.e. the MDBA Native Fish Strategy). However, The National Recovery Plan for the species was the catalyst in which this restoration work was conceptualised, and was written during the first stages of the Recovery Program. Fishery closures were implemented and resourced through state Fishery Agencies, and funds for population modelling, captive breeding and reintroduction were provided jointly by state and federal governments. The recovery planning process was essential to bringing together states under one banner and aligned to a common set of goals. It also gave authority for relevant parties to meet and make decisions as a group and on a regular basis (see Fig. 21.3). However, within this structure, the success of the program was primarily driven by the commitment of individuals in each state to drive the recovery actions. At the commencement of the formal Recovery Program (Douglas *et al.* 1994), which is now nearing completion of its third iteration (Trout Cod Recovery Team 2008), funding was available both to fill key knowledge gaps for the species, and importantly, allow the formation and regular meeting of a Recovery Team for the species.

The importance of the Recovery Team should not be underestimated. This was the first national recovery plan for an Australian freshwater fish and most state agencies at the time were poorly resourced for threatened fish management. Those who became team members were working in relative isolation within their own political borders, so the establishment of a recovery team to administer the plan was a key development because it provided a forum to discuss problems, new approaches, and foster shared understanding (Lintermans 2013). The recovery team had state/territory research, hatchery and management representatives, as well as Commonwealth representation and recreational anglers through

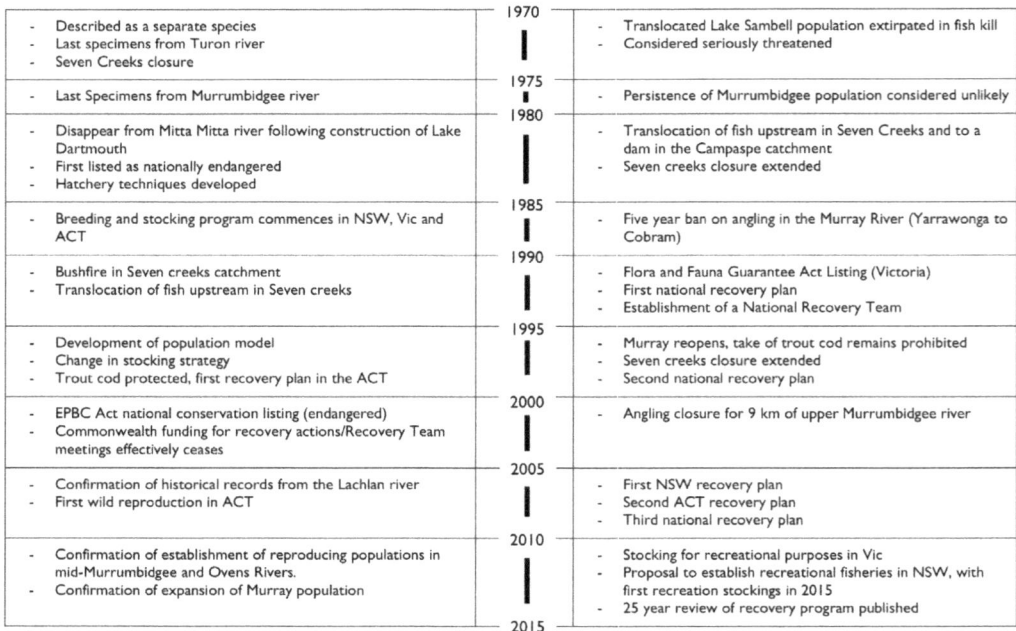

Fig. 21.3. Important events in the recovery of Trout Cod (modified from Koehn *et al.* 2013).

groups such as Native Fish Australia. As such, learnings could be quickly translated into revised management or policy arrangements. Without passionate people driving the process, it is unlikely that the effort would have been a success. In this case, state agency research staff made the most significant contributions in driving the program forward. Most of these staff are still working in research positions today and continue to be strong advocates, albeit mostly without dedicated resources for the tasks needed.

Monitoring

As a tri-state Recovery Program (Victoria, New South Wales, Australian Capital Territory), monitoring of the recovery following stocking was done at various intensities and with differing designs in each jurisdiction, given that funding usually came from a variety of different sources and as such was generally 'piggybacked' on other programs. However, across jurisdictions, monitoring was primarily directed at initially detecting survival of stocked fish, with the intention of eventually being able to detect wild-bred individuals (an indicator of successful population establishment). Monitoring of animals that live underwater is difficult. Regular monitoring at many sites demonstrated survival and growth of stocked individuals, but no recruitment or establishment. This was thought to be related to either dispersal of stocked individuals away from release sites, or simply the combination of low numbers of fish released and ongoing natural mortality. A secondary, but important, source of information was angler reports, which provided evidence both of initial survival to adulthood, but which also gave an indication of range expansion in the Murray. Having a committed angling fraternity provides an abundance of eyes and ears to help spread the message and educate other anglers in why trout conservation is important.

A key component of the monitoring work that commenced early in the Recovery Program was that on the Murray River population (the only 'natural' wild population remaining). Early concerns were that the Murray population was declining, but targeted survey work has revealed that this was not the case, and that in fact the population is growing and expanding in distribution (see Fig. 21.1). The relative lack of success of the stocking program in the early days (as evidenced by low capture rates in subsequent monitoring) was a flag that highlighted that a thorough review of the stocking approach was required, and the development of a population model was a key outcome (Todd *et al.* 2004). Use of the model changed the approach to stocking and resulted in larger numbers of fish being released over longer time spans (5–10 years for some populations) (see Lyon *et al.* 2012). This gave a greater probability of population re-establishment over defined time periods. Indeed, subsequent monitoring of populations (e.g. Lyon *et al.* 2012) confirmed the success of this modified approach.

Money

Initially, funds were provided by the federal and state governments. Commonwealth funding effectively ceased for recovery actions, and also for Recovery Team operation, in the early 2000s. However, an important ally was found in the form of recreational anglers, with funding from their recreational fishing licence fees becoming critical to the ongoing recovery effort. Although over-fishing (illegal or otherwise) was one of the many drivers in the original decline of the species, the increasing social conscience of the angling community has led not only to increased investment in both captive breeding and reintroduction of animals, but also into catchment restoration in the forms of co-funding fishways, instream habitat restoration and riparian revegetation.

The future

In the current National Recovery Plan For Trout Cod (Trout Cod Recovery Team 2008), a stated objective is the creation of a recreational fishery for trout cod in Victoria, and a similar program has recently commenced in NSW. Stockings for recreational fisheries have now been implemented in two closed lakes in north-east Victoria and one in NSW. Although to some this may seem controversial, given that it involved the harvest of a threatened species, the captive breeding and stocking of these fish is paid for by fishing licence fees, and over half of the fish bred are reintroduced to rivers for conservation purposes. In addition, the branding of the species as something that recreational anglers can 'value' and relate to maintains the species in the public eye (Lintermans et al. 2005) and brings an important stakeholder group to the table with a vested interest in supporting recovery efforts. The sentiments of a local angler on one of the rivers where trout cod have been re-introduced and are now self-sustaining are generally reflective of community attitudes: 'We hadn't seen a trout cod in the river for a couple of decades, but now my grandkids are down the fishing hole catching them. They release everything they catch, and now they know what they are, they feel a real passion for them – and they tell their mates at school and the community to take ownership'. Such support and growing social conscience from anglers is key in further recovery of both this species, but also other threatened fishes. As such, recovery programs for large bodied, threatened species are increasingly being driven by the recreational angling sector, rather than the conservation sector.

Further refinement and use of the population model will assist recovery actions. Additional hatchery capacity and dedicated funding is needed for stocking programs for the establishment of further populations. A recommitment of angler education (signage, publicity) is also needed to reduce mortalities from accidental or illegal take of trout cod.

As an interesting side-note, the increase in awareness around trout cod conservation within the angling community relates not only to angling species, but to all fishes. A great example of this is the lead role that the Australian Trout Foundation has taken in promoting the recovery of several small, threatened galaxiid species. Realising that their target species was impacting on threatened fish populations, this group has formed strategic partnerships with conservation managers and have now taken a lead to undertake works such as maintenance of barriers to keep trout out of critical habitats and replanting of riparian zones to support recovery of threatened galaxiids.

Conclusion

Further refinement and use of the population model will assist recovery actions. Additional hatchery capacity and dedicated funding is needed for stocking programs for the establishment of further populations. A recommitment of angler education (signage, publicity) is also needed to reduce mortalities from accidental or illegal take of trout cod.

With considerable success of many key recovery actions, especially the successful establishment of new populations and the education and support of anglers, the future of trout cod is much brighter than two decades ago. There is a continued need for conservation efforts, however, before its status in the wild is secured.

Acknowledgements

Many people have contributed to the recovery of this species. Key people to be thanked for their contributions to trout cod recovery include: staff from government hatcheries (Brett

Ingram, Steve Thurston, Matthew McLellan and Geoff Gooley); researchers and fisheries/conservation managers (Matthew Beitzel, Alistair Brown, Andrew Bruce, John Douglas, Dean Gilligan, Simon Nicol, John Pursey, Wayne Tennant Charles Todd, Steve Saddlier, Glen Johnson, Joy Sloan and Mark Jekabsons); and members of community groups such as Native Fish Australia (Will Trueman, Ron Lewis and Tim Curmi) and various staff from state and federal conservation policy groups.

References

Berra TM (1974) The trout cod, *Maccullochella macquariensis*, a rare freshwater fish of eastern Australia. *Biological Conservation* **6**, 53–56. doi:10.1016/0006-3207(74)90042-1

Cadwallader PL, Gooley GJ (1984) Past and present distributions and translocations of Murray cod *Maccullochella peeli* and trout cod *M. macquariensis* (Pisces: Percichthyidae) in Victoria. *Proceedings of the Royal Society of Victoria* **96**, 33–43.

Douglas JW, Gooley GJ, Ingram BA (1994) 'Trout cod, *Maccullochella macquariensis* (Cuvier) (Pisces: Percicthyidae) Resource Handbook and Research and Recovery Plan'. Department of Conservation and Natural Resources, Victoria.

Ebner BC, Thiem JD, Lintermans M (2007) Fate of 2 year-old hatchery-reared trout cod *Maccullochella macquariensis* (Percichthyidae) stocked into two upland rivers. *Journal of Fish Biology* **71**, 182–199. doi:10.1111/j.1095-8649.2007.01481.x

Ebner BC, Johnston L, Lintermans M (2009) Radio-tagging and tracking of translocated trout cod (*Maccullochella macquariensis*: Percichthyidae) in an upland river. *Marine and Freshwater Research* **60**, 346–355. doi:10.1071/MF08257

Harris JH, Rowland SJ (1996) Family Percichthyidae: Australian freshwater cods and basses. In *Freshwater Fishes of South-eastern Australia*. (Ed. R McDowall) pp. 150–163. Reed Books, Sydney.

Ingram BA, Rimmer MA (1992) Induced breeding and larval rearing of the endangered Australian freshwater fish trout cod, *Maccullochella macquariensis* (Cuvier) (Percichthyidae). *Aquaculture and Fisheries Management* **24**, 7–17.

Koehn JD, Harrington DJ (2006) Environmental conditions and timing for the spawning of Murray cod (*Maccullochella peelii peelii*) and the endangered trout cod (*M. macquariensis*) in southeastern Australian rivers. *River Research and Applications* **22**, 327–342. doi:10.1002/rra.897

Koehn JD, Lintermans M (2012) A strategy to rehabilitate fishes of the Murray-Darling Basin, south-eastern Australia. *Endangered Species Research* **16**, 165–181. doi:10.3354/esr00398

Koehn JD, Nicol SJ (2014) Comparative habitat use by large riverine fishes. *Marine and Freshwater Research* **65**, 164–174. doi:10.1071/MF13011

Koehn JD, Lintermans M, Lyon JP, Ingram BA, Gilligan DM, *et al.* (2013) Recovery of the endangered trout cod, *Maccullochella macquariensis*: what have we achieved in more than 25 years? *Marine and Freshwater Research* **64**, 822–837. doi:10.1071/MF12262

Kondolf GM, Angermeier PL, Cummins K, Dunne T, Healey M, Kimmerer W, *et al.* (2008) Projecting cumulative benefits of multiple river restoration projects: an example from the Sacramento-San Joaquin river system in California. *Environmental Management* **42**, 933–945. doi:10.1007/s00267-008-9162-y

Lintermans M (2013) Conservation and management. In *The Ecology of Australian Freshwater Fishes*. (Eds P Humphries, K Walker) pp. 283–316. CSIRO Publishing, Melbourne.

Lintermans M, Rowland SJ, Koehn J, Butler G, Simpson R, Wooden I (2005) The status, threats and management of freshwater cod species in Australia. In *Proceedings of the*

Management of Murray Cod in the Murray-Darling Basin: Statement, Recommendations and Supporting Papers. (Eds M Lintermans, B Phillips) pp. 15–29. Murray-Darling Basin Commission, Canberra.

Lyon JP, Todd C, Nicol SJ, Macdonald A, Stoessel D, Ingram BA, *et al.* (2012) Reintroduction success of threatened Australian trout cod (*Maccullochella macquariensis*) based on growth and reproduction. *Marine and Freshwater Research* **63**, 598–605. doi:10.1071/MF12034

Nicol SJ, Barker RJ, Koehn JD, Burgman MA (2007) Structural habitat selection by the critically endangered trout cod, *Maccullochella macquariensis* Cuvier. *Biological Conservation* **138**, 30–37. doi:10.1016/j.biocon.2007.03.022

Todd CR, Nicol SJ, Koehn JD (2004) Density-dependence uncertainty in population models for the conservation of trout cod, *Maccullochella macquariensis*. *Ecological Modelling* **171**, 359–380. doi:10.1016/j.ecolmodel.2003.06.002

Trout Cod Recovery Team (2008) 'National Recovery Plan for Trout cod *Maccullochella macquariensis*'. Trout Cod Recovery Team, <www.environment.gov.au/biodiversity/threatened>.

22

Underbelly: the tale of the threatened white-bellied frog

Manda Page, Kay Bradfield and Kim Williams

Summary

The problem
1. The frog has a small fragmented distribution, and is impacted by habitat disturbance and alterations.
2. It has specific habitat requirements and low breeding outputs.
3. Three-quarters of the suitable habitat is on private land.
4. The frog is vulnerable to future climate change.

Actions taken to manage the problem
1. 'Head-starting' program to establish new subpopulations and supplement existing subpopulations.
2. Habitat protection incorporating 1570 ha of land into the conservation estate, construction of conservation fences on private land.
3. Extensive survey and increased monitoring of riparian habitat.

Markers of success
1. Likely extinction averted.
2. Recovery Plan developed and implemented.
3. Recovery Team enduring and well managed.

Reasons for success
1. Biological features enabled effective conservation strategies to be implemented (the head-starting program).
2. Small core of dedicated and passionate members of the Recovery Team.
3. Coordinated action between Department of Parks and Wildlife, Perth Zoo and researchers, with direction from the Recovery Plan.
4. Habitat protection by incorporating land into conservation estate, and amended clearing regulations.

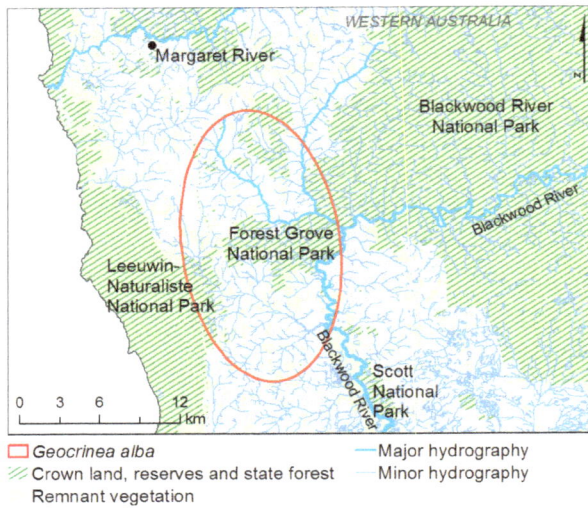

Fig. 22.1. The small distribution of the white-bellied frog in south-western Western Australia.

Introduction

The white-bellied frog (*Geocrinia alba*) was discovered in 1983 and formally described in 1989 (Wardell-Johnson and Roberts 1989). The species has an extremely restricted and fragmented distribution between the Margaret River and Augusta in the south-west of Western Australia (Fig. 22.1).

The small size of the white-bellied frog, being no bigger than an Australian 10 cent coin (~20–25 mm), limited distribution and specific habitat requirements probably account for why the species remained undiscovered for so long. But these features also put them at risk of extinction: a risk that was evident from the time of discovery.

The white-bellied frog has a white (or very faint yellow) ventral surface, or underbelly. It has a light brown to grey dorsal (upper) surface, which blends in well with its muddy environment and makes it hard to see from above (Fig. 22.2). The males' mating call is a series of 11–18 loud pulses repeated irregularly, akin to the sounds of a toy machine gun.

The white-bellied frog is listed as Critically Endangered under the Western Australian *Wildlife Conservation Act 1950* using the IUCN Red List criteria (2004), and listed as Endangered nationally under the Commonwealth (federal) *Environment Protection and Biodiversity Conservation Act 1999*. It was upgraded from Endangered to Critically Endangered in Western Australia in 1999, just 1 year after it was first assessed, because further surveys found that it inhabited only a very small geographic range. The area of occupancy is only 1.9 km^2. Subpopulations are severely fragmented and considered to be undergoing continuing decline. Although the formal conservation status of the white-bellied frog has not changed since it was listed as Critically Endangered in 1999, the actions undertaken as part of the Recovery Program have better secured the species by protecting habitat, supplementing existing small subpopulations, and establishing new subpopulations.

Conservation management

Biology and habitat requirements

White-bellied frogs breed in spring. They have fully terrestrial breeding biology: that is, the breeding cycle occurs on land, not in the water. The males create small burrows in moist soil from which they call to attract females. Once the female has chosen a male, she enters his

Fig. 22.2. The white-bellied frog has been known for only 30 years and could easily have been lost before it was even known. (A) First year metamorph. (B) Lateral view. (C) Oblique front view (photos: Manda Page).

burrow to mate. Although the male and female physically interlock (called amplexus), fertilisation is fully external: as the female lays eggs, the male fertilises them. The eggs are deposited in the burrow and then left unattended. The average number of eggs per clutch size is 10–12 (maximum recorded being 26), which is relatively small for frogs. The eggs are part of an egg mass that is held together by a jelly-like substance. Eggs hatch and the larvae develop and metamorphose, all within the burrow and inside this jelly mass. There is no free-swimming or feeding tadpole stage; the larvae have a yolk sac that is used as a food source during development. The eggs and larvae are highly susceptible to predators such as flatworms, spiders, insects and burrowing crayfish. Those that survive leave the egg mass following metamorphosis into froglets, but don't reach sexual maturity in the wild until around 2–3 years of age. Although they can live for 10 or more years in captivity, most adults survive long enough to breed only for a single season in the wild.

The white-bellied frog is highly sedentary (Driscoll 1997; Conroy 2001). Over small distances, the subpopulations can be genetically distinct, which means that they rarely move between subpopulations, even though these may be only a few kilometres apart. One study demonstrated that 95% of the adult male frogs moved less than 5 m in a year, and no more than 20 m between years (Driscoll 1997). This means that if a subpopulation goes locally extinct, possibly due to a failed wet season or a fire, there is little chance that the site will be recolonised and thus the local extinction may be permanent. Given the very low likelihood of immigration, subpopulations appear to be almost completely reliant on breeding for recruitment.

Distribution and land tenure

White-bellied frogs have a patchy distribution. Fragmentation is likely a natural phenomenon due to dependence on specific breeding habitat. Although they occur in drainage lines that are extensive, they don't seem to use all apparently suitable available habitat. We are yet to completely understand what limits them to small sites along the drainage lines, but it most likely relates to the level of moisture maintained throughout the year, food availability and temperature, which may be dictated by the type and amount of vegetation cover.

The habitat of the white-bellied frog is in the vicinity of Margaret River: the famous wine producing area of Western Australia. Unfortunately, much of this area has been cleared for agriculture such as beef and dairy production, with a more recent change in land use towards vineyards and tree plantations, resulting in the creation of large dams on the natural creek systems. It has been estimated that up to 70% of potentially suitable riparian habitat for this species has been cleared and, albeit on a smaller scale, clearing continues in the region. Most (77%) of the total known subpopulation of white-bellied frogs occurs on private land along narrow corridors of riparian vegetation among extensive areas of cleared farmland (Fig. 22.1).

As part of the Recovery Program, 1560 ha of white-bellied frog habitat was acquired for addition to the Blackwood River National Park in 2000. This area contained ~30% of the total number of known calling males. Later, some other crown lands where they occurred were incorporated in Blackwood River and Forest Grove National Parks. Despite this, most of the subpopulation still resides on private land.

In the early 1990s, extensive surveys were undertaken to locate existing subpopulations. This increased the known number of subpopulations from 26 to 72 by 1996 and by 2007 there were 102 recognised subpopulations. However, no new subpopulations have been found since 2007, and of the 102 known subpopulations, 39 are now considered extinct. Unfortunately, there has been a trend towards smaller subpopulations and increased extinctions since 2007.

Governance and planning

The first Recovery Plan for the white-bellied frog was prepared in 1995 (Wardell-Johnson *et al.* 1995), just a few years after it was formally described. A new Recovery Plan was developed and approved in 2015 (DPW 2015). Since 1995, there has been a formal Recovery Team that provides advice on the implementation of the Recovery Program for both the white-bellied and orange-bellied frogs. The organisational membership of the Recovery Team has changed little over the years, with Parks and Wildlife, Perth Zoo and the University of Western Australia providing the core members.

Evolution of a captive program

It was recognised early that the establishment of new subpopulations would be a vital tool for the recovery of the white-bellied frog. The 1995 Recovery Plan provided four options for this: (1) translocation of egg masses in the field; (2) translocation of egg masses into a laboratory where these egg masses are raised to adult stage for release; (2) translocation of adult individuals to new sites from existing subpopulations; and (4) captive breeding. A successful captive breeding program for the karri frog (*Geocrinia rosea*) – a closely related species with a similar breeding biology (i.e. direct-developing eggs) – had previously been established in New Zealand (Bell 1985). Given the precarious situation of the white-bellied frog, initial work involved breeding trials with the widespread and abundant karri frog (Mantellato *et al.* 2013). Following success with this species, the captive program focussed on using the orange-bellied frogs (*Geocrinia vitellina*) because they are more closely related to the white-bellied frogs than the karri frog, but less threatened.

However, the breeding techniques that were successful for the karri frog proved not to be transferable to the orange-bellied nor white-bellied frogs. Individuals of both threatened species were successfully maintained in captivity and juvenile orange-bellied frogs were successfully reared to maturity, which indicated that a captive program may still be a viable option.

Small subpopulation sizes precluded testing the translocation of adults; hence, the initial focus was on translocating egg masses, either to the field or to a captive facility to allow the eggs to develop to metamorphosis to increase survivorship and avoid the mortalities experienced in the field through predation and fungal infections. Early trials translocating orange-bellied frog egg masses in the field in 2000 had mixed success, and only a few adult frogs persisted at the translocation sites.

Given limited success from direct translocation of egg masses and the difficulty in breeding the threatened species in captivity, the focus soon turned to creating a 'head-starting' program, where egg masses could be collected in the field, transferred to a captive facility where they were maintained through metamorphosis, some 1–2 months later, and then for another 10–11 months before being released back to the wild as juveniles the following spring. Given how small these frogs are when they metamorphose (they are only a few mm long and weigh ~0.034 g – about the same as a single rice bubble), a specialist skillset is required to work with them in captivity. Perth Zoo took on the challenge and established a captive conservation project for the two threatened *Geocrinia* species. In 2008, the first two clutches of white-bellied frog egg masses collected from the wild by zoo staff were successfully maintained through metamorphosis and subsequently reared for release or the ongoing breeding trials.

To date, the head-starting program at the Perth Zoo has released over 500 juveniles. There are indications that at least two new subpopulations of the white-bellied frog have been established and other small subpopulations have been successfully supplemented. The head-starting program now achieves nearly 100% success of egg clutches to juvenile stage, whereas mortality rates are ~80% if they are left in the wild. This means that harvesting the egg masses will have a very small impact on the breeding success at the site, but potentially a dramatic positive impact at another site. The impact of collecting egg clutches on the source population is reduced by releasing a proportion of each clutch back into the subpopulation it was collected from.

Captive breeding also has been attempted with up to five pairs per year, but with less success. Breeding was successful in 2011 and 2016, but was unsuccessful in the intervening years. Understanding the cues to stimulate breeding in captivity is far from complete, but Perth Zoo staff will continue to trial different breeding enclosure designs and environmental conditions, with insights from Parks and Wildlife staff working in the field.

In-situ management

The main threats to the subpopulations of white-bellied frog include physical disturbance from feral and domestic animals and humans, alterations in hydrology (ranging from the effects of climate change to the impacts of herbicides, pesticides and fertilisers), vegetation clearing, inappropriate fire events and disease (DPW 2015).

The white-bellied frog is particularly susceptible to many of these threats because it occurs primarily on privately managed land. One of the key recovery actions is to protect occupied habitat from damage by livestock. To achieve this, conservation fencing has been constructed and maintained across 19 private properties, protecting ~17 ha of riparian habitat from livestock disturbance. In partnership with Parks and Wildlife, the local Natural Resource Management group (South West Catchments Council) has coordinated, funded and delivered on-ground projects such as feral pig control and ecosystem protection and rehabilitation. In addition, amendments to the Western Australian

Environmental Protection Act 1986 in 2004 provided a greater capacity to regulate vegetation clearing and acknowledge the principle that native vegetation should not be cleared if it comprises the whole or part of, or is necessary for the maintenance of, a significant habitat for fauna indigenous to Western Australia.

The conservation of the white-bellied frog is a shared responsibility when it occurs on private land, thus community awareness and participation is important. A 'Frog Recovery Kit' was first produced in the late 1990s, which outlined information about the species and the recovery process. This was disseminated to landholders and the wider community. Furthermore, the Department of Parks and Wildlife has attempted to keep track of changes in landowners for properties on which this species occurs, and has conducted field visits and engaged in discussions with landowners about habitat management. Many property owners, however, exhibit a lack of interest or are indifferent to the fact they have this threatened species on their property. Two private landholders established conservation covenants, in part, due to having habitat of white-bellied frogs on their land. However, at both properties, subpopulations of this species have not persisted, and this legal option for protection has not been taken up by any other landowners.

Monitoring

Accurate monitoring of rare species can be challenging, but male white-bellied frogs call loudly to attract a mate and this calling provides a robust basis for monitoring. However, monitoring based on such calling can only occur when males are calling reliably, which occurs only between September and late November. Females do not call. We know that the male to female ratio is approximately even, so counting calling males gives an estimate of the number of animals in a subpopulation.

Calling is loudest at night, and four different types of monitoring techniques have been developed to provide accurate and repeatable counts. Techniques include simple point counts where a person stands at a marked site and counts every male they can hear and linear transect counts where trained observers slowly move along the drainage line counting calls. Monitoring can be challenging, because surveys must be undertaken at night time and in wet, muddy, and often cold, conditions. This takes both skill and dedication. In addition to counts, the extent of occupied habitat at a site can be captured using GPS technology, allowing changes in the area of habitat used to be tracked over time.

Using these monitoring techniques, we have been able to analyse the data and track subpopulation trends. This monitoring now extends over 20 years and the results have consistently shown that:

- Small subpopulations with fewer than 10 calling males have declined, with many of them going extinct.
- Subpopulations with more than 50 calling males have persisted and in a few instances increased to over 100 calling males.
- The majority of subpopulations in the northern and southern extents of the distribution have declined or become extinct.

Unfortunately, the monitoring results indicate that subpopulations in the upper headwaters of creek systems, regardless of whether they occur on private lands or conservation estate, have declined or are declining to the point of no return. This seems to be a direct consequence of declining rainfall and climate change modifying their habitat. The core habitat for this species is now primarily restricted to one creek system and associated tributaries, the majority of which is in private ownership.

This monitoring not only provides information on the trends of the wild subpopulations but also informs aspects of the captive program, including where to source egg masses, which subpopulations may benefit from augmentation, and post-release success. Three subpopulations have been, or are in the process of being, established via translocations from the head-starting program, into creek systems where natural subpopulations have become extinct. Results have been promising. The best example is a site that has now sustained a stable subpopulation of >50 calling males 4 years after the last release.

Money

To date, the Recovery Program has been funded from a variety of sources and in-kind support. The most significant investment to date is A$5.5 million to purchase land. This is followed by the resources invested in the captive program, which has, to date, amounted to over A$800 000. Funding solely dedicated to the recovery of this species has been limited during the recovery project and has almost disappeared over the past 5 years. The voluntary contributions of individuals beyond their job requirements have been, and continue to be, necessary to continue the program. The captive program has dedicated support until 2019, but the Recovery Team continues to seek funds to support the entire program.

The future

The ability to successfully establish new subpopulations, and to monitor trends, places the white-bellied frog in a better situation that it might otherwise have been without the dedicated Recovery Program. However, they are not out of danger yet, because multiple threats continue, particularly climate change and competing demands on wetland habitat. This could have disastrous consequences for this species unless the Recovery Program can create or maintain suitable habitat.

The white-bellied frog's reliance on its wetland habitats means the species faces many future challenges that relate to water, its use and management. As the type of agricultural land use being practiced within its geographic range intensifies, so does the demand for ground and surface water. Coupled with the continuing rainfall reduction occurring in the south-west of Western Australia, reported to be as much as a 19% decline since 1970, resulting in a 50% decline in stream flow (CSIRO and BOM 2016), the natural habitat and hydrology on which this species relies is diminishing.

The changing climate is also expected to result in increased bushfire frequency and intensity, which threatens not only the habitat of the white-bellied frog, but human habitat also (Hughes and Steffen 2013). As society demands increased protection from fire, authorities are pushed to increase the scale and coverage of fuel-reduction programs to protect life and property and this also threatens a significant portion of the frog's habitat.

Resourcing threatened species recovery programs presents many challenges. Many species respond slowly, and it often needs 5 or more years to determine the success or failure of an action. Over the past decade, there has been an overall reduction in funding directed to implementing recovery programs, which has required alterations to the type and quantity of conservation actions that can be implemented. For this species, fundamental decisions relating to long-term strategies must be made on a year-to-year basis, despite the success achieved so far. Furthermore, these little frogs do not capture community empathy and support in the same manner as charismatic mammal fauna, such as koalas. Nor does white-bellied frog habitat have the same appeal as colourful tropical reefs or rainforest, but that does not mean that they are any less deserving of a fair share of the available funding.

Conclusion

Were it not for the fortune of its recent discovery, the white-bellied frog may well have gone extinct without anyone even realising that it had ever existed. Since its recent discovery, its extinction has been averted through the efforts of a small but dedicated team. The main drivers of success to date have been the ability to understand the breeding traits of the species and transfer this knowledge into an effective conservation strategy and, more broadly, improved legislation constraining broad-scale land clearing and additions to the conservation estate. The small Recovery Team combines the skills and experience from a range of disciplines and this has been fundamental to the Program. Scientists, conservation land managers, amphibian-breeding specialists and natural resource management groups have collectively progressed recovery actions to date, but now we may need to expand to include specialists in climate change and hydrology, and increased engagement with private land managers, community groups and funding bodies.

Acknowledgements

Current and past recovery team members have provided direction, innovation and passion to this Recovery Program and include Kim Williams (Chair, Parks and Wildlife), Dale Roberts (University of Western Australia), Chris Fleay (Parks and Wildlife), Manda Page (Parks and Wildlife), Ben Lullfitz (Parks and Wildlife), Kay Bradfield (Perth Zoo), Cathy Lambert (Perth Zoo), Peter Mawson (Perth Zoo), Helen Robertson (Perth Zoo) and Megan Flowers (Parks and Wildlife). Funding and other support has come from the South West Conservation Council, Shire of Augusta-Margaret River, Lower Blackwood Land Conservation District Committee, National Threatened Species Network, South West Catchment Council, WA State Natural Resource Management, Perth Zoo and the Department of Parks and Wildlife. Scientific input and advice has been driven by academic researchers such as Don Driscoll, Simon Conroy, Grant Wardell-Johnson and Nicky Mitchell. Staff in Perth Zoo's Native Species Breeding Program continue to be adaptive and innovative in the delivery of the breeding program.

References

Bell BD (1985) Conservation status of the endemic New Zealand frogs. In *Biology of Australasian Frogs and Reptiles*. (Eds G Grigg, R Shine, H Ehmann) pp. 449–458. Surrey Beatty, Chipping Norton, NSW.

Conroy SDS (2001) Population biology and reproductive ecology of *Geocrinia alba* and *G. vitellina*, two threatened frogs from southwestern Australia. PhD thesis, The University of Western Australia, Perth, Australia.

CSIRO and BOM (2016) State of the climate. CSIRO and Bureau of Meteorology, <http://www.bom.gov.au/state-of-the-climate/State-of-the-Climate-2016.pdf>.

DPW (2015) White-bellied and orange-bellied frogs (*Geocrinia alba* and *Geocrinia vitellina*) Recovery Plan. Wildlife Management Program No. 59. Department of Parks and Wildlife, Perth, WA.

Driscoll DA (1997) Mobility and metapopulation structure of *Geocrinia alba* and *Geocrinia vitellina*, two endangered frog species from southwestern Australia. *Australian Journal of Ecology* **22**, 185–195. doi:10.1111/j.1442-9993.1997.tb00658.x

Hughes L, Steffen W (2013) Be prepared: climate change and the Australian bushfire threat. Climate Council of Australia, <http://www.ufuq.com.au/files/9313/8655/5677/Dec_2013_Be_Prepared_report.pdf>.

Mantellato L, Gaikhorst G, Kruger R, Vitali S, Robertson H (2013) Growth and development of captive *Geocrinia rosea* (Myobatrachidae): a rare species analogue. *Zoo Biology* **32**, 374–380.

Wardell-Johnson G, Roberts JD (1989) Endangered! Yellow-bellied and white-bellied frog. *Landscope* **5**, 17.

Wardell-Johnson G, Roberts JD, Driscoll D, Williams K (1995) *Orange-bellied and White-bellied Frogs Recovery Plan*. 2nd edn. Department of Conservation and Land Management, Perth, WA.

23

Western swamp tortoise: slow and steady wins the race

Gerald Kuchling, Andrew Burbidge, Manda Page and Craig Olejnik

Summary

The problem
1. Only two tiny populations survived extensive habitat loss and fragmentation through agriculture, urbanisation and clay mining, and predation by the European red fox.
2. Current habitat is becoming drier and hotter due to climate change.

Actions taken to manage the problem
1. Protection of remaining habitat through creation of nature reserves specifically for the conservation of the species and habitat purchase and restoration to increase their size.
2. Research into their biology and long-term population monitoring.
3. Habitat management including predator control, drainage alteration and water supplementation in dry years.
4. Captive breeding, reintroduction of captive-bred progeny and assisted colonisation to establish new populations.

Markers of success
1. Extinction was probably averted and the population is now increasing, including in captivity.
2. The causes of decline have been identified and are now being managed effectively.
3. A Recovery Plan was developed and updated in four editions and is being implemented under guidance of an extremely experienced and committed Recovery Team.
4. Community involvement and support has been sustained over many decades.

Reasons for success
1. An iconic species with a long history of awareness and lobbying for its protection from the public and scientific community.

2. Strong public support driving significant fundraising together with long-term government commitment.
3. Strong partnerships via a Recovery Team between government and non-government agencies and scientists, allowing timely research and adaptive management once problems are recognised

Introduction

In 1839, Ludwig Preiss, a collector of plants and animals in the Swan River Colony of Western Australia, sold a freshwater turtle specimen to the Museum of Natural History in Vienna, Austria. It was described as a new species (*Pseudemydura umbrina*), in a monotypic genus by Friedrich Siebenrock in 1901. The next specimen known to science was schoolboy Robert Boyd's locally found pet freshwater turtle exhibited at the annual Wild Life Show of the Western Australian Naturalists' Club in Perth in 1953. This was joined by another specimen in 1954. These two 20th century specimens were described by the then Director of the Western Australian Museum, Ludwig Glauert, as a new species (*Emydura inspectata*), which in 1958 was shown to be a synonym of *Pseudemydura umbrina*. Searches by Western Australian naturalists during the 1950s found the western swamp tortoise (*P. umbrina*) (freshwater turtles are called 'tortoises' in Western Australia) only in small patches of winter-wet clay swamps in the Upper Swan suburb of Perth – apparently the stronghold of the western swamp tortoise had been the clay soils of the Swan Valley. However, these were the only fertile soils in the coastal plain and were the first part of Western Australia cleared and developed for agriculture during the 19th century. By the early 1960s, almost all this land had been urbanised, used for intensive agriculture or mined for clay to manufacture bricks and tiles.

A captive colony was established in 1959, but until 1988 survival and offspring production was erratic and low. Research by students of The University of Western Australia began in 1963 and one of them, Andrew Burbidge, went on to research the western swamp tortoise for his PhD. Cutting-edge turtle research techniques for that time were developed and applied, including radio-tracking and the use of radiography for the non-lethal assessment of egg and clutch size in gravid females.

The rarity and the restricted distribution of the western swamp tortoise was confirmed: in the mid-1960s, Burbidge estimated the wild populations to number over 200 individuals and concluded 'that *P. umbrina* should persist, at least at Twin Swamps Reserve' (Burbidge 1967). By 1978, the estimates were ~20 at Twin Swamps Nature Reserve and 15–20 at Ellen Brook Nature Reserve. Burbidge (1981) revised his earlier assessment to: 'if the downward trend in numbers of *P. umbrina* continues, the long-term survival of the species in the wild seems unlikely'. The downward trend did continue: by 1987 'they have apparently disappeared entirely from one of the two reserves set up for the species, and were reduced to a very few dozen individuals in the other … The captive breeding program also seems to have reached a standstill, with no reproductive activity (courtship or oviposition) over the last three seasons' (Pritchard 1987). During the 1980s the IUCN-Species Survival Commission Freshwater Chelonian Specialist Group considered *P. umbrina* the single most endangered freshwater chelonian in the world (Rhodin and Mittermeier 1983).

A rescue operation started in 1988 (Kuchling and DeJose 1989) and the Recovery Program has been operating since 1991 (Kuchling *et al.* 1992). Although the western swamp tortoise's IUCN Red List status of Critically Endangered has not changed since 1996, nor its Critically Endangered status under the *Environment Protection and Biodiversity Con-*

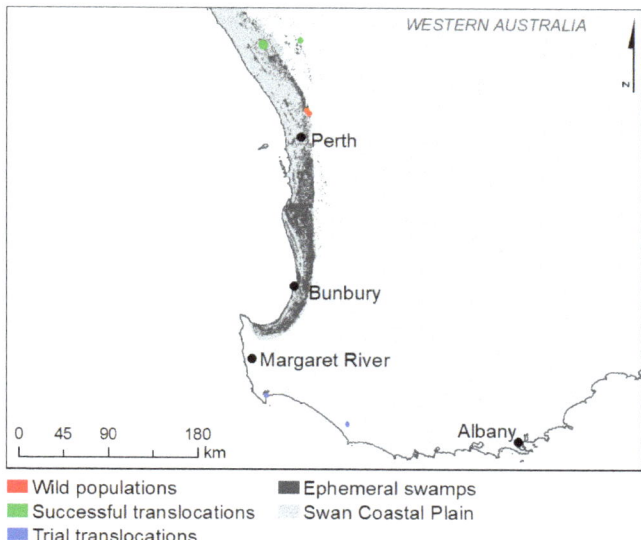

Fig. 23.1. Once the wild western swamp tortoise populations at the fringes of Perth recovered during the 1990s through habitat management and reintroduction, captive stock was, and still is, used to establish new populations further away from the city and across a broader climate gradient.

servation Act 1999 since 2004, major progress has been made. By 2016 the non-hatchling population was ~200 individuals in four sites with a captive population of 180 (Fig. 23.1).

Conservation management

There have been three phases of the conservation history of the western swamp tortoise.

Mid-1950s to mid-1960s

Public interest and what we now call 'citizen science' drove conservation efforts. The captive insurance colony was founded in 1959 with 25 individual tortoises taken into captivity in the backyard of the private residence of Dr D. Ride, Director of the Western Australian Museum. Two of those wild-collected tortoises are still today the best breeding females in the captive colony at Perth Zoo. By the early 1960s, the extensive loss and fragmentation of the clay swamp habitat of the western swamp tortoise was recognised as the single major threatening factor. The rediscovery of this species on the verge of extinction in the outskirts of Perth captured the imagination of the public and the response to the public appeal 'Operation Tortoise' for donations to secure their last remaining habitat was enthusiastic. Thus, in 1962, the Ellen Brook Nature Reserve (65 ha) and the Twin Swamps Nature Reserve (155 ha) were declared by the Western Australian government as Class A Nature Reserves specifically for the conservation of the western swamp tortoise.

Mid-1960s to mid-1980s

The public considered the job done and the momentum for further conservation actions was lost. The 13 surviving tortoises of Dr Ride's captive colony were transferred to Perth Zoo and produced 26 hatchlings between 1963 and 1978, with 14 of these alive in April 1978 (Spence *et al.* 1979). Four of those remain alive today. The then Department of

Fisheries and Fauna employed a research scientist in 1968, who maintained a low-level monitoring program of the wild western swamp tortoise populations. Monitoring demonstrated low recruitment and decreasing numbers throughout the 1970s and 1980s. Fewer than 30 individuals survived in the wild by the mid-1980s.

In 1979, the three remaining adult females and two males of Perth Zoo were transferred to the WA Wildlife Research Centre. Two females, two males and one juvenile found during the early 1980s in Twin Swamps Nature Reserve were also taken into captivity. However, artificial reproductive techniques used at the WA Wildlife Research Centre from 1979 to 1986 failed to boost hatchling production, and hatchlings could not be reared successfully. The five females produced 22 eggs in 1979 and 1980, eight of which hatched. Seven of the hatchlings died during their first year and the last one in its second year. No eggs were produced in captivity from 1981 to 1986. Of the 30 wild founders taken into captivity between 1959 and 1981, plus their 35 offspring (i.e. 65 individuals in total), only 17 survived by 1987, including one male, one female and one juvenile of the Twin Swamps population taken into captivity in the early 1980s.

By 1987, the total population of the western swamp tortoise (wild and captive combined) amounted to fewer than 50 individuals, of which only 25 were adults. The population at Twin Swamps Nature Reserve was functionally extinct. The species was slipping towards becoming a well-documented turtle extinction.

1987 to present

In 1987, a rescue operation for the western swamp tortoise was proposed to the Western Australian Department of Conservation and Land Management (Kuchling 1987) and changes were made to captive management. This resulted in two females laying seven eggs but none hatched due to long-term inadequate nutrition of the females (Kuchling 2006). The first action in 1988 was to rejuvenate the captive breeding operation. The western swamp tortoise Captive Breeding Management Committee was formed with members from the Department of Conservation and Land Management, the Perth Zoo and UWA. This first formal threatened species recovery group in Western Australia marked the starting point of the successful, still ongoing, conservation program. Based on thorough investigation of the reproductive cycle of western swamp tortoises (Kuchling and Bradshaw 1993), a successful captive breeding program was established at the WA Wildlife Research Centre and at Perth Zoo. With 44 surviving juveniles this rescue operation had doubled the population of the western swamp tortoise by 1991. In the winter of 1991 all captive tortoises were transferred to Perth Zoo who took over responsibility for the captive breeding project. Since 1988 more than 1000 western swamp tortoises have been bred in captivity.

Due to the steady production of hatchlings since 1988, releases of captive-bred tortoises commenced in 1994 to re-establish a former or establish new populations:

- 272 juvenile tortoises have been reintroduced to Twin Swamps Nature Reserve since 1994.
- 180 juvenile tortoises have been introduced (assisted colonisation) to Mogumber Nature Reserve since 2000. Assisted colonisation is translocation where a species is intentionally released into a site outside of its indigenous range (IUCN/SSC 2013). This is undertaken primarily because there is a recognised threat in its current range, in this case habitat loss, which could increase the risk of the species becoming extinct.
- 174 juvenile tortoises have been introduced (assisted colonisation) to Moore River Nature Reserve since 2007 (Fig. 23.2).

Fig. 23.2. Juvenile captive-bred western swamp tortoise introduced to Moore River Nature Reserve (photo: Gerald Kuchling).

At all those translocation sites, some released tortoises reached maturity and started to reproduce, with successful recruitment into the juvenile population having been recorded at all sites. At Twin Swamps Nature Reserve, however, recruitment only became successful over the last decade, probably due to recently longer lasting swamp supplementation with groundwater in winter and spring. Due to the long life cycle of the species, it is too early to demonstrate success in establishing self-sustaining populations. Continuing monitoring at least for another decade will be necessary for this evaluation.

Research, biology and key threats

The western swamp tortoise inhabits only shallow, ephemeral, winter-wet swamps on clay or sand over clay soils (Burbidge 1967, 1981). These swamps have a high richness and biomass of invertebrate species, and a diverse flora. After the swamps fill in early winter, the tortoises can be found in water. The western swamp tortoise is carnivorous, eating only living food such as insect larvae, small crustaceans, worms and tadpoles. Food is taken only under water and feeding is restricted to winter and spring. Females take 8–15 years to reach maturity and lay one clutch of three to five eggs in late spring. When the swamps dry out in late spring, the western swamp tortoise moves from the low swamp areas into slightly elevated bushland to aestivate during summer and autumn in naturally occurring holes or under leaf litter (Burbidge 1967, 1981).

Research in the 1960s led to the conclusion that, under then implemented management actions, the population at Twin Swamps Nature Reserve (180–200 individuals) should increase in numbers and persist indefinitely, while the future of the small population (20 individuals) at Ellen Brook Nature Reserve was considered less certain (Burbidge 1967).

However, the tiny population at Ellen Brook Nature Reserve remained viable and persisted over more than half a century, while the originally larger population at Twin Swamps Nature Reserve had nearly disappeared by the mid-1980s (Burbidge *et al*. 1990). Due to the species' biological traits of high longevity, high adult survival and a very long individual reproductive span, even a very small population like that at Ellen Brook Nature Reserve can remain viable and persist in suitable habitat. Largely based on anecdotal information, the introduced European red fox (*Vulpes vulpes*) was identified as important predator of tortoises and their eggs (Burbidge 1967, 1981) and as a key threat: the rapid decline in tortoise numbers at Twin Swamps Nature Reserve, where aestivation refuges were mostly on the surface, was primarily attributed to fox predation, whereas underground aestivation at Ellen Brook Nature Reserve better protected the species from foxes (Burbidge *et al*. 1990). Increasing aridity was seen as another contributing factor for decline that affected Twin Swamps Nature Reserve more heavily than Ellen Brook Nature Reserve (Burbidge *et al*. 1990; Fig. 23.3).

With fox predation largely under control, the drying climate, in combination with the very fragmented habitat, is now the major threat for the survival of the western swamp tortoise. Swamp life depends on rainfall and hatchlings have to reach a certain body mass during their first winter and spring to be able to survive their first aestivation period. Progressively shorter wet periods and longer dry seasons reduce recruitment without the pos-

Fig. 23.3. The worst enemy of western swamp tortoise: habitat loss and fragmentation. Aerial view of the Upper Swan area towards north-west (January 2015) with Twin Swamps Nature Reserve and Ellen Brook Nature Reserve, the locations in which the last wild populations persist and the years of completion of fox-proof fences and additions of rehabilitated areas (yellow, orange, pink) (photo: Gerald Kuchling).

sibility for tortoises to move to wetter habitats – individual home ranges of western swamp tortoises can be larger than any of the existing nature reserves.

Planning and policy

Due to their critically low numbers, the conservation response for the western swamp tortoise had to be planned and had to commence in a context where there was too little material and no time for exhaustive data collection and field studies, in particular on aspects such as predation, which may be a rare event but significant for population persistence. Due to their highly specialised life style, there were also no clear-cut, by-the-book captive breeding or conservation strategies to follow. Despite these challenges, the problems with captive breeding and rearing of western swamp tortoise had been solved by 1991 and the Captive Breeding Management Committee was transformed into the formal western swamp tortoise Recovery Team with expanded membership. The Western Swamp Tortoise Management Program was published in 1990 (Burbidge *et al.* 1990), and was launched by the World President for WWF, Prince Phillip. The Management Program was converted to a Recovery Plan and published in 1994, with the main objective to ensure the survival of the western swamp tortoise by creating at least two viable wild populations (Burbidge *et al.* 1990). This was the first recovery plan in Western Australia; it has undergone several revisions and is now in its 4th edition (Burbidge *et al.* 2010).

Every revision reviewed the success in reaching the recovery criteria of the previous edition and defined new criteria and objectives. An independent review of the Plan's implementation was conducted by Dr Hal Cogger (Australian Museum, Sydney), dated 14 April 1999. Dr Cogger stated *inter alia* 'In summary, the Recovery Plan was well planned and has been implemented so successfully that many of its 2002 goals have already been achieved. Logistical and ecological problems have arisen in the course of the project, and most have been appropriately and successfully addressed. The Recovery Team has operated effectively and has provided expertise critical to the success of the program.'

People, agencies, governance and accountability

The Department of Biodiversity, Conservation and Attractions is the statutory authority responsible for the conservation of the species. Perth Zoo runs the breeding program, and university academics support these organisations by conducting relevant research. However, in this context, the enthusiasm and the creativity of dedicated individuals appear to have been more important drivers to implement the recovery actions than bureaucratic mechanisms *per se*. The success of the western swamp tortoise recovery project caused enthusiasm for the species to re-emerge in the wider community. This also generated political attention: the release of the 500th juvenile captive-bred western swamp tortoise was celebrated with the WA Minister for Environment in 2014.

Monitoring

Monitoring of the two western swamp tortoise populations at Ellen Brook and Twin Swamps Nature Reserves commenced in 1963, with a first intensive phase until 1966 (Burbidge 1967), continued then in low-level mode until the late 1980s (Burbidge 1981; Burbidge *et al.* 1990) and became again more intensive from 1991 onwards. Due to the cryptic nature of western swamp tortoises and the difficulty associated with luring them into traps, the proportion of individuals recorded annually is relatively low. This causes a temporal delay for reliable population estimates of about 4–5 years – we can never be sure about total population numbers (Burbidge *et al.* 2010). Still, this is one of the world's longest running freshwater tortoise monitoring programs and it provides reliable trends for the Recovery Team to act on. Examples include prolongation of the groundwater pumping protocol at Twin Swamps Nature Reserve in response to low juvenile recruitment, provision of tortoise passages through the fence at Ellen Brook Nature Reserve, and an agreement with a private neighbour of Mogumber Nature Reserve to manage a swamp on his property for western swamp tortoises that use it seasonally.

In 2013, a celebration marking 50 years of scientific monitoring of the western swamp tortoise took place at Ellen Brook Nature Reserve. A wild female, 'Tortoise Number 4', who was first marked in 1963 and who is now more than 65 years old, was briefly on display.

Money

Since the 1960s, funds raised by sponsorships and public appeals have been a feature of the research into and management of the western swamp tortoise. In addition to the provision of funds and staff time from Parks and Wildlife and Perth Zoo, grants from the Australian National Parks and Wildlife Services, Natural Heritage Trust (via the Swan Catchment Council) and Community Conservation – money, equipment, substantial discounts on purchases and direct assistance have come from universities, non-government organisations and private sector organisations. Recovery actions are costed and budgeted in the Recovery Plan and includes pro rata salaries of a chief investigator, zoo keepers, technical staff for reserve management, and administration in addition to capital maintenance. The annual costs increased from about A$300 000 in 1992 to about A$650 000 over the last few years. This does not include major capital costs such as a new breeding facility at Perth Zoo.

With support from industry and the wider community, in tandem with collaboration between government and non-government organisations, funding for the project has been maintained, albeit at varying levels. The formation of a Friends Group further raised the profile of the species and it developed innovative ways to raise funds for the program while assisting with field projects. One example is the partnership with the Margaret River Chocolate Factory, resulting in the sale of chocolate tortoises called 'Swampy' to help this Endangered reptile in a delicious way (see http://www.westernswamptortoise.com.au/news/fowst-news/item/54-margaret-river-choc-launches-swampy.html).

The future

One of the growing threats to this species is the impacts of a changing, drying climate. The current reserves where the western swamp tortoise resides are becoming drier and hotter. Habitat modelling based on current climate change predictions has indicated that, without remedial actions, these locations are in danger of becoming unsuitable for the tortoise by 2050. The modelling suggested areas to the south of Perth may in the future offer better conditions of swamp life during winter and spring than those further north (Mitchell *et al.* 2013). Through modelling how the different scenarios may affect released tortoises and future egg development, scientists and managers have identified new sites south of Perth that may become suitable under future climate change scenarios during the lifetime of now juvenile western swamp tortoise individuals (Mitchell *et al.* 2013, 2016). This work, undertaken at the University of Western Australia, has identified that the most suitable location and sized reserves are along the south coast of Western Australia. A 12-month trial was initiated in 2016 when 24 captive-bred western swamp tortoises from the Perth Zoo were released into two sites 300 km south of their current distribution. One trial site is a small swamp area easily accessible for monitoring, while the second site is in a large national park with extended wetland systems for which monitoring is more challenging.

This trial will determine if the tortoise can survive in a cooler climate, while identifying which food types are consumed and how individuals behave. This translocation represents the first known assisted colonisation of a vertebrate species specifically in response to climate change impacts.

New challenges associated with managing and maintaining the Recovery Program continue to emerge 29 years on. One such example impacting on funding includes government-dependent political influence of conservation academics promoting ecosystem conservation at the cost of the recovery of Critically Endangered species. There is no institutionalised or long-term funding security for the Western Swamp Tortoise Recovery Program apart from general reserve management, zoo operations and some basic population monitoring. The enthusiasm and ingenuity of individuals ensures that many parts of the Program keep going, and that grants are raised from external sources by members of the Recovery Team.

Conclusion

The western swamp tortoise is a classic example of a long-lived, late maturing and slow reproducing species with highly specialised habitat requirements that occurs in a small area subject to agriculture, mining and development. Recovering such species with low intrinsic population growth rates from the brink of extinction in the wild is slow and difficult in the face of habitat loss, climate change, predation and other threats. Captive breeding has proved to be instrumental for the western swamp tortoise Recovery Program, in combination with hands-on, adaptive management of the remaining habitat based on long-term research and monitoring data. After 29 years implementing recovery actions, there are now more than 50 mature western swamp tortoises in the wild, but the race isn't won yet. Due to the impacts of progressing climate change, it is not yet possible to downlist the species from Critically Endangered and many challenges still lie ahead. Continued support of the Australian public, government and non-government organisations will be necessary to help this little smiling tortoise keep its head above water.

Acknowledgements

We thank past and present Recovery Team members: Jan Bant, Don Bradshaw, Dean Burford, John DeJose, Brady Durell, Phil Fuller, David Groth, Graham Hall, Cathy Lambert, Jacqui Maguire, Tanya Marwood, Peter Mawson, Sandra McKenzie, Darryl Miller, Nicki Mitchell, Lyndon Mutter, Ray Nias, Helen Robertson, Sally Stephens, Gordon Wyre. For major funding we acknowledge: the Natural Heritage Trust, the former Australian National Parks and Wildlife Service and Australian Nature Conservation Agency, the Swan Catchment Council, and the World Wide Fund for Nature Australia. Help was provided by: the Martyn brothers, Joe Glucina, Alex Errington, the Kuiper family, Rod Martyn, Glyn Hughes, Graham Elsdon, Swan Coastal District staff, the Friends of the Western Swamp Tortoise, Trevor Evans, Paul Gioia, Norm Hall, John Ingram, Guundie Kuchling-Fesser, Bert Main, Floyd Riffey, Perth Zoo WST Program staff, East–west Veterinary Supplies, Aherns Pty Ltd, Kailis Brothers, Murdoch University Mineral Science Program, Beth and Peter Sproxton, the Smith family, the White family, the Purser family, the Lefroy family and the Rowe family of the Benalong Grazing Co. Financial support was also provided by: Bundesverband fur fachgerechten Natur-und Artenschutz, British Chelonia Group, School of Biomedical Sciences, Curtin University of Technology, The Natural Heritage Trust via the Swan Catchment Council, World Wide Fund for Nature-Australia,

Threatened Species Network, TiWest, ANZ Bank, Lotterywest, Midland Brick Co Ltd, Minerva Airconditioning, WA Nature Conservation and National Parks Trust Account, Western Australian Water Corporation, UWA Zoology Department, Toshiba Medical, and Gull Service Station Upper Swan.

References

Burbidge AA (1967) The biology of south-western Australian tortoises. PhD. thesis. The University of Western Australia, Perth, Australia.

Burbidge AA (1981) The ecology of the western swamp tortoise, *Pseudemydura umbrina* (Testudines, Chelidae). *Australian Wildlife Research* **8**, 203–222. doi:10.1071/WR9810203

Burbidge AA, Kuchling G, Fuller PJ, Graham G, Miller D (1990) 'The western swamp tortoise'. Western Australian Wildlife Management Program No. 6. Department of Conservation and Land Management, Perth, WA.

Burbidge AA, Kuchling G, Olejnik C, Mutter L (2010) 'Western Swamp Tortoise *(Pseudemydura umbrina)* Recovery Plan, 4th Edition'. Western Australian Wildlife Management Program No. 50. Department of Conservation and Land Management, Perth, WA.

IUCN/SSC (2013) *Guidelines for Reintroductions and Other Conservation Translocations.* Version 1.0. IUCN Species Survival Commission, Gland, Switzerland.

Kuchling G (1987) Proposal for a program to improve the captive breeding situation of the western swamp turtle *Pseudemydura umbrina*. *IUCN/SSC Tortoise and Freshwater Turtle Specialist Group Newsletter* **2**, 9–11.

Kuchling G (2006) An ecophysiological approach to captive breeding of the western swamp turtle *Pseudemydura umbrina*. In *Turtles. Proceedings of the International Turtle and Tortoise Symposium Vienna 2002*. (Eds H Artner, B Farkas and V Loehr) pp. 196–225. Edition Chimaira, Frankfurt, Germany.

Kuchling G, Bradshaw SD (1993) Ovarian cycle and egg production of the western swamp tortoise *Pseudemydura umbrina* in the wild and in captivity. *Journal of Zoology* **229**, 405–419. doi:10.1111/j.1469-7998.1993.tb02645.x

Kuchling G, DeJose JP (1989) A captive breeding operation to rescue the critically endangered western swamp turtle *Pseudemydura umbrina* from extinction. *International Zoo Yearbook* **28**, 103–109. doi:10.1111/j.1748-1090.1989.tb03262.x

Kuchling G, DeJose JP, Burbidge AA, Bradshaw SD (1992) Beyond captive breeding: the western swamp tortoise *Pseudemydura umbrina* recovery programme. *International Zoo Yearbook* **31**, 37–41. doi:10.1111/j.1748-1090.1991.tb02358.x

Mitchell NJ, Hipsey MR, Arnall S, McGrath G, Bin Tareque H, Kuchling G, et al. (2013) Linking eco-energetics and eco-hydrology to select sites for the assisted colonization of Australia's rarest reptile. *Biology (Basel)* **2**, 1–25. doi:10.3390/biology2010001

Mitchell NJ, Rodriguez N, Kuchling G, Arnall S, Kearney M (2016) Reptile embryos and climate change: modelling limits of viability to inform translocation decisions. *Biological Conservation* **204**, 134–147. doi:10.1016/j.biocon.2016.04.004

Pritchard PCH (1987) News from Australia. *IUCN/SSC Tortoise and Freshwater Turtle Specialist Group Newsletter* **1**, 9–10.

Rhodin AGJ, Mittermeier RA (1983) Conservation and status of chelid turtles. *Hamadryad – Newsletter of the Madras Snake Park Trust* **8**, 43–44.

Spence T, Fairfax R, Loach I (1979) The Western Australian swamp tortoise in captivity. *International Zoo Yearbook* **19**, 58–60. doi:10.1111/j.1748-1090.1979.tb00527.x

24

Twenty-five years of helmeted honeyeater conservation: a government–community partnership poised for recovery success

Dan Harley, Peter Menkhorst, Bruce Quin, Robert Anderson, Sue Tardif, Karina Cartwright, Neil Murray and Merryn Kelly

Summary

The problem
1. Helmeted honeyeater habitat has largely been cleared and the population is now restricted to a single forest patch.
2. Remaining habitat is in poor condition and difficult to restore due to altered hydrology and browsing by herbivores.
3. The honeyeater population has fallen to critically low numbers.

Actions taken to manage the problem
1. Assiduous monitoring is conducted in association with research on the biology of the honeyeater and ecology of its habitat.
2. Captive breeding has been employed to supplement the wild population.
3. Supplementary food is provided to increase recruitment and survival.
4. Significant effort and resources have been directed at restoring honeyeater habitat.

Markers of success
1. Extinction has almost certainly been averted and the population trend is now positive.
2. The causes of decline have been identified and efforts are under way to manage them.
3. For almost three decades, a cohesive Recovery Team has been guiding management actions.

Reasons for success
1. Dedicated champions, sustained volunteer input and a well-managed recovery process.
2. Monitoring, research and management have always been closely linked.
3. Funding has been sustained at an adequate level for many decades.
4. Management of the captive and wild populations are closely integrated.

Introduction

The helmeted honeyeater (*Lichenostomus melanops cassidix*), Victoria's Bird Emblem since 1971, is a distinctive subspecies of the more widespread yellow-tufted honeyeater (Pavlova *et al*. 2014; Fig. 24.1). The subspecies is listed as Critically Endangered under the *Environment Protection and Biodiversity Conservation Act 1999* owing to the small population size (until recently as low as 50 individuals) and restriction of the wild population to a single locality. Concern for the subspecies is long held. The honeyeater's restricted distribution and apparent decline prompted the formation of 'Survey *cassidix*' in 1952, and this group undertook surveys during the 1950s and 1960s. Lobbying from the former Bird Observers Club of Australia, the Victorian Ornithological Research Group (VORG) and Survey *cassidix* led to establishment, in 1965, of what is now Yellingbo Nature Conservation Reserve (hereafter referred to as 'Yellingbo') ~50 km east of Melbourne. In 1989, continuing concern led to the formation of one of the most active friends groups in Australia, the Friends of the Helmeted Honeyeater, with 200–300 members, four part-time employees, and an indigenous plant nursery that sells plants for revegetation undertaken by government agencies, community groups and private landholders.

Although always quite localised, the distribution of the helmeted honeyeater once extended ~20 km north of Yellingbo and more than 60 km to the south (Fig. 24.2). By the early 1900s, extensive areas of floodplain forest inhabited by the taxon had been selectively

Fig. 24.1. After a long decline, helmeted honeyeaters have benefitted greatly in recent years from supplementary feeding (photo: Merryn Kelly).

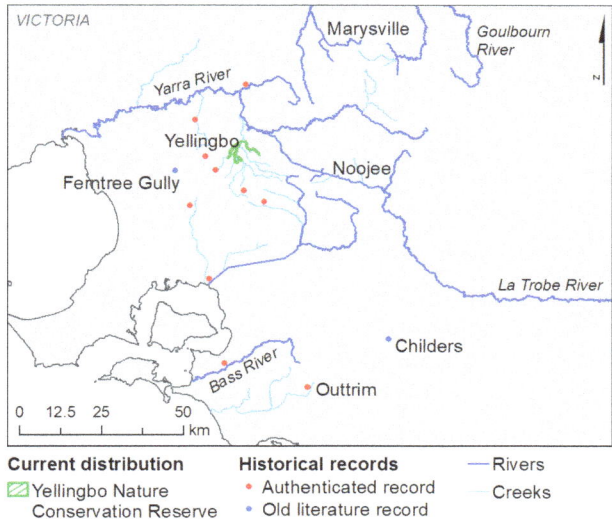

Fig. 24.2. Helmeted honeyeaters always had a restricted range in eastern Victoria and most of their habitat was rapidly alienated for agriculture.

cleared for agriculture. Since 1983, when isolated populations at Upper Beaconsfield and Gembrook were lost in the Ash Wednesday bushfires, the last wild population of helmeted honeyeater has been restricted to Yellingbo. Here the honeyeaters are currently confined to less than 80 ha of riparian and floodplain forest along two separate creeks.

A coordinated recovery program for the helmeted honeyeater began in 1989 following formation of a Recovery Team, the Friends group, preparation of the first Recovery Plan (Menkhorst and Middleton 1991) and establishment of a captive breeding program at Healesville Sanctuary. The latest (third) Recovery Plan for the helmeted honeyeater was prepared in 2008 (Menkhorst 2008a, 2008b), with substantial progress on securing the taxon and laying the foundations for recovery having occurred since that time.

Conservation management

The key threats to the helmeted honeyeater are well understood and documented (Menkhorst and Middleton 1991; Smales *et al*. 1995; Menkhorst 2008a, 2008b; Menkhorst *et al*. 2010; Harrisson *et al*. 2016). The primary reason is the limited area of suitable habitat, particularly large habitat patches, with most of the remaining habitat being in poor condition due to altered hydrology and a lack of regeneration. The patches are highly vulnerable to fire and invading bell miners (*Manorina melanophrys*) that exclude helmeted honeyeaters. Until recently, juvenile recruitment and rates of establishment for captive-bred birds released to the wild were low, and genetic diversity declining.

The recovery model for the taxon is based around: sustaining and increasing the wild population with supplementary feeding while habitat conditions at Yellingbo are restored; supporting the wild population through releases of captive-bred honeyeaters drawn from a small, genetically diverse captive population of honeyeaters maintained as insurance until multiple wild populations are established; restoring habitat and establishing new populations beyond Yellingbo to increase total population size and reduce risk from bushfire, disease and drought; and increasing community engagement and active participation in threatened species conservation.

Based on these objectives, the Recovery Team has developed explicit recovery targets that include:

- a total population size of ≥500 individuals to maintain genetic diversity
- five sites supporting wild populations, as a risk-spreading mechanism
- each wild site capable of supporting ≥100 individuals so that populations are demographically stable and self-sustaining.

These targets focus on creating the minimum viable populations of the taxon needed to secure a sustainable, wild meta-population where captive intervention is no longer required. However, *in-situ* management is expected to continue beyond this. The recovery objectives are necessarily modest, given the small area of remnant habitat suitable for reintroduction, and the high cost of land acquisition and habitat restoration. Establishment of multiple populations of the helmeted honeyeater is the highest priority, and is only achievable through effective habitat management.

Habitat protection and restoration – an essential foundation

Lack of quality habitat (owing to past draining and clearing of floodplains) is the major factor limiting the restoration of helmeted honeyeater populations. Less than 10% of Yellingbo Nature Conservation Reserve currently provides high-quality habitat for helmeted honeyeaters, with sections of the reserve subject to severe vegetation dieback. Some parts of the floodplain now experience protracted water inundation and siltation; other sites receive too little water.

Beyond the vegetation dieback zone, some habitat is becoming less dense as it matures to an older successional stage, reducing its suitability for helmeted honeyeaters. There has been poor recruitment of key helmeted honeyeater food and nesting plants, including mountain swamp gum (*Eucalyptus camphora*), scented paperbark (*Melaleuca squarrosa*) and woolly tea-tree (*Leptospermum lanigerum*). One objective for a current Australian Research Council research project is to identify the triggers to stimulate natural regeneration in the floodplain vegetation communities that provide critical habitat.

The extent of suitable habitat is being increased through revegetation. Between 2014 and 2016, ~750 000 trees and shrubs were planted at Yellingbo as part of state and Commonwealth (federal) tree-planting programs, complementing the long-standing revegetation program of the Friends of the Helmeted Honeyeater, which involves the local community and schools. Melbourne Water and Parks Victoria have also invested substantial resources in hydrological and habitat restoration at Yellingbo over recent years. If habitat restoration initiatives are successful, the reserve could support more than 300 honeyeaters without the need for supplementary feeding.

Low rate of juvenile recruitment

Smales *et al.* (2009) reported that helmeted honeyeaters produced just 0.18 fledglings per egg laid compared with an average of 0.45 for southern hemisphere passerines, with this low rate largely due to predation. Accordingly, there have been extensive attempts to protect helmeted honeyeater nests from predators. Two nest-protection techniques have been used: (1) perspex wrapped around tree trunks to impede climbing predators such as tiger snakes (*Notechis scutatus*), and (2) cages providing a barrier around nests to exclude larger predators such as the laughing kookaburra (*Dacelo novaeguineae*). One or both of these protection measures were applied to 111 nests. Nest protection was found to be a high effort strategy with mixed results and, ultimately, achieving sufficient annual recruitment is reliant upon having many pairs undertaking multiple nesting attempts.

Juvenile recruitment at Yellingbo may also be low due to poor habitat quality. Recruitment appears to have improved since supplementary feeding was expanded (see below) and the end of the drought in 2010. In 2015, 79 nesting attempts were recorded, resulting in 68 fledglings, the most recorded since the Recovery Program began, and equivalent to 0.62 fledglings per egg laid. Supplementary feeding may also help offspring survival: combining the 2013–14 and 2014–15 breeding seasons, 60 of the 65 fledglings (92%) whose survivorship could accurately be determined survived their first year compared with a first-year survival rate of 56% between 1984 and 1997 (Smales *et al.* 2009).

Supplementary feeding

At various times over the past 25 years, the Recovery Program has used supplementary feeding to support site-specific objectives for the helmeted honeyeater, such as supporting captive-bred birds following their release to the wild. This was expanded in 2013 because numbers were so low and habitat quality so poor and now most wild birds receive 50–70% of their daily dietary requirements from food supplementation. It takes a lot of work: in the first 6 months of 2016, 95 regular volunteers contributed 4037 volunteer hours to the supplementary feeding program and the Victorian Government-funded a volunteer coordinator 2 days a week. Such investment is delivering results. The site currently supporting the most helmeted honeyeaters (~85 birds) was twice abandoned by wild birds before the commencement of supplementary feeding. Conversely, a reduction in supplementary feeding at Bunyip State Park may have contributed to failure of this reintroduction attempt.

Captive breeding and release

Healesville Sanctuary's captive breeding program began in 1989, providing some insurance against extinction in the wild. Approximately 16 adult breeding pairs are maintained at Healesville Sanctuary and more than 370 chicks have been raised. Wild and captive populations are managed as one 'captive-wild metapopulation' through the transfer of eggs and/or nestlings between the wild and captivity during most breeding seasons.

More than 250 captive-bred honeyeaters have been released to the wild, with an average release of 16 birds per year since 2001. Captive-bred honeyeaters have successfully augmented the extant population at Yellingbo, with two of the four occupied sites established with birds released from captivity. Captive-bred birds were also used in an unsuccessful attempt to establish a new population at Bunyip State Park (Menkhorst *et al.* 2010). This failed after 10 years of releases, partly because drought reduced the vegetation condition and some creeks dried. A contributing reason for failure may be that too few birds were released at once, so Healesville Sanctuary is attempting to increase the number of young available for release annually by using a closely related subspecies of yellow-tufted honeyeater (*Lichenostomus melanops gippslandicus*) to foster helmeted honeyeater eggs. Other innovations may also help: a predator-awareness training regime has been introduced for the captive-bred honeyeaters before release, and the season of release has shifted from autumn to spring (when food is more abundant).

Competition with bell miners

Colonisation of helmeted honeyeater habitat by bell miners increased competition between the two species, ultimately resulting in the decline of helmeted honeyeater 'neighbourhoods' (Pearce *et al.* 1995). In the early 1990s, removal of bell miner colonies at Yellingbo resulted in the formation of at least 16 new helmeted honeyeater breeding territories (Pearce *et al.* 1995; Menkhorst 2008b). Parks Victoria continues to manage bell miner reinvasion close to helmeted honeyeater habitat.

Genetic management

The use of DNA-based techniques was initiated in 1989 to clarify parentage of wild birds and to characterise wild and captive gene pools, with an archive of DNA samples being established. Harrisson *et al.* (2016) concluded that by 2013 allelic richness and expected heterozygosity had declined to ~84% and 90% of 1990 levels, respectively, with 11 previously recorded alleles missing, and levels of relatedness increasing. Importantly, pre-1993 *cassidix* were more *gippslandicus*-like than subsequent wild samples, suggesting that careful introduction of some *gippslandicus* individuals into the breeding program may restore aspects of the gene pool and improve fitness (genetic rescue).

Population size and trend

In the early 1990s, and again from 2005–2013, the total wild population of helmeted honeyeaters numbered ≤70 birds (Fig. 24.3). Since 2013, there has been dramatic improvement in response to new recovery measures (e.g. expanded supplementary feeding), and the population has increased from a low of 50 to ~190 individuals. This is the highest population size recorded since the Recovery Program began in 1989.

Wider biodiversity benefits arising from the Recovery Program

The protection and restoration of habitat for the helmeted honeyeater, on private and public land, has resulted in broader benefits for biodiversity. The recovery of the honeyeater is directly linked with the conservation of the genetically distinct lowland population of the Critically Endangered Leadbeater's possum (*Gymnobelideus leadbeateri*), given their overlapping habitat requirements. Both taxa are highly dependent upon a threatened vegetation community, sedge-rich *Eucalyptus camphora* woodland, more than 95% of which has been cleared for agriculture due to the suitability of floodplains for farming. Thus, the investment in habitat restoration for the helmeted honeyeater provides significant benefits to another highly threatened species and a threatened vegetation community.

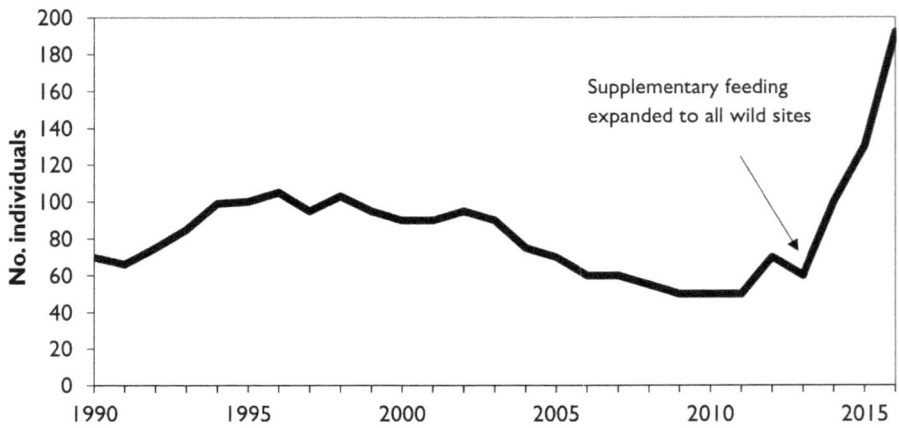

Fig. 24.3. The helmeted honeyeater population trend at Yellingbo Nature Conservation Reserve from 1990 to 2016.

Key features in success

Never stop learning – the importance of research and monitoring

Owing to extensive research and monitoring, the threats to the helmeted honeyeater are well understood. It is the most intensive, long-term population monitoring program in place for a Victorian bird. There has been a full-time field ornithologist working on the Recovery Program since 1990, and a comprehensive colour-banding program has been maintained since the early 1990s. From the outset, the program placed considerable value on the important insights and progress that can be driven by research (Smales *et al*. 1995). Significant lessons have emerged from an unsuccessful attempt to establish the honeyeater in Bunyip State Park, and this experience has played an important part in shaping the current recovery model and release strategy. The current research focus is on habitat restoration and genetic rescue, both of which involve Australian Research Council Linkage grants. Monitoring and measuring the effectiveness of actions continues to be a key part of the recovery effort, the results informing the refinement of methods and strategies within an adaptive framework.

Strong working partnerships – the key to success

Smales *et al*. (1995) provided a detailed outline of the structure of the Helmeted Honeyeater Recovery Team and operation of the Recovery Program in its early stages. Most of these elements remain. The Recovery Team has been in operation for 27 years, continues to meet at least three times per year (subgroups more frequently), and provides a successful model of long-standing partnerships between multiple government agencies, non-government organisations, community groups and tertiary institutions. Defining a clear vision of success has been important for effective coordination among project partners. With an increased focus on habitat restoration since 2014, hydrological management and vegetation management have been more strongly linked to helmeted honeyeater conservation and release priorities at Yellingbo.

Valuing volunteers

The strength of this program is in the people it attracts. Dedicated naturalists first reported the honeyeater's decline, eventually resulting in the establishment of the Friends of the Helmeted Honeyeater and calls for a recovery program. The Friends of the Helmeted Honeyeater continue to be extremely successful in their ability to attract, inspire and retain a critical mass of committed individuals and now supports four paid part-time positions to manage their local nursery, undertake education and revegetation activities with local schools, and coordinate the supplementary feeding program. The current contribution of volunteers to supplementary feeding alone amounts to more than 152 h per week (excluding their travel time). They remain the heartbeat of this recovery effort.

Engaging politicians

Maintaining political support for threatened species recovery is as important as tackling the biological drivers of decline. To this end, the Friends of the Helmeted Honeyeater have worked tirelessly to engage politicians, which has helped keep helmeted honeyeater conservation a high priority for Victoria. The Governor of Victoria is a Patron of the Friends group.

Sustained funding

The sustained political and public support for helmeted honeyeater conservation, in large part due to the efforts of the Friends of the Helmeted Honeyeater, has been one of the most important factors in maintaining financial support for this Recovery Program. This is reflected in the continuous employment of a field ornithologist for 27 years, which is central to the recovery effort. The Friends of the Helmeted Honeyeater have also built strong relationships with certain donors, such as the Judith Eardley Save Wildlife Association.

The future
The extinction of the helmeted honeyeater has been prevented, but there is some ongoing vulnerability while the wild population is confined to Yellingbo. However, there is better spatial spread of honeyeaters there due to the past 3 years of releases of captive-bred birds.

Funding – the basis of sustained action
The resources required to promote population recovery are far greater than those required to prevent extinction. The helmeted honeyeater program is entering the recovery phase, with greater resource requirements ahead, particularly relating to the extensive habitat restoration required at Yellingbo and new release sites elsewhere. Other high-cost/effort activities include captive breeding, genetic management of the metapopulation, supplementary feeding, and maintaining political and community support.

Habitat quality
The limited area of high-quality habitat remains the key factor restricting the expansion of helmeted honeyeater populations at Yellingbo and beyond. Efforts to remedy this have been scaled up since 2014, and will continue over the coming decade if funds can be secured. Research on vegetation dynamics and regeneration ecology of helmeted honeyeater habitat will be essential.

Improving hydrological conditions in the floodplain at Yellingbo is a necessary precursor to improving vegetation condition. After several years of data collection and modelling commissioned by Melbourne Water, works dealing with part of the problem should start in 2017, and will be linked to an Australian Research Council Linkage Project examining the influence of water on vegetation condition and dynamics.

Population expansion beyond Yellingbo
The confinement of helmeted honeyeaters to a single reserve places the subspecies at risk from bushfire. Based on recent habitat suitability modelling, six priority areas will form the basis of the release plan for the coming decade. Community engagement will need to be expanded to encompass new localities. The translocation program requires more birds for release, and now that population size at Yellingbo has exceeded 150 individuals, translocation of wild-born young to new sites is possible. Given the decline in genetic variation since 1990, genetic rescue of the helmeted honeyeater is also being explored as part of an Australian Research Council Linkage project.

Conclusions
The major outcome from this Recovery Program is that the extinction of the helmeted honeyeater – one of Victoria's Faunal Emblems – has been averted. Since 2013, the population has tripled, demonstrating that we have the required management tools and a strong platform in place to recover helmeted honeyeater populations at Yellingbo and beyond.

The supplementary feeding program has proven critical while habitat quality improves, but long-term recovery of self-sustaining helmeted honeyeater populations will rely upon improving habitat, maintaining effective vegetation management and ongoing community support.

Capitalising on the momentum created by the 'successes' has been an important feature of the Helmeted Honeyeater Recovery Program, and the Friends of the Helmeted Honeyeater have been at the forefront of communicating the 'wins' widely. However, providing that programs are structured in a manner that results in continual learning from each endeavour, **failed** initiatives also contribute to subsequent success. The path to success includes a willingness to take calculated risks and absorb some failures along the journey.

An inspiring vision of success has focused and united the Recovery Team and the community, allowing us to continue to attract funding. The conservation of threatened species is a multidisciplinary task, and success has relied on the commitment, cooperation and coordination of multiple partners. The Friends of the Helmeted Honeyeater have been pivotal to maintaining community and political support.

Threatening processes typically operate for decades before amelioration begins, so recovering species can be slow. In the case of the helmeted honeyeater, habitat restoration takes time and considerable funding. Although the conservation status of the helmeted honeyeater remains Critically Endangered, substantial progress over the past 3 years means that the program could now be described as 'successful', albeit with much work ahead. A strong vision of success has been developed, the foundations are in place, and there is currently considerable momentum in realising this vision as the Recovery Program expands into its third decade.

Acknowledgements

During the past 27 years, numerous people have made critical contributions to the survival of the helmeted honeyeater. They include members of the Recovery Team, Friends of the Helmeted Honeyeater and staff from Zoos Victoria, Parks Victoria and Melbourne Water. Ian Smales and Don Franklin have made substantial contributions to our biological understanding of the bird. Vital financial support for recovery initiatives has been provided by the Australian and Victorian Governments, and several donors including the Judith Eardley Save Wildlife Association, Merrin Foundation and Myer Foundation. Melbourne Water have committed substantial resources to hydrological restoration at Yellingbo, and Greening Australia have contributed greatly to habitat restoration there. Trust for Nature has been instrumental in the acquisition and protection of habitat.

The list of volunteers working tirelessly at the Friends of the Helmeted Honeyeater nursery, planting trees, controlling weeds, providing supplementary food to support the wild birds, and working behind the scenes on committees, reference groups and at community events, is far too long to list here, but your contribution and dedication has been, and continues to be, critical to the success of this program.

This chapter was improved with comments from Stephen Garnett, John Woinarski, Peter Latch and Josephine MacHunter.

References

Harrisson KA, Pavlova A, da Silva AG, Rose R, Bull JK, Lancaster ML, *et al.* (2016) Scope for genetic rescue of an endangered subspecies through re-establishing natural gene flow with another subspecies. *Molecular Ecology* **25**, 1242–1258. doi:10.1111/mec.13547

Menkhorst P (2008a) 'National Recovery Plan for the Helmeted Honeyeater *Lichenostomus melanops cassidix*'. Department of Sustainability and Environment, Melbourne.

Menkhorst P (2008b) 'Background and implementation information for the Helmeted Honeyeater *Lichenostomus melanops cassidix* National Recovery Plan'. Department of Sustainability and Environment, Melbourne.

Menkhorst P, Middleton D (1991) 'Helmeted Honeyeater Recovery Plan'. Department of Conservation and Environment, Melbourne.

Menkhorst P, Quin B, Cartwright K, Smales I (2010) Using captive-bred helmeted honeyeaters to establish a new population and supplement an existing population. In *Global Re-introduction Perspectives: Additional Case-studies from around the Globe*. (Ed. PS Soorae) pp. 165–170. SSC Re-introduction Specialist Group, Abu Dhabi, United Arab Emirates.

Pavlova A, Selwood P, Harrison KA, Murray N, Quin B, Menkhorst P, et al. (2014) Integrating phylogeography and morphometrics to assess conservation merits and inform conservation strategies for an endangered subspecies of a common bird species. *Biological Conservation* **174**, 136–146. doi:10.1016/j.biocon.2014.04.005

Pearce J, Menkhorst P, Burgman MA (1995) Niche overlap and competition for habitat between the helmeted honeyeater and the bell miner. *Wildlife Research* **22**, 633–646. doi:10.1071/WR9950633

Smales I, Menkhorst P, Horrocks G (1995) The Helmeted Honeyeater Recovery Programme: a view on its organization and operation. In *People and Nature Conservation: Perspectives on Private Land Use and Endangered Species Recovery*. (Eds A Bennett, G Backhouse and T Clark) pp. 35–44. Royal Zoological Society of NSW and Surrey Beatty & Sons, Chipping Norton, NSW.

Smales IJ, Quin B, Menkhorst PW, Franklin DC (2009) Demography of the helmeted honeyeater (*Lichenostomus melanops cassidix*). *Emu* **109**, 352–359. doi:10.1071/MU09002

25

Bringing back warru: return of the black-footed rock-wallaby to the APY Lands

John Read, Peter Copley, Matt Ward, Ethan Dagg, Liberty Olds, David Taggart and Rebecca West

Summary

The problem
1. By the 1980s, ongoing declines led to the fragile persistence of only two small known populations of warru (*Petrogale lateralis* MacDonnell Ranges race) in the APY Lands of northern South Australia.
2. Introduced predators and invasive buffel grass (*Cenchrus ciliaris*) are ongoing threats to remaining warru populations.

Actions taken to manage the problem
1. A sustained and collaborative recovery effort involving traditional landowners, researchers and state agencies was developed with explicit employment opportunities for Indigenous landowners.
2. Adaptive management was undertaken based on management trials, monitoring and refinement of management actions.
3. An insurance captive breeding program was established through cross-fostering.
4. A range of targeted feral cat control methods have been trialled or proposed.

Markers of success
1. Likely extirpation was averted.
2. Causes of decline have been identified and others have become the focus of adaptive management.
3. A reintroduction is now planned.
4. Enhanced community involvement and support has occurred since project inception.

Reasons for success
1. Indigenous landowners are highly engaged in, and have been providing direction to, the Recovery Program.

2. There are close links between monitoring, research and management.
3. Funding has been sustained for over a decade.
4. There has been a preparedness to adopt and adapt novel and potentially risky approaches.

Introduction

Warru (black-footed rock-wallaby *Petrogale lateralis*, MacDonnell Ranges race) are small macropods inhabiting arid rocky ranges of northern South Australia, central Western Australia and the southern Northern Territory (Pearson 2010). They were formerly widespread and locally abundant, with records from early explorers noting that they 'inhabited all the ranges and granite hills along the northern margin of the Great Victoria Desert' and 'the Musgrave, Mann and Tomkinson Ranges swarmed with rock-wallabies' (Finlayson 1961).

Unfortunately, by the 1960s, they had declined markedly: '… today, although it still persists at scattered points … [the rock-wallaby] is a comparatively rare form' (Finlayson 1961). This decline was particularly severe in northern South Australia, where extensive surveys yielded rock-wallabies at only one site (Copley *et al.* 1989) (Fig. 25.1).

Warru therefore became a 'target species' during Anangu Pitjantjatjara Yangkunytjatjara (APY) Lands surveys from 1991 to 2001. These surveys, informed by local Aboriginal landowners, recorded warru from only two sites in the Musgrave Ranges – near New Well and at Sentinel Hill (Wamitjara), north and north-east of Pukatja (Nesbitt and Wikilyiri 1994). Old faecal pellets were found in caves at many sites, supporting the earlier accounts of the species' former widespread distribution and of their more recent serious decline.

The warru was therefore listed as Endangered in South Australia (schedule 7, *National Parks and Wildlife Act 1972*) and assigned the ominous recognition as South Australia's most endangered mammal (Copley and Alexander 1997). Nationally, it is listed as Vulnerable, with this lower conservation status primarily because there is still a sizable and relatively stable population in the MacDonnell Ranges, Northern Territory, although elsewhere there are ongoing declines and local extinctions. This chapter focuses particularly on the conservation recovery effort in northern South Australia.

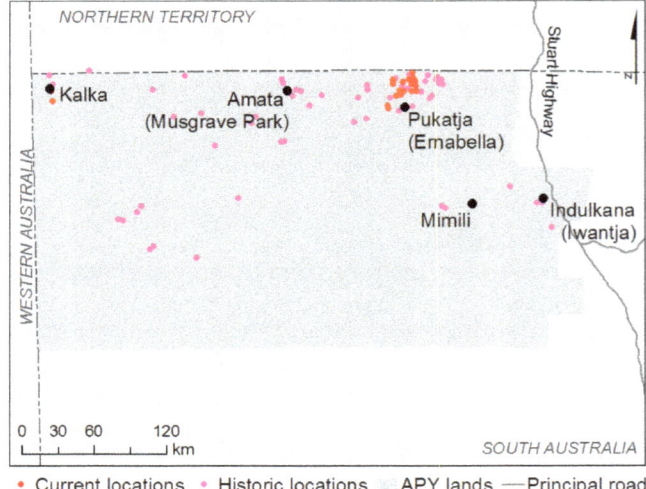

Fig. 25.1. Warru were once distributed widely across northern South Australia but are now confined to a few sites near Kalka and Pukatja.

Conservation management
Early conservation focus
Because red fox (*Vulpes* vulpes) control had been shown to lead to rock-wallaby recovery at other sites (Kinnear et al. 1988, 2010; Chapter 15), the South Australian environment department and APY Land Management initiated discussions about fox control with the traditional owners of the hills where the two known remaining South Australian warru colonies occurred. Fox-baiting using dried kangaroo meat baits impregnated with 1080-poison commenced around perimeter tracks at New Well and Wamitjara in 1996 (Geelen 1999). This baiting was intensified, including aerial baiting from a helicopter, after it was confirmed the only known warru colony in South Australia outside of the APY Lands – a recently discovered population in the Davenport Ranges – had recently become extinct (Moseby et al. 1998).

Unfortunately, despite the regular baiting program, monitoring of the colonies at New Well and Wamitjara through the early 2000s indicated a continued decline and, by 2006, the Wamitjara colony also became extinct (Read 2010). However, further opportunistic and systematic surveys located another metapopulation in the far north-west of the APY Lands in the Tomkinson and Hinkley Ranges near Kalka and Pipalyatjara (Ward et al. 2011).

Establishment of the South Australian Warru Recovery Team
In response to the recent population extinctions, the South Australian environment department and the Alinytjara Wilurara (AW) Natural Resource Management Board in 2005 secured funding to increase warru recovery efforts. Along with the employment of two ecologists and initiation of a trap–mark–release–recapture program, the provision of additional operational funding provided the impetus for the establishment of a formal Warru Recovery Team (WRT) in early 2007. This Recovery Team comprised, and still comprises, representatives of APY Land Management (APY LM), traditional owners, Ecological Horizons Pty Ltd, Department of Environment Water and Natural Resources (DEWNR), the Australian Government's Working on Country and Indigenous Ranger programs, Zoos SA, the Australian Museum and research students and graduates from the University of Adelaide.

The team reviewed learnings from previous management and research and then prepared a detailed South Australian Warru Recovery Plan (Read and Ward 2010) outlining an ambitious program to improve the status of warru by reversing decline of *in-situ* populations and undertaking reintroductions into their former range. The cornerstones of the plan are ownership of the recovery process and creation of employment opportunities for Anangu, good science and regular monitoring and reporting of management effectiveness, and the maintenance of adequate funding levels to accomplish its goals.

Identification and management of key threats
Red foxes and feral cats (*Felis catus*) are small and agile enough to penetrate into wallaby den-sites, which are inaccessible to the larger dingoes (*Canis familiaris*), and have been targeted by the Recovery Team. However, the continued suppression of warru recruitment, despite very low fox numbers, led in June 2012 to the initiation of a regular spotlight-shooting program to target feral cats. Historically occurring under the guidance of experienced contractors, the Warru Rangers have since gained the necessary skills in conducting regular spotlight searches, shooting and data collection and have now assumed responsibility for this management activity. The importance of this shooting program was highlighted when the rangers found a freshly killed warru at the site where they shot a large feral cat (Fig. 25.2).

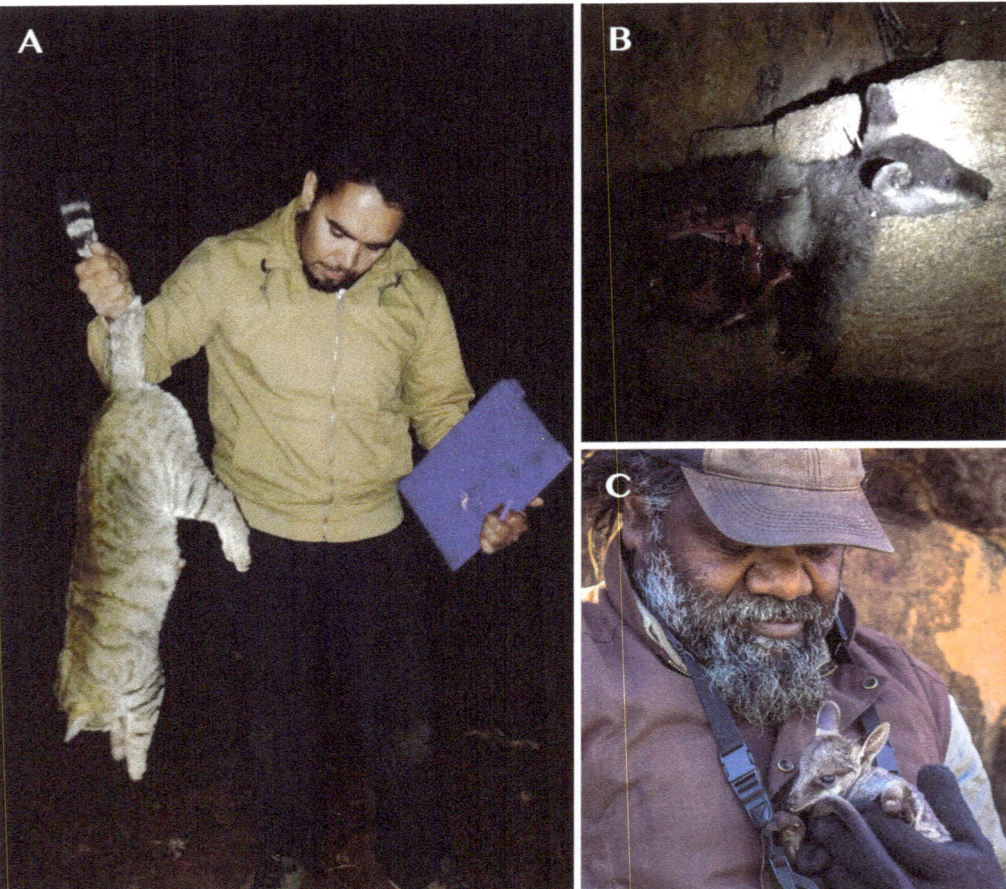

Fig. 25.2. (A) Warru Ranger Ethan Dagg holding a cat killed in warru habitat. (B) Remains of a warru killed and partly eaten by a cat. (C) Warru ranger Jacob McKenzie holding a warru (photos: A, APY Land Management; B, Ethan Dagg; C, Ellen Ryan-Colton).

Over 80 cats have been shot, and the 2016 trapping results suggest warru numbers have either remained stable or increased where cat shooting has occurred, and declined where shooting has not been possible. On three occasions since 2012, ~4500 Eradicat® baits were dropped over ~200 km^2. Camera-trap monitoring of complementary ground-based bait stations showed that few baits were taken by cats. Consequently, there was little evidence of reduction in cat numbers due to this baiting (WRT 2016).

Kumanara Haggie, one of the traditional owners instrumental in the establishment of the recovery team, explained the reason for this work conducted by Warru Rangers:

> 'We don't want to eat them anymore because we looking after now, today. We're working for rock wallaby, looking after. Some fox might come and eat him, that's why we're looking after them.'

Habitat-changing processes: buffel grass and fire

Buffel grass (*Cenchrus ciliaris*) was introduced to the APY Lands as a dust suppressant around communities and later as forage for the cattle industry. Over the past few decades, it has formed dense swards of near monoculture grassland throughout the northern APY Lands. Buffel grass poses a threat to warru populations by outcompeting important forage species, such as native grasses, and by increasing the spread and intensity of wildfires that can damage or destroy important fire-sensitive food plants such as native fig (*Ficus platypoda*) and spear bush (*Pandorea doratoxylon*).

Consequently, control of buffel grass within warru habitat and strategic reduction of buffel grass in areas surrounding warru habitats have become high priority actions for the Recovery Team.

> 'Rangers have been working tirelessly to get rid of buffel grass in areas where there is not much of it. Protecting ili [native fig] and other food sources for warru and other animals. If we do not do anything about it, it will destroy this beautiful country.' – Ethan Dagg, Warru Field Officer

To date, buffel management on the APY Lands has been opportunistic, using a combination of burning, herbicide application and manual removal. Management has targeted well-defined patches such as the warru pintji (fenced predator exclosure) and access tracks to the colonies, enabling infestations to be controlled before they spread further into vital warru habitat. The flow-on effects are also significant. For instance, the pintji efforts will protect 100 ha of diverse vegetation and small animal communities amidst the increasing sea of buffel grass.

Supplementary feeding and water

Research was undertaken to determine whether supplementary water could be used to alleviate resource pressure during dry periods, and hence whether provision of water could be used as a tool to increase survival of warru (West *et al.* 2017b). Wild and reintroduced warru both used supplementary water points within 10 days of installation with no significant increase in visits by predators or competitors (West *et al.* 2017b). Drinking rates at these water points were significantly higher during dry winter months, suggesting the provision of supplementary water during drier months or after dramatic events such as wildfire, may increase warru survival rates.

To reduce the need for vulnerable young-at-foot to leave the refuge of their den sites when food becomes scarce, a supplementary feeding program was also initiated at the Hinckley Range warru colonies following extensive fires in 2013. Marsupial food pellets and water have been provided in feeders in prime refuge sites and topped up monthly by the rangers. In the Tomkinson Ranges, Warru Rangers are now using scat plots to assess the value of adding supplementary feeders and water points.

Insurance population and captive management

The Recovery Team decided the establishment of a captive insurance population of warru was required to manage the risk of ongoing colony declines and extinctions. Fortunately, the technique of cross-fostering of macropod pouch-young in pouches of related species had recently been demonstrated to accelerate reproductive outputs of captive rock-wallaby species (Taggart *et al.* 2010). This technique had been developed at Zoos SA, so with their involvement in the Recovery Team, a proposal to cross-foster pouch-young warru in the pouches of captive yellow-footed rock-wallabies (*Petrogale xanthopus*) at Monarto Zoo was prepared. Because rock-wallaby females with pouch-young usually have a second embryo in suspended animation (or embryonic diapause) that will re-commence development once lactation for the existing pouch-young ceases, taking pouch-young from the wild should have limited impact on the productivity of the wild population, while enabling the establishment of an insurance population.

Taking the young to a place so far away from their 'homeland' was initially viewed with considerable alarm by Anangu because it equated to the establishment of a 'stolen generation' of rock-wallabies. Although concern for the welfare of the pouch-young remained, the proposal was agreed to on the understanding the warru would be returned to the APY Lands to establish a new colony. Explicit approval for a Warru Translocation Proposal (Ward and Clarke 2007) was gained from the APY Executive, as well as permission from traditional owners on whose land the remaining populations were present.

Anangu not only engaged with this novel conservation technique: they also directed how both conservation and cultural outcomes could be satisfied. Perhaps even more significantly, senior Anangu women then under-pinned the cultural significance of this proposal with the production of a painting of the recovery story (now the logo for the Recovery Program) and followed this with the development of a contemporary *tjukurpa inma* (cultural story) involving song and dance. As the project's story has progressed, both the song and the dance have been elaborated.

> '*Warru has been here for a long time, before us, and they should be in our lands because they are traditional owners too … We want to see them back in all places where they used to be. And not just warru, we should bring back Ninu, Mala, Tjuwalpi and Wayuta too. All of them.*' – Frank Young, former APY LM coordinator

A total of 22 pouch-young warru were flown to Monarto Zoo between 2007 and 2009 to establish the insurance population. Zoos SA continues to manage a small breeding population at Monarto, which has also proven invaluable for research and raising awareness of the Warru Recovery Program. Twenty-two captive-raised/captive-bred warru were returned to the warru pintji between 2011 and 2015. By 2016, the pintji had an estimated 40 animals, which is a significant milestone for the Recovery Team because these animals form a strong population ready for wild release onto the formerly occupied hills of the APY Lands.

Population genetics

Initial genetic studies using microsatellite markers showed significant differentiation between three populations; New Well and Alalka in the Eastern Musgraves Ranges and Kalka in the Tomkinson Ranges (Ruykys and Lancaster 2015). Consequently, the New Well/Alalka and Kalka populations were treated as two separate population management units both in the wild and at Monarto Zoo. However, re-examination of the genetic relationships using mitochondrial, as well as microsatellite, markers across the warru range indicated there was no pattern in the genetic and geographic distance, providing evidence of much greater historical connectivity between warru populations (West 2014). The Recovery Team determined the risk of outbreeding was small and therefore decided to interbreed animals from all three populations in captivity and within the pintji. This allowed the team to overcome genetic hurdles of maintaining a small captive Kalka group with few founders and the logistical hurdles of housing the groups separately. Based on this experience, we recommend, where possible, to use both contemporary and historical markers when determining best-practice genetic management.

Monitoring

Due to their cryptic nature, the status of low-abundance rock-wallaby colonies is more readily monitored by scat counts than direct observations or trapping (Telfer *et al.* 2006). Scat counts on marked quadrats, along with spotlight surveys from the Wamitjara and

New Well colonies (Geelen 1999; Read 2010), form the longest-running South Australian warru monitoring dataset that now also includes several other colonies, including New Well North and five Tomkinson Range colonies. These counts have proven invaluable as a coarse measure of warru decline, including the extirpation of warru at Wamitjara in 2006 (Read 2010) and also successful recruitment and recovery.

Ground and helicopter-based scat surveys were also implemented throughout the major ranges and at historic warru sites throughout the APY Lands from 2001 to 2010 (Ward *et al.* 2011). Although these surveys increased the known extent of both warru metapopulations, no additional colonies were located.

Since 2005, the Recovery Team has undertaken trapping at sites across Alalka, Kalka and New Well to monitor the survival, fecundity and recruitment of warru. A mark–recapture analysis of these data suggested that, although adult survival rates were reasonable (annual average 73%), juvenile survival rates were only 41% and strongly correlated with winter rainfall (Ward *et al.* 2011). These low survival rates of juveniles led to increased prioritisation for more intensive management of feral cats. In 2010–2011, two new sites were added to the trapping schedule after small populations of warru were located nearby. Since 2011, these populations have grown, which is a promising sign for warru recovery and reinforces the importance of the management actions being undertaken.

Radio-collars (VHF and GPS) have been fitted to adult warru and radio-tracking has been undertaken to monitor survival and to better understand warru ecology, especially home range dynamics and movement patterns. Warru rangers conducted 'kulini' (radio-tracking) on adult warru at New Well and Kalka from 2009 to 2014, and found adult survival was generally high, with some adults surviving across the 5-year study period.

'It's really good working for land management because you get to work in the country and see amazing scenery. It makes you feel great that you are doing a positive job, looking after our ancestor's country' – Ethan Dagg, Warru Field Officer

GPS collar data have shown warru at New Well have home ranges of ~23 ha, with male home ranges larger than for females, and multiple overlapping female home ranges, supporting the genetic evidence for a polygamous mating system (West 2014).

Ongoing monitoring of warru and predators by camera trapping, track counts and radio-tracking confirmed low recruitment success but good adult survivorship, indicative of heavy predation on juveniles and relatively high cat numbers (West *et al.* 2017a).

Recovery partners

APY landowners and agencies, the AW NRM Board and state agencies have been pivotal and long-term partners in the conservation effort for warru. We note here also other crucial contributions.

Support from Zoos SA for warru recovery began in 2006 when establishing a captive population became a priority action. Funding for purpose-built warru enclosures at Monarto Zoo saw significant investment from DEWNR, AW NRM region and philanthropists, along with Zoos SA core funding. Since then, Monarto Zoo has housed and bred warru as part of Zoos SA's investment in this species' recovery. Support is also provided for the program through veterinary, keeping and conservation staff, as well as provision of equipment for trapping surveys. Over 40 staff at Monarto Zoo have been directly involved in warru recovery through trapping, hand-rearing and everyday husbandry.

Research students have brought funds to support their projects and, more importantly, they have brought passion and provided exposure for Anangu to experience a range of wildlife research techniques. Two PhD students, Laura Ruykys (2007–2011) and Rebecca

West (2010–2013), secured grants to provide casual ranger salaries to assist with data collection and engage with community members during their multiple trips to the APY Lands. Both students also filled the secretarial role for the Recovery Team during their PhDs and assisted with community engagement events such as NAIDOC week and meetings in Adelaide. Independent researcher John Read has also provided long-term monitoring and research data for, and is the Chair of, the WRT.

National recognition
The South Australian Warru Recovery Program has received national recognition including:
- establishment of the Kalka – Pipalyatjara Indigenous Protected Area, encompassing the warru populations in the far west of the APY Lands
- listing of warru in the list of top 20 priority threatened mammals for focused attention under the National Threatened Species Strategy 2015–2020
- the 2011 National NAIDOC LandCare Award.

The South Australian Warru Recovery Plan covers all key areas in, and provided long-term datasets and other input to, the National Recovery Plan while also providing more detailed site-specific targets and actions. The plan for the proposed 2017 re-establishment of the Wamitjara colony benefited greatly from experience gained from Western Australian translocations and may include warru from the MacDonnell Ranges (Northern Territory), as well as from the pintji.

Money
From 2000 to 2007, monitoring and management through baiting, received A$10 000–$25 000 per year through South Australian Government environment department and WWF Threatened Species programs, managed through APY LM. Investment in the warru program increased from 2007 to 2012 through co-investment from the Natural Heritage Trust through the AW NRM Board (A$130 000 per year) and the South Australian Government (A$130 000 per year). This investment in governance, partnerships, Aboriginal employment and on-ground outcomes saw confidence in the team rise, leading to investment by the Australian Government's Working on Country program of approximately A$600 000 per year from 2008 to 2012, which included construction of the 100 ha warru pintji completed in July 2010. A second Working on Country grant of over A$3.5 million (2013–17), received and managed by APY LM on behalf of the Team, guaranteed continued empowerment for Anangu traditional owners and employment of four part-time warru rangers at both the Musgrave and Tomkinson Ranges regions. Furthermore, between 2009 and 2012, the state invested A$200 000 per year to support reintroduction efforts, while the AW NRM Board has subsequently invested smaller amounts (A$20 000–50 000 per year) from the South Australian and Australian Governments to support governance of the Recovery Team.

The future
Our long-term adaptive management suggests that, even with effective fox control, warru populations do not necessarily recover. This outcome suggests that predation by feral cats, dramatic habitat modification through invasion by buffel grass, and potentially also climate change-mediated increase in droughts, may also be significant threats.

Feral cat shooting is unlikely to provide a sustainable option in inaccessible or particularly remote locations and trials of special cat baits at the warru colonies have not been promising, so now the Warru Recovery Team is considering trials of both cat specific Felixer grooming traps (Read *et al.* 2014) and rendering individual warru toxic through use of population protection implants (Read *et al.* 2015) so individual catastrophic predators do not extirpate entire colonies.

Buffel grass represents probably an even more insidious and intractable threat to warru. Despite the widely accepted and broad-scale devastating effect buffel is having on threatened species, ecosystems and Indigenous culture, there is a lack of political will to devote appropriate resources to the management of buffel infestations. Advocating for effective control mechanisms for this transformer weed species remains a focus for the Recovery Team and traditional landowners in APY Lands.

Other aspirations of the Recovery Team involve *in-situ* management of extant colonies, and reintroductions resulting in the re-establishment of colonies in formerly occupied areas. Increasing recruitment rates by provisioning of food and/or water and planting key food plants may provide an important opportunity for Warru Rangers to increase the colonies' resilience against climate change.

Direct transfers of adults or pouch young between warru colonies, both within and outside South Australia, may be required to compensate for the current predator-induced low levels of natural dispersal and limited genetic transfer among colonies. Greater collaboration between landowner groups and agencies in other states should also assist in sharing research outcomes and lessons learned.

Conclusion

The Warru Recovery Team recognises that, unless we are progressing reintroductions and adaptive management, stochastic or cumulative threats will ultimately force warru to their extinction within South Australia and possibly beyond. Our success will ultimately be measured by sustainable warru reintroductions, initially within the APY Lands and then eventually to other formerly occupied areas. These successes will only be possible through continued close collaboration between traditional owners and scientists, proactive adaptive management and maintaining access to governance and operational funding.

Acknowledgements

We are indebted to the many other dedicated members of the Warru Recovery Team who have been integral to our progress. Significant contributions have been made by the Warru Minyma: Inpiti Winton, Ninguta Edwards, Tjariya Stanley, Dora Haggie, Molly Miller, Yarichi Connelly, Milyka Paddy and many Warru Rangers including long-term permanent rangers Bronson Bennet, Matthew Miller, Kym Nelson, Sherada Stanley, Thomas Tjila, Margaret Winton, Eric Abbott, Grant Nyaningu and Elisha Roesch. Special thanks to Jacob McKenzie who now co-chairs the Warru Recovery Team and leads the western rangers as Ethan Dagg leads the eastern rangers. Warru Recovery Team members including APY LM, AWNRM, SA DEH and ZoosSA staff who have played an important role in the project include Jasmina Muhic, Simon Booth, Rachel Barr, Belinda Cooke, Mark Eldridge, Wendy Foster, Laura Ruykys, Amber Clarke, Anika Dent, Luke Ireland, Anna Miller, Helen Palmer, Sara Weir, Magda Zabek, Lexi Knight, Kate Holmes, Beth Pohl, Mick Post, Vicki-Jo Russell, Jason van Weenan, Eric De Smit, Ian Smith and Thalie Partridge.

References

Copley PB, Alexander PJ (1997) Overview of the status of rock-wallabies in South Australia. *Australian Mammalogy* **19**, 153–161.

Copley PB, Kemper CM, Medlin GC (1989) The mammals of north-western South Australia. *Records of the South Australian Museum* **23**, 75–88.

Finlayson HH (1961) On central Australian mammals. IV. The distribution and status of Central Australian species. *Records of the South Australian Museum* **14**, 141–191.

Geelen LJ (1999) A preliminary study of the black-footed rock-wallaby (*Petrogale lateralis* MacDonnell Ranges race) in the Anangu Pitjantjatjara lands, South Australia. BSc. (Hons) Thesis, University of Adelaide, Adelaide.

Kinnear JE, Onus ML, Bromilow RN (1988) Fox control and rock-wallaby population dynamics. *Australian Wildlife Research* **15**, 435–450. doi:10.1071/WR9880435

Kinnear JE, Krebs CJ, Pentland C, Orell P, Holme C, Karvinen R (2010) Predator-baiting experiments for the conservation of rock-wallabies in Western Australia: a 25-year review with recent advances. *Wildlife Research* **37**, 57–67. doi:10.1071/WR09046

Moseby K, Read J, Gee P, Gee I (1998) 'A study of the Davenport Range black-footed rock wallaby colony and possible threatening processes'. Department for Environment and Heritage, Adelaide.

Nesbitt BJ, Wikilyiri D (1994) 'Waru (black-footed rock wallaby) survey of the Anangu Pitjantjatjara Lands'. Anangu Pitjantjatjara Council and South Australian National Parks and Wildlife Service, Adelaide.

Pearson DJ (2010) 'Recovery Plan for five species of rock-wallabies: black-flanked Rock-wallaby (*Petrogale lateralis*), Rothschild Rock-wallaby (*Petrogale rothschildi*), short-eared Rock-wallaby (*Petrogale brachyotis*), monjon (*Petrogale burbidgei*) and nabarlek (*Petrogale concinna*)'. Department of Environment and Conservation, Perth, WA.

Read J (2010) 'Status of the known Waru (black-footed Rock-wallaby) populations in the Anangu Pitjantjatjara Yankunytjatjara Lands. A report for APY Land Management and the Department for Environment and Heritage'. Ecological Horizons, Adelaide.

Read J, Ward MJ (2010) 'Warru Recovery Plan: recovery of *Petrogale lateralis* MacDonnell Ranges race in South Australia, 2010–2020'. Department of Environment and Natural Resources, Adelaide.

Read JL, Gigliotti F, Darby S, Lapidge S (2014) Dying to be clean: pen trials of novel cat and fox control devices. *International Journal of Pest Management* **60**, 166–172. doi:10.1080/09670874.2014.951100

Read JL, Peacock D, Wayne AF, Moseby KE (2015) Toxic trojans: can feral cat predation be mitigated by making their prey poisonous? *Wildlife Research* **42**, 689–696. doi:10.1071/WR15125

Ruykys L, Lancaster ML (2015) Population structure and genetic diversity of the black-footed rock-wallaby (*Petrogale lateralis* MacDonnell Ranges race). *Australian Journal of Zoology* **63**, 91–100. doi:10.1071/ZO14009

Taggart DA, Schultz DJ, Fletcher TP, Friend JA, Smith IG, Breed WG, *et al.* (2010) Cross-fostering and short term isolation in macropodid marsupials: implications for conservation. In *Macropods*. (Eds G Coulsen and M Eldridge). pp. 263–278. CSIRO Publishing, Melbourne.

Telfer WR, Griffiths AD, Bowman DMJS (2006) Scats can reveal the presence and habitat use of cryptic rock-dwelling macropods. *Australian Journal of Zoology* **54**, 325–334. doi:10.1071/ZO05074

Ward MJ, Clarke A (2007) 'Proposed translocation of Warru pouch young (black-flanked rock-wallaby MacDonnell Ranges race *Petrogale lateralis*) from the Anangu Pitjantjatjara Yankunytjatjara Lands'. Department of Environment and Heritage, Adelaide.

Ward MJ, Urban R, Read JL, Dent A, Partridge T, Clarke A, vanWeenen J (2011) Status of warru (*Petrogale lateralis* MacDonnell Ranges race) in the Anangu Pitjantjatjara Yankunytjatjara Lands of South Australia. 1. Distribution and decline. *Australian Mammalogy* **33**, 135–141. doi:10.1071/AM10047

West RS (2014) Reintroduction as a tool for the recovery of warru (*Petrogale lateralis* MacDonnell Ranges race) on the Anangu Pitjantjatjara Yankunytjatjara Lands of South Australia. PhD. thesis, University of Adelaide, Australia.

West R, Read JL, Ward MJ, Foster WK, Taggart DA (2017a) Monitoring for adaptive management in a trial reintroduction of the black-footed rock-wallaby. *Oryx* **51**, 554–563. doi:10.1017/S0030605315001490

West R, Ward MJ, Foster WK, Taggart DA (2017b) Testing the potential for supplementary water to support the recovery and reintroduction of the black-footed rock-wallaby. *Wildlife Research* (online early) doi:10.1071/WR16181

WRT (2016) 'Recovery of *Petrogale lateralis* MacDonnell Ranges race in South Australia'. Progress Report. Warru Recovery Team, APY Land Management, Umuwa, South Australia.

26

Recovery of the mainland subspecies of eastern barred bandicoot

Richard Hill, Amy Coetsee (nee Winnard) and Duncan Sutherland

Summary

The problem

1. By the late 1980s, the mainland subspecies of the eastern barred bandicoot (*Perameles gunnii* unnamed subspecies) was functionally extinct in the wild, mainly due to habitat loss and fox predation.
2. Active management of the last wild population (at Hamilton in south-western Victoria) ceased after 1988. In the following 16 years, attempts to establish reintroduced populations failed.

Actions taken to manage the problem

1. A review in 2008 followed by a new Recovery Plan in 2010 re-focused recovery on fenced sites that excluded introduced predators. By 2016, three fenced sites (totalling 805 ha) supported more than 1000 bandicoots.
2. Trial introductions were undertaken on French Island in 2012 and on Churchill Island in 2015 to determine the suitability of islands for bandicoots in anticipation of introductions to large fox-free islands.
3. Gene pool mixing trials are investigating introducing new genetic diversity to mitigate past losses and restore adaptive potential of the taxon.
4. Trials are underway on the use of guardian dogs to allow reintroductions of bandicoots to unfenced sites.

Markers of success

1. Probable extinction was averted and the population size has increased markedly.
2. Causes of decline have been identified and are now being managed effectively.
3. A Recovery Plan is guiding management under direction from a well-managed Recovery Team.
4. The broader community has become increasingly aware of, involved in, and supportive of the program.

Reasons for success
1. A renewed Recovery Plan with achievable targets and objectives that recognised the intensity of the threat from introduced predators.
2. Substantial financial support for the creation of reintroduction sites at which introduced predators are excluded.
3. Persistence and determination to achieve good conservation outcomes including a willingness to trial new strategies such as genetic mixing and guardian dogs.
4. Strong links between science and on-the-ground management actions.

Introduction

The eastern barred bandicoot (*Perameles gunnii*) is a small (<1 kg) insectivorous marsupial. Tasmanian and mainland populations are distinct at the subspecies level (Robinson *et al.* 1991). Unless otherwise specified, this chapter focuses on the mainland subspecies of eastern barred bandicoot. The mainland subspecies is currently (2016) listed as Endangered under the *Environment Protection and Biodiversity Conservation Act 1999*, Threatened under the Victorian *Flora and Fauna Guarantee Act 1988* and Extinct in the Wild under the Victorian Advisory List of Threatened Vertebrate Fauna 2013.

This taxon was once widespread and locally common across the grasslands and grassy woodlands of the far south-east corner of South Australia and south-western Victoria (Fig. 26.1), particularly on the basalt soils of the Victorian Volcanic Plains (Menkhorst and Seebeck 1990; Backhouse 1992). From the time of European settlement, this area quickly became the most productive and intensively farmed area of Victoria, and grazing and cropping remain the main land uses over the entire original mainland range of the eastern barred bandicoot. The native grasslands of this area have been extensively cleared or degraded, with less than 1% of the original area remaining (Scarlett *et al.* 1992). The 'natural temperate grassland of the Victorian Volcanic Plain', the ecological community that encompasses the Victorian distribution and habitat of the eastern barred bandicoot, is itself Critically Endangered (EPBC 1999).

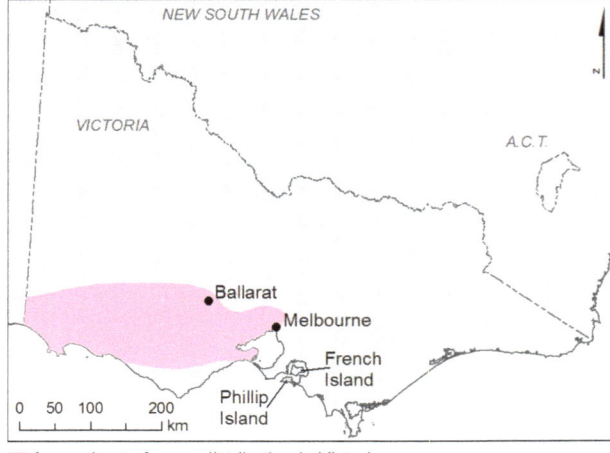

Fig. 26.1. Eastern barred bandicoots were once spread throughout western Victoria.

Fig. 26.2. (A) Eastern barred bandicoots are nocturnal, camping during the day in grassland in nests. (B) Captive breeding and release programs have been essential to restoring the wild population. (C) Research has driven recent successful conservation management, identifying the major threats so they can be ameliorated. (D) Fox-proof fencing has been the single most effective tool for maintaining bandicoots in the landscape (photos: A, Annette Rypaski; B, Richard Hill; C and D, Kim Volk)

By 1960, the mainland eastern barred bandicoot was reduced to three small populations. By 1972, it was extinct throughout its range, except for a small population at Hamilton in western Victoria (Seebeck 1979). By the late 1980s, this last population was considered to be rapidly trending to extinction (Backhouse 1992).

The principal threats to the eastern barred bandicoot are past habitat loss and predation, primarily by the red fox (*Vulpes vulpes*) (Hill *et al*. 2010). Foxes became established in western Victoria around the turn of the 19th century, and anecdotal reports suggested that they killed many bandicoots (Menkhorst and Seebeck 1990). Observations of the last wild population at Hamilton suggested that predation, particularly of juveniles (<500 g), by domestic and feral cats (*Felis catus*) also contributed to their decline (Maguire *et al*. 1990).

In the past 10 years, the status of the mainland eastern barred bandicoot has improved markedly (Fig. 26.2). Recent recovery efforts have focused on re-establishing populations within fenced sites that exclude introduced predators, and the total population size has increased from around 100 individuals in 2006 to more than 1000 in 2016. Populations currently occur in three fenced sites: Mt Rothwell Biodiversity Interpretation Centre (~600 animals), Woodlands Historic Park (~500 animals) and Hamilton Community Parklands (~50 animals). It also has been released on Churchill Island (~100 animals), which lacks introduced predators. A fourth fenced site (Tiverton) should be ready to receive animals in 2018.

Although fenced sites have prevented the extinction of the mainland eastern barred bandicoot and are currently the only successful model for maintaining the taxon, it is uncertain whether fenced populations can be sustained in the long term (i.e. for many human generations). To mitigate this risk, we are attempting to establish populations of mainland eastern barred bandicoots on Victoria's two largest islands. French Island is fox-free, and a recent fox-control program on Phillip Island has made it essentially fox-free. Both islands support around 9000 ha of farmland that should provide suitable grassland habitats. However, both islands also have populations of feral cats, which, if not controlled, may prevent successful and sustained establishment of bandicoots.

Conservation management

1982 to 1989

A Recovery Program for the mainland eastern barred bandicoot commenced in 1982 with the production of the first management prescriptions for the remaining population around and within Hamilton (Brown 1989). This focused on protection and enhancement of habitat through the provision of habitat strips, artificial shelters (from predators) and land purchase. Community groups were formed to promote bandicoot conservation and to assist in recovery actions. Predation was tackled through fox control and the promotion of responsible cat ownership (Brown 1989). Road warning signs and reflectors were installed in attempts to reduce road mortality. Community workshops were held and information displays and brochures were prepared to assist in community education. Despite these considerable efforts, bandicoot abundance and range declined quite rapidly, and by 1991 the population had declined to probably fewer than 50 individuals, and monitoring and modelling indicated that extinction in the wild appeared inevitable (Clark *et al*. 1995).

1989 to 2005

Between 1989 and 1991, as many animals as possible from the sole surviving wild population were captured to establish a captive breeding program. The newly formed Recovery Team developed a captive breed and release strategy aimed at establishing new populations at eight locations within the original range. A review of the approach and outcomes over this period (Winnard and Coulson 2008) noted that success across these sites varied and all sites ultimately failed, primarily due to fox predation. The most successful reintroduction site was at Mooramong, where a small population persisted for ~17 years before becoming extinct.

2006 to present

Following the Winnard and Coulson (2008) review, a new Recovery Plan was developed in 2010 and now guides recovery effort (Hill *et al*. 2010). The new Recovery Plan adopted a 'fox-free recovery model', focused on increasing population size to 1000 individuals within 5 years and setting a long-term recovery target of 2500 individuals. This would be achieved through establishment of ~2500 ha of sites that exclude introduced predators, based on an assumption of long-term average density of 1 animal/ha. In 2006, one site was already established at Mt Rothwell Biodiversity Interpretation Centre, near Geelong

(470 ha). In 2007, we reintroduced animals into a 100 ha site at Hamilton after fitting a 'floppy top' and repairing the 'apron' on an existing 1.8 m high fence, and in 2013 we reintroduced animals into Woodlands Historic Park, again after enhancements to an existing fence (235 ha). In 2016, we commenced fencing Tiverton, a working 1000-ha farm near Mortlake, which is anticipated to be ready for a release of bandicoots in 2018. This will take the total fenced, fox-free area available to the mainland eastern barred bandicoot to 1805 ha, capable of holding at least 1800 individuals, but probably many more in favourable years.

In 2012, we conducted a trial release of 18 non-breeding eastern barred bandicoots onto French Island. The island has extensive areas of largely cleared farmland (around 9000 ha), previously grassy woodland, which was judged potential habitat based on habitat preferences within the natural range, and habitats occupied by eastern barred bandicoots in Tasmania. We found that the impact of cats (through toxoplasmosis and direct predation) was the largest identifiable source of mortality, but the fate of this trial did not rule out the possibility of a subsequent successful establishment. In 2015, we introduced a breeding population to Churchill Island, a 57 ha fox- and cat-free island connected to Phillip Island via a 100 m concrete bridge. This release is to test breeding success and maximum likely densities of island introductions and to demonstrate to the communities of French and Phillip Islands what to expect from an eastern barred bandicoot release. We are currently seeking approval to release the species onto Phillip Island in 2017. Approvals for these island releases have required extensive evaluation of the potential environmental, economic and social impacts and benefits of these proposals. In a project proposed for 2018, we will investigate whether guardian dogs can reduce feral predator activity sufficiently to allow reintroduction to unfenced sites.

Captive breeding

A breed-for-release program started in 1988 at Woodlands Historic Park, but was transferred to Zoos Victoria facilities in 1991. The program started with 42 individuals captured from Hamilton, although only 19 of these are thought to have successfully contributed to the gene pool: all animals in the extant subpopulations are descendants of these 19 founders. The eastern barred bandicoot is largely solitary and animals in captive facilities need to be housed separately, which ultimately limits captive holding capacity. Based on a review of the genetic diversity in the captive population (Weeks *et al.* 2013), the captive and reintroduced populations are now managed as a single metapopulation to maximise overall effective population size. The captive program is currently based on 15 breeding pairs annually, with animals sourced from reintroduction sites and typically bred once before being re-released. The primary function of the captive program is now to provide additional insurance against catastrophic loss of reintroduced populations, as well as supplying small numbers of suitable individuals for establishing new sites.

Although population numbers have increased, genetic diversity among the populations has dropped by 35–40% since 1990 (Weeks *et al.* 2013). As a consequence, the Recovery Team has commenced a gene-pool mixing trial to investigate whether genetic diversity and the long-term adaptive potential of mainland eastern barred bandicoots can be increased by introducing additional genetic variation from the Tasmanian subspecies.

Research, biology, identification of key threats

The eastern barred bandicoot rests above ground during the day in nests concealed within grass tussocks or fallen timber (Brown 1989). The species was originally considered a habitat specialist preferring dense grassland cover. However, at fox-free reintroduction sites, eastern barred bandicoots use a wide range of grasslands and grassy woodlands for feeding, and construct above-ground nest sites, as well as using rabbit warrens for daytime refuges (Brown 1989). The eastern barred bandicoot was believed to be severely impacted by drought and loss of grass cover, but, in the absence of foxes, it can persist where grassland cover is markedly reduced by drought or by overgrazing (Winnard et al. 2013). It has a very high potential rate of increase, with a 12.5 day gestation, multiple litters (up to five) in a single year, and time to sexual maturity of only 3 months.

Winnard and Coulson (2008) reviewed the detailed population monitoring data available for all eight reintroductions of the eastern barred bandicoot from 1989 to 2005. They found that, although the population monitoring data were detailed and consistently collected, no potential explanatory variables had been measured, so it was not possible to explain the population drivers of these reintroduction failures. Nevertheless, they considered that predation (by foxes) and drought were the most likely causes of loss in each case. Winnard (2010) monitored the establishment of a reintroduced population at a fox-free site at Hamilton Community Parklands and the nascent population at Mt Rothwell. At Hamilton, she showed that the eastern barred bandicoot could establish in fox-free sites, even where cats were present at low densities.

People, agencies, governance and accountability

The Recovery Team commenced in 1989 and has been chaired by the state government agency throughout this time (Richard Hill since 2005). Implementing the Recovery Plan (Hill et al. 2010) has been the primary focus of the Recovery Team. The Plan has two interrelated targets: (1) to minimise further loss of genetic diversity; and (2) to rapidly increase population size to at least 1000 individuals within the life of the 5-year plan, and 2500 individuals for long-term recovery.

Since 2011, the Recovery Team has worked through four sub-groups: science, strategy, operations and communications. These sub-groups meet at least twice yearly, while the full Recovery Team meets annually. Most of the decision making, although provisional until endorsed by the Recovery Team, tends to be done in the sub-groups. Using the Recovery Plan actions as a framework, the work of the teams follows an annual action plan, which identifies the detail and priority of each recovery plan action for the coming year. The full team meeting reviews and endorses works undertaken over the preceding year and then reworks the action plan details and priorities for the coming year.

Most communication within the Recovery Team is within sub-groups via email and teleconference. Sub-groups have worked more effectively when they are small, so we now limit the membership of each sub-group to no more than six. Annual reviews are important in demonstrating to all team members and to other stakeholders that we have a clear plan, that we regularly evaluate and review our actions, and thus are focused on clear goals that we are achieving. This structure also motivates members to foster support for the Recovery Team's activities within their own organisations.

Volunteers are a vital part of the program and play a valuable role at all sites. Direct community support for the program varies from site to site. Mt Rothwell has a Landcare group comprising committed volunteers who contribute many thousands of hours annually to fence maintenance, weed and rabbit control and other activities. At Hamilton and Woodlands, Conservation Volunteers Australia offers volunteer opportunities to companies as well as individuals. The volunteers are involved in fence maintenance and weed control, as well as spotlight walks and the popular trapping program where they get to see and touch an animal. Media is done through member organisations rather than as a team, although an agreed media protocol aims to ensure consistency of messages and acknowledgement of team members.

Metapopulation management

New populations of the eastern barred bandicoot are established using animals sourced from captive breeding and from existing reintroduction sites. Genetic guidelines are used to maximise the variability of the founder group and to restrict the release of siblings and half-siblings from the captive breeding program.

We aim to manage the entire population as a single metapopulation, but have limited data on the breeding success of translocated animals, and so are unsure of the rate of population movement needed to achieve this. A research priority is to determine the level of breeding success of translocated individuals.

Our main threat mitigation strategy is predator-proof fencing. Such fences are expensive to establish and maintain, but have been far more successful than previous programs where fences were not deployed. These fences require regular (near-daily) inspections to maintain, and at current sites this is done by paid staff and volunteers. The eastern barred bandicoot is not a social animal, and fenced populations do not reach densities where they impact negatively on individual animal welfare or habitat conditions. Rather, population density ultimately appears to be regulated by social intolerance. In contrast, all fenced sites have populations of native eastern grey kangaroos *Macropus giganteus*, which are controlled to agreed maximum population densities. Rabbits (*Oryctolagus cuniculus*) require ongoing control at all sites, and we are still developing effective rabbit control methods that can be used in the presence of the eastern barred bandicoot. At new fenced sites, we plan to eradicate rabbits altogether before establishing the eastern barred bandicoot. We mosaic-burn sites for fire protection and ecological reasons, with lighting patterns designed to minimise the risk of animals being entrapped by fire.

Monitoring

Our two key monitoring measures are total population size and genetic diversity. At most sites we use well-established trapping methods, which provide population estimates through mark–recapture models. At some sites, trap interference by abundant non-target species reduces the robustness of this approach. For this reason, distance sampling on fixed transects, which models spotlight sighting data, is being trialled at Churchill Island, Mt Rothwell and Woodlands; early results validated against a site with known population size are promising.

Genetic diversity is measured through genotyping of ear biopsies or hair samples taken from all bandicoots captured in the trapping program, bred in the captive breeding program, and/or used as founders. These data are used to track overall genetic diversity, as well as to select suitable pairings for captive breeding.

Monitoring for fox and cat activity occurs at all sites to ensure they remain free of these predators. Methods vary from site to site, but all include one or more of the following: free feeds of liver baits, scat searches, sand pads, cameras and spotlighting. In the rare event a fox is detected, a range of control methods are available including poison-baiting, spotlight shooting, trapping, fox drives using hunters, and trained dogs.

Infrastructure

Predator-barrier fences have been essential in preventing the extinction of the eastern barred bandicoot, but cost around A$30 000 per km (2016 dollars). The three current release sites have 23 km of fence, with another 18 km currently being built at Tiverton. The oldest fences are 25 years old, but are still serviceable. The Recovery Team does not currently have a plan, or the funds, for replacement or rebuilding of aging fences, but this is an emerging priority.

There are 69 pens within the captive breeding program. Four other facilities have small (4–8 ha) enclosures or 'retirement homes' that exclude introduced predators and currently hold single-sex, post-reproductive individuals that are too old to be released at reintroduction sites. Together, these provide space for up to 96 bandicoots.

Money

Over the past 10 years in particular, this program has benefitted from substantial public and private investment. Philanthropic investors purchased and manage two properties totalling 1470 ha (Mount Rothwell and Tiverton), and currently maintain the largest population of the eastern barred bandicoot at Mt Rothwell. Since 2012, Zoos Victoria has made a multi-million dollar investment in their threatened species masterplan, including funding a full-time project officer. Direct public investment has come from the Australian and Victorian Governments. The Australian Government's Caring for Country program has been a strong supporter of this project. Considerable investments supporting staff and on-ground management also have been made by Parks Victoria, Conservation Volunteers and Phillip Island Nature Parks. The Victorian Department of Environment, Land, Water and Planning funds the Recovery Team chair role, and have provided well over $1 million of project funding in the past 20 years. Volunteers donate thousands of hours annually to the operation at Mount Rothwell.

A key challenge is to make the fenced sites as cost-effective as possible. Nigel Sharp, co-owner of the two largest sites, is developing a sustainable long-term funding model. For example, the entire property of Tiverton is being predator-fenced, and will continue as an operating sheep farm, as well as deriving funds through the sale of native vegetation offsets. At Tiverton, we are trialling an automated fence-checking camera which, if successful, will markedly reduce the amount of time spent patrolling fence boundaries.

The future

By 2020 we hope to reach our target population of 2500 bandicoots. We expect to have an established reintroduced population at Tiverton by 2020 and also hope to have populations on French Island and/or Phillip Island. Results of the guardian dog trials also should be available and, if successful, will provide the possibility of having unfenced populations of eastern barred bandicoot within the historic range. This set of measures should provide sufficient conservation security to allow closure of the captive insurance program.

Currently, several thousand hectares of grassland reserves are now being established immediately west of Melbourne. Fencing or guardian dogs may provide the possibility of re-establishing the eastern barred bandicoot within those large parks.

Conclusion

Extinction of mainland populations of the eastern barred bandicoot was averted by salvaging the last wild population and establishing a captive breeding program. For the following 16 years, the main recovery effort focused on a captive insurance population and re-establishing animals within their historic range. The inability to sufficiently reduce fox predation was probably the main reason these attempts ultimately failed. These failures prompted a comprehensive review and thence a major change in program direction. Conservation of eastern barred bandicoots is now being secured at fenced, fox-free sites. This has been a very successful model, and that demonstrated success has helped to attract considerable additional funding.

The main ingredients of this recovery have been a clearly articulated set of actions in the Recovery Plan, a focus on the most intense threats, a Recovery team focused on results and regularly reviewing its activity, and integration of science and on-ground management. Thirty years after the bandicoot's near extinction in the wild, more than 1000 individuals are now established in three fenced fox-free areas. With fencing of a fourth site underway, trials of island translocations progressing, and plans to trial reintroductions into unfenced areas, the conservation future of the eastern barred bandicoot is looking bright as we progress to our recovery target population of 2500 animals by 2020.

Acknowledgements

The recovery of the eastern barred bandicoot has had a long history, and been coordinated by one of the nation's most long-lasting recovery teams. The late John Seebeck, Peter Brown and Mandy Watson (Victorian Government) were key people earlier in the program, as were members of the Hamilton Field Naturalists club, and Peter Courtney (ex Zoos Victoria).

The current Recovery Team has many participants, organisations and supporters. These include: Conservation Volunteers Australia (Travis Scicchitano, Jonathon Lee, Kim Volk), Department of Environment, Land, Water and Planning, Victoria (Jim O'Brien, Michelle Basset), French Island Landcare group, Mt Rothwell Biodiversity and Interpretation Centre and Tiverton Property Partners (Nigel Sharp, Annette Rypalski, Tim Hill), the National Trust (Val Lang), Parks Victoria (Phil Pegler, Fiona Smith, Tristan Factor, Tamara Karner, Brendan Sullivan, David Collins), Phillip Island Nature Parks (Duncan Sutherland, Peter Dann, Anthony Rendall), University of Melbourne (Graeme Coulson, Andrew Weeks), and Zoos Victoria (Amy Coetsee, Dan Harley, Michael Magrath, Marissa Parrot, Dave Williams).

References

Backhouse B (1992) 'Recovery Plan for the Eastern Barred Bandicoot *Perameles gunnii*'. Department of Conservation, Forests and Lands, Victoria and Zoological Board of Victoria, Melbourne.

Brown PR (1989) 'Management Plan for the Conservation of the Eastern Barred Bandicoot *Perameles gunnii* in Victoria'. ARI Technical Report 63. Department of Conservation, Forests and Lands, Melbourne.

Clark TW, Gibbs JP, Goldstraw PW (1995) Some demographics of the extirpation from the wild of eastern barred bandicoots (*Perameles gunnii*) in 1988–91, near Hamilton, Victoria, Australia. *Australian Wildlife Research* **22**, 289–297. doi:10.1071/WR9950289

Hill R, Winnard A, Watson M (2010) 'National Recovery Plan for the Eastern Barred Bandicoot (mainland) *Perameles gunnii* unnamed subspecies'. Department of Sustainability and Environment, Melbourne.

Maguire LA, Lacy RC, Begg RJ, Clark TW (1990) An analysis of alternative strategies for recovering the eastern barred bandicoot in Victoria. In *Management and Conservation of Small Populations*. (Eds TW Clark and JW Seebeck) pp. 147–164. Chicago Zoological Society, Brookfield IL, USA.

Menkhorst P, Seebeck JS (1990) Distribution and conservation status of bandicoots in Victoria. In *Bandicoots and Bilbies*. (Eds JH Seebeck, PR Brown, RL Wallis and CM Kemper) pp. 51–60. Surrey Beatty & Sons, Chipping Norton, NSW.

Robinson NA, Sherwin WB, Brown PR (1991) A note on the status of the eastern barred bandicoot, *Perameles gunnii*, in Tasmania. *Australian Wildlife Research* **18**, 451–457. doi:10.1071/WR9910451

Scarlett NH, Wallbrink SJ, McDougall K (1992) *Field Guide to Victoria's Native Grasslands*. Victoria Press, Melbourne.

Seebeck JH (1979) Status of the barred bandicoot, *Perameles gunnii* in Victoria: with a note on husbandry of a captive colony. *Australian Wildlife Research* **6**, 255–264. doi:10.1071/WR9790255

Weeks A, Van Rooyen A, Mitrovski P, Heinze D, Winnard A, Miller A (2013) A species in decline: genetic diversity and conservation of the Victorian eastern barred bandicoot, *Perameles gunnii*. *Conservation Genetics* **14**, 1243–1254. doi:10.1007/s10592-013-0512-9

Winnard A (2010) Reintroduction biology of the eastern barred bandicoot. PhD thesis. University of Melbourne, Australia.

Winnard AL, Coulson G (2008) Sixteen years of eastern barred bandicoot *Perameles gunnii* reintroductions in Victoria: a review. *Pacific Conservation Biology* **14**, 34–53. doi:10.1071/PC080034

Winnard AL, Di Stefano J, Coulson G (2013) Habitat use of a critically-endangered species in a predator-free but degraded reserve in Australia. *Wildlife Biology* **19**, 429–438. doi:10.2981/12-116

27

Arid Recovery: a successful conservation partnership

Katherine Moseby, Peter Copley, David C. Paton and John L. Read

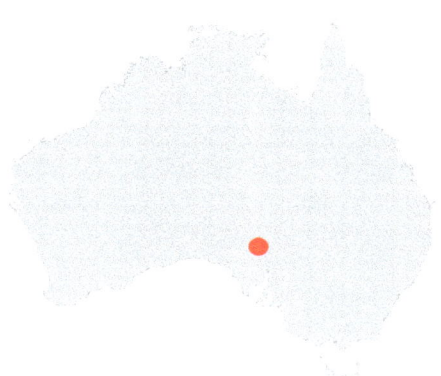

Summary

The problem

1. Many native mammal species disappeared from arid Australia as a result of predation by introduced cats and foxes and competition with introduced herbivores.
2. Introduced predators and competitors are extremely difficult to control at a landscape scale.

Actions taken to manage the problem

1. 123 km^2 of arid shrubland were fenced to exclude rabbits and introduced predators, with this fencing continually maintained.
2. Rabbits, cats and foxes were removed from within 60 km^2 of the fenced area.
3. Threatened arid zone mammal species were reintroduced.
4. There has been a commitment to intensive monitoring, and long-term research including into methods to re-establish native species outside of fenced reserves.

Markers of success

1. Causes of decline were identified and are now effectively managed, such that populations of threatened mammals are stable or increasing.
2. There has been substantial and ongoing community involvement and support.
3. High-quality research outputs have led to improvement in the control of introduced species and an increased understanding of threats and pre-European faunal assemblages.

Reasons for success

1. Dedicated champions and staff, some of whom have been with the project since its inception.
2. A reliable and long term source of funding, in this case from a mining company.

3. A dedicated team of volunteers, many of whom have been with the project for decades.
4. High-quality research and adaptive management, often as part of multi-stakeholder partnerships

Introduction

More than 60% of the mammal species around Roxby Downs in arid South Australia have become locally extinct since European settlement. These mostly medium-sized species included the greater bilby (*Macrotis lagotis*), burrowing bettong (*Bettongia lesueur*), pig-footed bandicoot (*Chaeropus ecaudatus*), crest-tailed mulgara (*Dasycercus cristicauda*), golden bandicoot (*Isoodon auratus*), greater stick-nest rat (*Leporillus conditor*), Gould's mouse (*Pseudomys gouldii*) and long-tailed hopping-mouse (*Notomys longicaudatus*). Many are now globally extinct; others still hang on in refuges such as offshore islands. The main reason for the decline in mammals in arid areas is thought to be a combination of land degradation by introduced rabbits (*Oryctolagus cuniculus*) and domestic stock (primarily in the late 1800s and early 1900s), and predation by the introduced red fox *(Vulpes vulpes)* and feral cat (*Felis catus*) (Morton 1990). Overgrazing by introduced herbivores can reduce vegetation cover, leading to both a decline in food resources and an increased susceptibility to predation (Copley 1999). Introduced foxes and cats prey on native species and reach high densities in areas where introduced rabbits are present.

Arid Recovery was initiated in 1997 after the arrival of rabbit calicivirus disease led to the widespread decline of rabbits. Four co-founders – Katherine Moseby, John Read, Peter Copley and David Paton – worked together to submit an ambitious proposal to Western Mining Corporation (WMC) who owned the Olympic Dam (copper, gold, uranium) Mine in central South Australia (now owned by BHP Billiton). WMC agreed to co-fund the project and a partnership was developed between WMC, the South Australian Environment Department, the University of Adelaide and the local community. Our initial objective was to build on the temporary respite from rabbit overgrazing by building landscape-scale, rabbit-free exclosures to protect native vegetation, but this objective broadened to encompass

Fig. 27.1. Kaye Kessing's *Arid Recovery* portrays a vision for a future as it might be without feral animals.

reintroductions and restoration. The over-arching goal of Arid Recovery was to restore an area of arid land as close as possible to its pre-European state and to apply and disseminate the lessons from that restoration process to other arid areas (Fig. 27.1). In addition, the project was a pioneering example of a resource company committing to a long-term proactive conservation partnership. Therefore, Arid Recovery became a research, adaptive management and demonstration project, rather than simply a conservation reserve.

Conservation management

Fencing of the Arid Recovery Reserve began in 1998 and, over the next 5 years, a total of six contiguous exclosures were built enclosing 123 km^2 (Fig. 27.2). Eradication of rabbits was challenging (Read *et al.* 2011b), but deemed integral to the restoration project and has subsequently freed Arid Recovery from the challenges of co-managing rabbits and reintroduced native mammals. Between 1998 and 2001, rabbits, feral cats and foxes were removed from 60 km^2, and four nationally threatened mammal species were reintroduced. The greater stick-nest rat was formerly present across much of semi-arid southern Australia, but declined by more than 95% and became extinct on the mainland by the 1940s; it occurs naturally now only on two small offshore islands – East and West Franklin (Copley 1999). Arid Recovery reintroduced the greater stick-nest rat in 1998 – the first successful mainland reintroduction site for the species. They are now found throughout the Reserve and Arid Recovery supports the only viable mainland population of the species, with this population important for maintaining the species' genetic diversity for potential adaptation to future climate change.

The burrowing bettong was reintroduced in 1999 from a reintroduced population at Heirisson Prong, Western Australia. Western barred bandicoots (*Perameles bougainville*) were reintroduced 2 years later, also from Western Australia. Burrowing bettongs and western barred bandicoots are now found naturally on only three and two islands in Western Australia, respectively, and their natural populations number fewer than a few thousand individuals. Both species have adapted well to Arid Recovery, with population

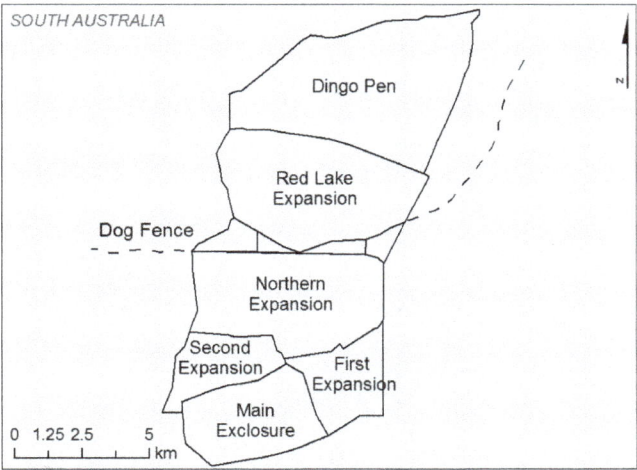

Fig. 27.2. The number of exclosures at Arid Recovery has expanded over time as funds have become available, allowing an increasing number of options to be tested.

trajectories still increasing inside the reserve. Indeed, burrowing bettongs have become so abundant that methods for reducing their population are now being implemented.

The greater bilby was reintroduced in April 2000. It is still extant in northern deserts of mainland Western Australia and the Northern Territory, and in south-western Queensland, but has been extinct in South Australia and New South Wales for nearly 100 years. The bilbies released at Arid Recovery were obtained from the Zoos SA Monarto Captive Breeding Facility and the first release comprised only nine individuals. Bilbies bred up to very high numbers, and were soon digging out of the reserve and self-dispersing. This had to be stopped with foot-netting on the inside of the fence to prevent rabbits entering the Reserve through bilby holes under the fence. Their numbers inside the Reserve now fluctuate in response to climatic conditions.

These four reintroduced mammal species are still extant within the Arid Recovery Reserve and have healthy populations (Moseby *et al.* 2011). Despite low founder size, all four species expanded their distribution throughout the Reserve over time and now occupy all of the areas from which rabbits, cats and foxes have been eradicated. One of the key measures of success was their ability to survive drought conditions without provision of supplementary food or water, and all four species have now met this criterion (Fig. 27.3). Population estimates are difficult to obtain but are thought to be over 500 for stick-nest rats, 6000 for bettongs, 1000 for western barred bandicoots, and 600 for bilbies (Fig. 27.4).

Two other species' reintroductions have been attempted. A trial release of five numbats (*Myrmecobius fasciatus*) was conducted but the three females were taken by birds of prey within 6 months (Bester and Rusten 2009). The two males remained within the reserve for over 12 months. A release of woma pythons (*Aspidites ramsayi*) was also attempted, but only captive-bred individuals could be obtained and these were preyed upon by mulga snakes (*Pseudechis australis*) soon after release (Read *et al.* 2011a).

Research, biology, identification of key threats

Arid Recovery is a research-based conservation project and more than 50 publications have been produced from research at the site. Honours and PhD students have conducted numerous studies into topics as diverse as bettong diet, bilby behaviour, termite abundance, scorpion burrows, bettong over-population, kangaroo carcass decomposition, mulga regeneration, foraging digs and thornbill ecology.

Although predation and competition from introduced herbivores were thought to be the key threats that led to local extinction of mammal species, it was only after attempting to reintroduce them outside the reserve that the devastating impacts of predators became apparent. We have attempted to reintroduce bilbies and bettongs outside the reserve on four separate occasions to date and, in all cases, predation by foxes, feral cats and, to a lesser extent, dingoes caused population extinction within a few months (Moseby *et al.* 2011; Bannister *et al.* 2016).

The significance of predator-exclusion fencing has also been demonstrated for other extant native species within the reserve with the threatened plains mouse (*Pseudomys australis*) increasing more than 10-fold and the spinifex hopping mouse (*Notomys alexis*) increasing by more than 15-fold inside the reserve compared with adjacent areas outside the fence (Moseby *et al.* 2009). Several other small mammals have also increased in abundance within the reserve.

Attempts to reduce the impacts of exotic predators outside the reserve using poison-baiting have had mixed results. Foxes could be controlled, but bait uptake by feral cats was highly variable and cats continued to cause extirpations of external releases of native mammals (Moseby and Hill 2011).

Planning and policy

Arid Recovery was an ambitious project: rabbits had never been removed from such a large area of prime rabbit habitat, and mining companies had not partnered with government, universities and conservation organisations at this scale. Receiving most funding from a mining company allowed Arid Recovery to plan reintroductions based on ecological recovery rather than having the tight timing constraints often imposed by government funding or philanthropic donations. This enabled a focus on fence design, rabbit eradication, monitoring and research to occur before reintroductions commenced.

Arid Recovery produced a research and reintroduction plan that guided on-ground work, but was also flexible enough to take advantage of opportunities as they arose because of its proactive and innovative team. Arid Recovery conducted its own fundraising through the local community group Friends of Arid Recovery. This provided valuable funding for materials such as radio-transmitters and trapping equipment and built community ownership of the project. Arid Recovery's unique partnership and governance arrangements allowed full advantage of resourcing, leveraging and technical input from partner organisations to be taken, while also allowing staff to operate somewhat autonomously to pursue proactive and innovative opportunities at short notice where necessary.

People, agencies, governance and accountability

Arid Recovery was born from the passion and dedication of four individuals who were committed to restoring the deserts around Roxby Downs. They were supported by their various organisations and given the freedom to make decisions. The four individuals formed a management committee that met approximately three times a year, usually with project staff, but were also in regular contact through emails and phone. As the project developed and success flowed, the project received more outside attention, the biodiversity assets became more valuable and the infrastructure investments and replacement liabilities increased. Consequently the project management team expanded and meetings and accountabilities were formalised. Although this expansion and change was inevitable, a more rigid corporate structure and board arrangement meant that some of the factors responsible for the initial success – including local knowledge, rapid inclusive adaptive management decision making, corporate memory, passion and ownership of challenges – were diminished.

Arid Recovery is now an incorporated not-for-profit organisation managed by a Board consisting of representatives nominated from three of the original partner organisations (BHP Billiton, DEWNR and the University of Adelaide) and up to three other members chosen by these organisations. Inception of the board saw a major change of BHP Billiton's involvement from partner to sponsor. Although ongoing funding from the mining company is obviously vital to the delivery of ongoing success and sustainability of Arid Recovery, diminished ownership through the new sponsorship and auditing role has seen a departure from the original collaborative 'hands-on' model (Read 2013). A scientific advisory group, including two Board members, now provides direct links between the Board and staff, founders and scientists. There is no longer a Friends of Arid Recovery community group and there is currently no community representation on the board, nor are there any elected members. However, the general public can, and do, still volunteer at the reserve.

Managing

During the initial stages, much time was dedicated to erecting predator exclusion fencing and removing rabbits, cats and foxes. Methods used have included 1080 poisoning, burrow fumigation, burrow destruction, shooting, cage trapping and leg-hold trapping. Removing rabbits was by far the most challenging activity and took thousands of hours (Read et al. 2011b). Unexpectedly, the standard 40 mm rabbit-proof fencing that was used was found to be breached by small rabbits. Reinforcing the fence with 30 mm-netting helped exclude them and enabled eradication to occur.

Reintroduction of threatened mammals was staged after eradication of rabbits, foxes and feral cats. An 8 ha release pen was used to allow animals to acclimatise to the arid conditions and facilitate intensive monitoring after release. Focus then turned to restoration activities beyond the reserve. A 20 km buffer-zone was established and poison baits were laid from an aircraft. In later years, rabbit incursions and fence maintenance have become significant management issues. Fence replacement is an ongoing concern because saline soils have seen sections of the fence corrode within 15 years. The large area of the reserve (123 km^2) and significant lengths of perimeter and internal fences has meant that a fence maintenance and replacement strategy is a high priority. Regular fence checks and track counts to detect holes and breaches are an ongoing and time-consuming, but critical, activity. Cat and fox traps are set continuously around the perimeter of the reserve to reduce incursions. There are occasional breaches of the fence by cats, foxes and rabbits, usually after large rainfall events when sections of corroded foot netting are washed away.

Recently, Arid Recovery has also started to manage over-abundance of the reintroduced bettongs, because predator-exclusion has allowed them to increase to levels where they are affecting plants and other native mammal species. One-way gates and the reintroduction of native predators such as quolls and dingoes are being trialled in the hope that a more balanced and sustainable ecosystem can be created within the reserve.

Monitoring

Arid Recovery has established more than 150 monitoring sites to measure changes in vegetation, small mammals, birds, reptiles, reintroduced species and soils. Sites are generally paired inside and outside the Reserve to enable changes associated with the exclusion of exotic species and introduction of native species to be differentiated from climatic influences. A control paddock inside the reserve was also established where there are no introduced or reintroduced species. Monitoring has demonstrated that following rabbit eradication, the vegetation started to recover, including improved recruitment of palatable perennials and dramatic increases in grass cover. The abundance of small mammals increased spectacularly, with more than 15 times more small mammals inside the Reserve compared with matched sites outside the fence (Fig. 27.4). The reintroduced mammal species also increased, establishing populations throughout the sections of the reserve that were free of foxes and cats. The threatened plains mouse, previously extirpated from the location, has re-established naturally and is now found at significantly higher densities inside the reserve than outside.

In recent years, there has been a decline in vegetation condition as burrowing bettongs have increased and overgrazing has occurred. Reptile trends are complex, with large reptiles, such as goannas, increasing inside the reserve, but geckoes and sandsliders (*Lerista* spp.) declining. This is probably because the absence of mammalian predators inside the reserve has led to increases in goannas, which may exert greater predation pressure on smaller reptiles (Read and Scoleri 2015). Bilbies and bandicoots may also be preying on smaller reptiles and their invertebrate prey. Abundances of birds have fluctuated depending on conditions within the Reserve and regionally. Generally the numbers of bird species and their total abundance have been higher within the Reserve. However, recent declines in frugivorous honeyeaters have occurred, with this trend probably due to overgrazing of key food plants by the large population of bettongs.

Fig. 27.3. Arid systems respond quickly to rainfall but droughts can reduce food availability for reintroduced species. Nevertheless all four species have survived drought without water being provided.

Infrastructure

The predator-proof fence has been the most important piece of infrastructure contributing to the success of Arid Recovery. Tests of the exclusion fence design were made with a small replica pen and captured foxes and cats, to assess whether these could breach the fence. The final fence design has since been published (Moseby and Read 2006) and widely

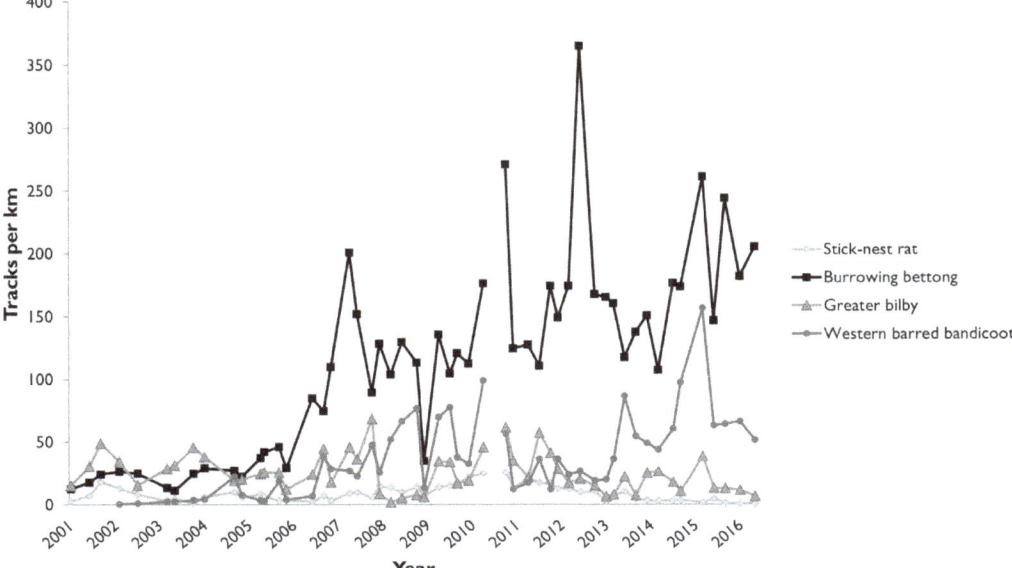

Fig. 27.4. Population trends for the four reintroduced species after release.

adopted by practitioners in Australia and overseas. Although the fence is vital for the protection of threatened species, it has also been responsible for issues of over-population and, potentially, for exacerbation of prey naivety of island- and zoo-sourced species. Total predator exclusion can lead to prey species losing their fear of predators and developing 'island syndrome' as anti-predator behaviour no longer affords individuals a survival advantage. Many threatened species that are reintroduced from islands are likely to have been isolated from predators for thousands of years, making them naive and extremely susceptible to novel predators. This hampers releases outside of fenced 'island' reserves, because individuals either do not recognise cats and foxes as predators, or they engage in inappropriate responses to their presence. Arid Recovery is attempting to remedy this issue by exposing some populations of threatened species within one section of the Reserve to low levels of cat predation in an attempt to improve anti-predator behaviour and accelerate natural selection for anti-predator traits. Initial results are promising and have led to plans to also reintroduce native predators (e.g. chuditch *Dasyurus geoffroii*) to the reserve in an attempt to reinstate anti-predator behaviour and an improved ecosystem balance.

Money

Arid Recovery has benefitted from a long-term funding commitment from WMC Resources and then BHP Billiton. This funding has allowed employment of full-time staff and to budget for fencing and equipment over the longer term. This base funding has also enabled the project to leverage other funds from grants and government funding schemes using the BHP Billiton funds as Arid Recovery contributions. For the first 5 years, Arid Recovery generated at least as many funds from external sources as it did from the mining company contribution. The large volunteer input also assisted with providing in-kind funding for grants. Any funding deficiencies were made up by increasing the volunteer commitment or by instigating imaginative fundraisers such as auctioning-off naming rights of reintroduced bandicoots, undertaking fundraising walks, organising quiz nights, merchandise sales, organising fundraising BBQs and spotlighting tours and inviting local businesses to sponsor particular activities. The Board is continually looking for innovative funding models and opportunities, to ensure that the built, natural and knowledge assets can be maintained and continually improved.

The future

In 2017, Arid Recovery celebrated its 20th year and this was a time for major reflection and planning for the future. The project has made significant progress towards understanding the causes and remedies for declines in the integrity of arid environments and has successfully established populations of five nationally threatened mammal species. These insurance populations will help to ensure that these species do not become globally extinct. A long-term funding mechanism is still not in place and until this is achieved Arid Recovery will be reliant on funding from an industry where access to discretionary spending fluctuates with global resource markets. The main challenge, however, is to facilitate widespread restoration by establishing these species back into the wild, outside of fenced reserves. Unfortunately, we do not yet have the tools to implement such widespread restoration for many species vulnerable to cat predation, but the research that is conducted at Arid Recovery is firmly focused on achieving this goal in the future. Innovative research into prey naivety (Moseby *et al*. 2016), prey switching, cat control via new technologies such as grooming traps and toxic Trojans (Read *et al*. 2015), and native predator reintroductions are being trialled at Arid Recovery and the project will continue to take risks and trial innovative techniques in an attempt to improve the future for threatened ecosystems and species in arid Australia.

Conclusion

The main drivers of initial success for Arid Recovery were a combination of passion and vision of the project champions, the long-term funding support from the local mining company, the opportune timing in terms of the spread of the rabbit calicivirus disease, the dedication and commitment of volunteers and staff, and the ability, at least in the formative stages of the project, for management to take calculated risks beyond what is typically endorsed by multi-partner project teams. One aspect of volunteer input that is often overlooked in conservation reporting is the large amount of volunteer hours contributed by project staff or co-founders who often worked long hours or on weekends to get various jobs done. Without this core volunteer contribution, coupled with the enthusiasm and commitment from long-term volunteers, it is unlikely that Arid Recovery could have achieved anywhere near what it has today. The project's strong research focus also allowed lessons to be learned and disseminated. Even our failures have been openly published so that others can learn about inadequacies of some fencing materials, over-ambitious rabbit eradication targets, contributing factors to failed reintroductions, and over-population issues. Arid Recovery has also been a successful training centre for students who receive training and mentoring in arid zone conservation ecology and learn through hands-on experience. For most of the research students, working at Arid Recovery was the first time they had visited the arid zone and many have gone on to be successful ecologists based in all parts of Australia.

Acknowledgements

Project partners DEWNR, BHP Billiton, the University of Adelaide and the Friend of Arid Recovery were instrumental to project success. A special mention needs to go to early supporters such as the Threatened Species Network (WWF), Nature Foundation SA, NHT, Envirofund, Green Corps and the Wildlife Conservation Fund (DEWNR). Steve Green has been a long-term supporter of the project and assisted with promoting the strategic value of Arid Recovery. Andrew Freeman, Greg Kammerman, Jackie Bice, Nicki Munro, Brydie Hill, Pete Paisley, Steve White, Frank Bernhardt and Katie Oxenham worked tirelessly to help establish the project. Adam Bester, Helen Crisp, Bree Galbraith, Reece Pedler, Jeff Turpin, Jenny Stott, Clint Taylor, Geoff Axford, Yvette Mooney, Travis Gotch, Chris McGoldrick, Jason Briffa, Adam Kilpatrick, Heather Neilly, Hugh McGregor and Rebecca West helped take the project forward to another level. A special mention to Rusty who was a long-term volunteer. In recent years, Kylie Piper, KJ Kovac and Kath Tuft have managed the project. Finally, a big thanks to all the students and researchers who have conducted research at the site: you have helped increase our knowledge of arid zone ecology and contributed to improved conservation management at the reserve.

References

Bannister H, Lynch C, Moseby KE (2016) Predator swamping and supplementary feeding do not improve reintroduction success for a threatened Australian mammal, *Bettongia lesueur*. *Australian Mammalogy* **38**, 177–187. doi:10.1071/AM15020

Bester AJ, Rusten K (2009) Trial translocation of the numbat (*Myrmecobius fasciatus*). *Australian Mammalogy* **31**, 9–16. doi:10.1071/AM08104

Copley P (1999) Natural histories of Australia's stick-nest rats, genus *Leporillus* (Rodentia: Muridae). *Wildlife Research* **26**, 513–539. doi:10.1071/WR97056

Morton SR (1990) The impact of European settlement on the vertebrate animals of arid Australia: a conceptual model. *Proceedings of the Ecological Society of Australia* **16**, 201–213.

Moseby KE, Hill BM (2011) The use of poison baits to control feral cats and red foxes in arid South Australia. 1. Aerial baiting trials. *Wildlife Research* **38**, 338–349. doi:10.1071/WR10235

Moseby KE, Read JL (2006) The efficacy of feral cat, fox and rabbit exclusion fence designs for threatened species protection. *Biological Conservation* **127**, 429–437. doi:10.1016/j.biocon.2005.09.002

Moseby KE, Hill BM, Read JL (2009) Arid Recovery – a comparison of reptile and small mammal populations inside and outside a large rabbit, cat and fox-proof exclosure in arid South Australia. *Austral Ecology* **34**, 156–169. doi:10.1111/j.1442-9993.2008.01916.x

Moseby KE, Read JL, Paton DC, Copley P, Hill BM, Crisp HM (2011) Predation determines the outcome of 11 reintroduction attempts in arid Australia. *Biological Conservation* **144**, 2863–2872. doi:10.1016/j.biocon.2011.08.003

Moseby KE, Blumstein D, Letnic M (2016) Harnessing natural selection to tackle the problem of prey naivety. *Evolutionary Applications* **9**, 334–343. doi:10.1111/eva.12332

Read JL (2013) *Red Sand, Green Heart: Ecological Adventures in the Outback*. Revised edn. PAGE Publishing, Adelaide.

Read JL, Scoleri V (2015) Ecological implications of reptile mesopredator release in arid South Australia. *Journal of Herpetology* **49**, 64–69. doi:10.1670/13-208

Read JL, Johnston GR, Morley TP (2011a) Snake predation thwarts trial reintroduction of threatened woma pythons, *Aspidites ramsayi*. *Oryx* **45**, 505–512. doi:10.1017/S0030605310001110

Read JL, Moseby KE, Briffa J, Kilpatrick AD, Freeman A (2011b) Eradication of rabbits from landscape scale exclosures: pipedream or possibility? *Ecological Management & Restoration* **12**, 46–53. doi:10.1111/j.1442-8903.2011.00567.x

Read JL, Peacock D, Wayne AF, Moseby KE (2015) Toxic Trojans: can feral cat predation be mitigated by making their prey poisonous? *Wildlife Research* **42**, 689–696. doi:10.1071/WR15125

28

Effective conservation of critical weight range mammals: reintroduction projects of the Australian Wildlife Conservancy

John Kanowski, David Roshier,
Michael Smith and Atticus Fleming

Summary

The problem
1. A third of Australia's mammals are extinct, or threatened with extinction, with predation by feral cats and foxes being the most significant factor in their decline. Extensive areas, particularly of arid and semi-arid Australia, now have only a vestigial representation of their pre-European mammal assemblages
2. Such extensive losses of the native mammal fauna have also led to continental-scale ecological changes and disruption of some ecological processes

Actions taken to manage the problem
1. The Australian Wildlife Conservancy (AWC) has established a network of fenced areas (and an island) from which introduced predators have been removed and into which threatened native mammal species have been reintroduced.
2. AWC measures and reports on the population of target species, actively manages threatening processes, and conducts research related to the restoration of ecosystems through mammal reintroductions.

Markers of success
1. The risk of extinction has been significantly reduced with populations of many threatened species increasing.
2. Causes of decline have been identified and are now being managed effectively at multiple sites.
3. The ecological return from implementing a fenced area strategy is relatively high.

4. Community involvement has been substantial, with ongoing support from the community and innovative public–private partnerships.

Reasons for success
1. Reasons for initial decline were clearly identified and are now controlled.
2. Managers have been prepared to take risks, based on robust evidence produced by dedicated, science-literate staff tightly linked to management.
3. Adequate resourcing has been available for long-term and sustainable management.
4. Projects are delivered in a framework that requires accountability for outcomes (increased populations) and expenditure.

Introduction

The Australian Wildlife Conservancy (AWC) is a not-for-profit conservation organisation. Our mission is the effective conservation of Australia's wildlife and their habitats. Currently, we manage over 3.8 million ha for conservation (Fig. 28.1). Our wildlife sanctuaries include representation of an estimated 86% of Australia's bird species, 71% of Australia's mammal species and over 50% of Australia's reptile and frog species.

AWC implements a range of conservation actions on its wildlife sanctuaries, including landscape-scale management of fire, feral animals and weeds. One of the major conservation actions undertaken by AWC involves the reintroduction of threatened mammals to wildlife sanctuaries from which introduced predators have been removed. Feral cats (*Felis catus*) and foxes (*Vulpes vulpes*) are the primary threat to most small to medium-sized Australian mammals (Woinarski *et al.* 2014; Doherty *et al.* 2017). Extensive areas of southern and central Australia now have only a vestigial small to medium-sized native mammal fauna (McKenzie *et al.* 2007), disrupting a suite of ecological processes in which these mammals participate, including pollination, seed dispersal, herbivory, nutrient and water retention (Fleming *et al.* 2014). These disruptions have knock-on consequences for other components of ecosystems and human welfare.

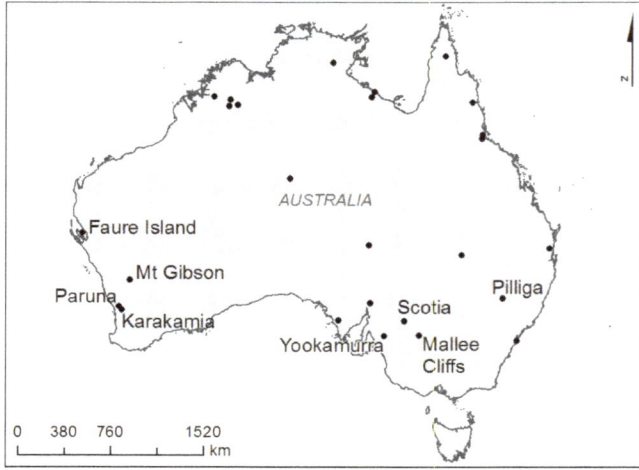

• Australian Wildlife Conservancy sanctuaries

Fig. 28.1. The Australian Wildlife Conservancy has sanctuaries across the country. The named sanctuaries exclude predators to help conserve threatened mammals.

Table 28.1. Threatened mammals protected by Australian Wildlife Conservancy's reintroduction program. Global population estimates from Woinarski et al. (2014)

Species		EPBC status	Global population size	Number of AWC sanctuaries in which present	Population size in AWC sanctuaries	Proportion of global population, AWC
Numbat	Myrmecobius fasciatus	Vulnerable	1000	2	>400	40%
Western barred bandicoot	Perameles bougainville	Endangered	~5000	1	600	12%
Greater bilby	Macrotis lagotis	Vulnerable	<10 000	2	1350	15%
Western ringtail possum	Pseudocheirus occidentalis	Vulnerable	<8000	1	40	<1%
Burrowing bettong	Bettongia lesueur	Vulnerable	~15 000	3	>7500	>50%
Woylie	Bettongia penicillata	Endangered	<18 000	3	1000	6%
Mala	Lagorchestes hirsutus	Endangered	~4000	1	70	2%
Banded hare-wallaby	Lagostrophus fasciatus	Vulnerable	~5000	1	350	7%
Bridled nailtail wallaby	Onychogalea fraenata	Endangered	~2500	1	2300	90%
Greater stick-nest rat	Leporillus conditor	Vulnerable	>3000	1	~60	2%
Shark Bay mouse	Pseudomys fieldi	Vulnerable	~10 000	1	900	9%

The reintroduction of threatened mammals to feral predator-free sanctuaries is one of the standout 'success stories' in Australian conservation. Currently, AWC supports 11 threatened species that have been reintroduced to wildlife sanctuaries in which introduced predators have been removed (Table 28.1), including one island and four fenced 'mainland islands'. Reintroductions have improved the conservation prognosis of these species by increasing both the number of secure populations and their global population size.

Conservation management

AWC's reintroduction program is based on a simple premise: feral cats and foxes are the primary threat to most of Australia's critical weight range mammals. Consequently, the removal of foxes and cats is often sufficient to re-establish viable populations of threatened mammals in areas of their former range. To date, AWC has removed introduced predators from Faure Island in Shark Bay (Western Australia, 5000 ha, predator removal completed in 2001), and four fenced 'mainland islands': Karakamia (Western Australia, 275 ha, 1994); Yookamurra (South Australia, 1100 ha, 2002); Scotia (New South Wales; 8000 ha, 2004); and, most recently (in 2015), Mt Gibson (Western Australia, 7832 ha). These sanctuaries support the 11 threatened mammals listed in Table 28.1.

Foxes and cats have been removed from these areas by a combination of baiting, shooting and trapping. Intensive monitoring was undertaken to determine where best to target control efforts and when control could cease. 'Mainland islands' are protected by a mesh fence ~2 m high, with a skirt where the fence meets the ground, a 'floppy top', and electrified wires to stop incursions of introduced predators and other pest animals (Fig. 28.2). Fences are patrolled every 2–3 days to maintain the integrity of the fence, and the area

Fig. 28.2. A Conservation fence, Mt Gibson wildlife sanctuary, WA.

inside the fence is also monitored regularly using sandplots, spotlighting and cameras to detect any incursions. There have only been a few incursions of introduced predators to AWC's fenced areas since establishment, mostly through human error; all incursions have been promptly eliminated.

Case studies of three species protected by AWC's reintroduction program – woylie (*Bettongia penicillata*), bridled nailtail wallaby (*Onychogalea fraenata*) and greater bilby (*Macrotis lagotis*) follow.

The woylie

The woylie or brush-tailed bettong (Fig. 28.3A) is a small (~1–2 kg) macropod, once distributed across most of southern and central Australia, from forests to semi-arid habitats. Woylies eat truffles, supplemented by tubers, bulbs, seeds, fruits and invertebrates. Woylies can produce up to three young per year (Claridge *et al.* 2007).

Woylies are highly susceptible to predation by foxes and feral cats. Populations collapsed across Australia in the late 19th and early 20th centuries; only a few populations in south-western Australia avoided extinction. Fox-baiting and a series of reintroductions helped the species recover in south-western Australia, but in the last two decades, populations in the wild have again crashed, largely due to predation by feral cats (Marlow *et al.* 2015; Wayne *et al.* 2015). The species is listed as Endangered nationally and Critically Endangered in Western Australia.

A Recovery Plan has been developed for the species (Yeatman and Groom 2012). Reintroductions are one of the main recovery actions listed in the plan.

The woylie has been subject to more reintroductions than any other Australian mammal (Short 2009). AWC has reintroduced the woylie to fenced areas at Karakamia, Scotia, Yookamurra and Mt Gibson wildlife sanctuaries. Populations in all these locations have established successfully; Karakamia has supported a stable population of around 400 woylies for over 20 years. At present, AWC sanctuaries support over 1000 woylies. Additional reintroductions are planned for Pilliga, Mallee Cliffs, Newhaven and Paruna in the next few years, with total population size predicted to exceed 10 000 individuals, which is more than 50% of the current global population size.

Bridled nailtail wallaby

The bridled nailtail wallaby (Fig. 28.3B) is a medium-sized (4–8 kg) macropod once widely distributed across inland eastern Australia from north Queensland to the Victorian mallee (Lundie-Jenkins and Lowry 2005). The species is well adapted to semi-arid conditions, capable of rapid population growth in favourable seasons. Populations of this species collapsed around the turn of the 20th century, such that, by 1937, this once common macropod was thought to be extinct. In 1973, a remnant population of bridled nailtail wallaby was discovered in central Queensland. This population, now in Taunton National Park, has been intensively managed; nevertheless, its persistence is tenuous due to predation and only 100–200 individuals remain (Woinarski *et al.* 2014).

A semi-wild population of the bridled nailtail wallaby was established at Scotia in 1998 by Earth Sanctuaries Ltd when 12 animals sourced from Queensland were released to a 250 ha breeding compound. Animals were provided with food and water and the population quickly expanded. Upon acquiring Scotia in 2004, AWC established a feral predator-free area of 8000 ha using conservation fencing. In 2004–2005, 162 bridled nailtail wallabies were released from the breeding population into Stage 1 (4000 ha). In 2008, an additional 267 individuals were released to Stage 2. By the end of 2011, following 2 years of good rainfall, Stages 1 and 2 supported more than 2000 animals; the population peaked at ~3600 animals in late 2012. The population subsequently declined to ~2300 following a run of dry years and is currently stable.

AWC's investment at Scotia has been critical in securing the future of the bridled nailtail wallaby. The reintroduction to Idalia NP in Queensland has failed (Bridled Nailtail Recovery Team *pers. comm.*) and the remnant population at Taunton has declined over the last 20 years. AWC also plans to reintroduce the species to fenced feral predator-free areas at Mallee Cliffs and Pilliga in NSW; the combined population size across these populations and Scotia is predicted to reach 6000. This is consistent with the Recovery Plan (Lundie-Jenkins and Lowry 2005), which identifies reintroductions as one of the main recovery actions. It remains difficult to re-establish the species outside fenced areas. Animals released outside the fence at Scotia in 2010 did not persist more than a few months (Hayward *et al.* 2012). Nonetheless, because of their moderately large size, the bridled nailtail wallaby remains a candidate for translocations 'outside the fence' if more effective strategies for feral predator control can be developed. AWC is currently conducting research on the ecology of feral cats and foxes at Scotia as part of a long-term strategy to establish a bridled nailtail wallaby population in the broader landscape.

Greater bilby

The greater bilby (Fig. 28.3C) is a small (1–2.5 kg) burrowing mammal once widely distributed across Australia (Pavey 2006). Bilbies are omnivorous, feeding on insects, bulbs and fungi dug from the ground. Like other small mammals, bilbies are susceptible to predation by foxes and feral cats. The lesser bilby (*Macrotis leucura*) is extinct, while the distribution of the greater bilby has contracted to south-west Queensland and part of the Northern Territory and Western Australia, with a population size estimated at 10 000 (Woinarski *et al.* 2014). An interim conservation plan recently developed for bilbies mostly dealt with the management of remnant populations (Bradley *et al.* 2015); nevertheless, reintroductions were recognised as one of the actions contributing to the species' conservation.

Bilbies have been reintroduced to AWC's Scotia, Yookamurra and Mt Gibson sanctuaries, as well as Arid Recovery (South Australia), Lorna Glen (Western Australia) and Thistle Island (WA). The total population size on AWC sanctuaries is currently 1350. The population at Scotia has increased from 400 in 2010 to around 1200 in 2016 (Fig. 28.4). AWC plans to reintroduce bilbies to feral predator-free areas at Pilliga, Mallee Cliffs and Newhaven in the next few years. The total population of bilbies on AWC sanctuaries is predicted to approach 8000 individuals, or 40% of the current global population size.

Research, biology, identification of key threats

The small to medium-sized mammals that are the focus of AWC's reintroduction program are primarily threatened by introduced predators. The re-establishment of reintroduced populations on islands and fenced 'mainland islands' free of cats and foxes has demonstrated that removal of introduced predators is often sufficient for the recovery of these species.

AWC conducts intensive research on various aspects of the ecology of reintroduced species (e.g. Hayward *et al.* 2015). At present, we are conducting research to better understand the ecology of feral predators, with the goal of suppressing cats and foxes sufficiently to allow reintroduction of at least some threatened mammals outside fenced areas. A technical challenge faced by all AWC's reintroduction programs is the need to optimise and maintain the genetic integrity of reintroduced populations; this is also the subject of active research by AWC.

Fig. 28.3. (A) Woylie, Karakamia wildlife sanctuary. (B) Bridled nailtail wallaby, Scotia wildlife sanctuary. (C) Greater bilby, Scotia wildlife sanctuary (photos: Wayne Lawler, AWC).

Planning and policy

AWC is one of the pioneers of reintroductions of threatened mammals to fenced feral predator-free areas in Australia. Our experience is that, where threatened species occur in jurisdictions where reintroductions are a well-established conservation measure (e.g. Western Australia, South Australia), recovery plans and government agencies are supportive of reintroduction programs.

Reintroductions to feral predator-free areas were identified in the Mammal Action Plan 2012 (Woinarski *et al.* 2014) as one of the few effective conservation actions undertaken for threatened mammals in Australia. This recognition, along with failure of many conservation programs reliant on management of feral predators in extensive landscapes, has helped in the broader acceptance of reintroductions as a conservation tool. This acceptance is demonstrated by the recent initiative of the New South Wales Government, where AWC has been contracted to establish feral predator-free areas and reintroduce regionally extinct mammals into two national parks. Reintroductions are currently one of the areas of research in the Threatened Species Recovery Hub: an initiative of the Australian Government.

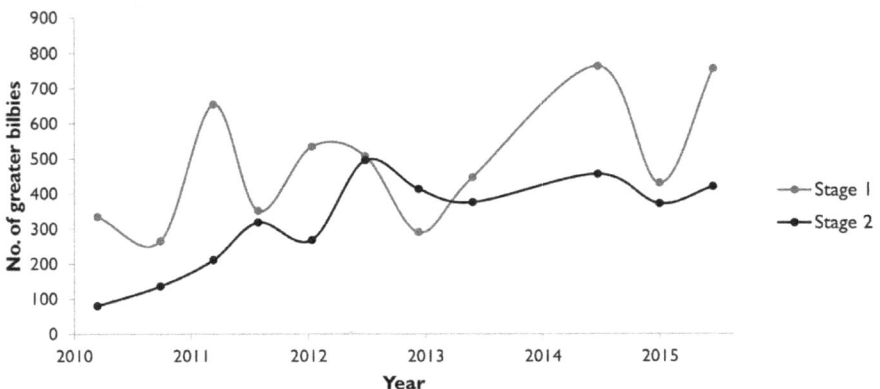

Fig. 28.4. Trends in numbers of Greater Bilby at AWC's Scotia Wildlife Sanctuary.

People, agencies, governance and accountability

As well as dealing directly with the primary threat to small to medium-sized mammals, the success of AWC's reintroduction program is due to several organisational factors. Most importantly, AWC is focused on delivering the highest possible ecological returns with available resources. Our reintroduction projects are delivered within a framework of accountability for outcomes (e.g. mammal populations) and expenditure. This promotes investor confidence, helping to secure resources. It also promotes innovation in relation to the establishment of large fenced areas and the removal of feral predators. Second, AWC invests substantially in 'in-house' scientific expertise. Currently, we have a staff of over 40 professional ecologists, who contribute to the design, implementation and monitoring of reintroduction projects. Third, our managers and many of our scientists are located on wildlife sanctuaries: an essential pre-requisite for successful execution of reintroduction programs. For example, our on-site managers are available to respond immediately to threats to the integrity of the fence.

AWC's reintroduction program has been supported by government agencies. For example, we have a good working relationship with the WA Department of Parks and Wildlife, which manages remnant populations of many threatened mammals and has substantial expertise in reintroductions; we are currently delivering two reintroduction projects in partnership with the New South Wales Government. Partner organisations such as Perth Zoo have assisted with captive breeding of source animals. Our experience is that organisations are generally willing to work together for the benefit of threatened species, although greater accountability within our sector for outcomes and costs would enhance the quality and productivity of collaborations.

Management

AWC is a not-for-profit organisation overseen by a board and accountable to supporters for conservation outcomes and expenditure. Our conservation priorities are informed by our expertise in ecology and land management, our scientific advisory network, and by extensive collaboration with external researchers and research organisations. Our reintroduction programs are informed by a structured decision-making prioritisation process.

Monitoring

Consistent with the need for accountability and adaptive management, AWC has devoted considerable resources to monitoring the outcomes of reintroduction projects. Typically, animals translocated to a new site are studied intensively, using telemetry and trapping, to determine survival, movement patterns and habitat preferences. For example, half the woylies reintroduced to Mt Gibson in 2015 and 2016 were fitted with radio-collars with a mortality sensor; the survival and location of collared animals was determined for a period of months. This information has been used to calculate home range, dispersal and habitat preferences and will, in turn, allow AWC to refine release protocols for future reintroductions of woylies to semi-arid habitats.

Once animals have established at a site, AWC ecologists use a variety of methods to estimate population size. Mark–recapture methodology, based on trapping animals, is used to estimate population size of readily trapped species such as burrowing bettong (*Bettongia lesueur*) and woylie. Spotlight transects, where animals are counted and their distance from a transect measured, are used to estimate population size of less readily trapped species such as bilby and bridled nailtail wallaby, using a strip-transect approach. A similar approach, based on diurnal counts, is used to estimate population size of the numbat.

In some cases, it is difficult to estimate population size robustly. For example, burrowing bettongs have become very common on AWC's Faure Island Wildlife Sanctuary, with an estimated 5000–10 000 individuals. The burrowing bettong readily enters traps, so that capturing sufficient individuals of other reintroduced species has become a challenge. AWC ecologists have trialled a range of alternative methods to estimate population size of these species, including spotlight transects, arrays of camera traps, ink-pads and scat counts. Genetic analysis of scats offers another potential solution.

Infrastructure

AWC's reintroduction programs to sites on mainland Australia are all dependent on a key piece of infrastructure: namely a feral predator-proof fence. The design of the fence has been refined over time and incorporates passive (high mesh fence with floppy top) and active (hot-wire) deterrents to incursions by feral predators. The ongoing success of reintroduction programs relies on maintenance of the integrity of the fence. AWC staff conduct fence patrols every 2–3 days to check for and fix potential problems, such as areas where animals have tried to burrow under the fence. The consequences of inadequate fence maintenance are demonstrated by the collapse of the bilby population at Currawinya National Park, Queensland, after a fence was breached by feral cats (Woinarski *et al.* 2014).

A second critical piece of infrastructure is accommodation for staff that allows ready access to the reintroduction site and fence. AWC has located workshops, office buildings and science laboratories at or near mainland reintroduction sites to facilitate staff working on site.

Money

AWC reintroduction projects have delivered an exceptional return on investment relative to other strategies: see, for example, the bridled nailtail wallaby case study. Establishing a feral predator-free fence involves a significant capital investment. Major reintroduction projects, such as AWC's Mt Gibson project – which involved construction of 43 km of fence and will entail reintroduction of 10 regionally extinct mammals – cost more than A$2 million to establish, with reintroductions costing in the order of A$100 000 per species, depending on complexity of sourcing, transporting and monitoring animals. However, running costs are modest (mainly staff time for checking the fence and monitoring reintroduced species) and the returns for many threatened mammal species have proven to be higher than alternative strategies. The Western Australian Government's 'Western Shield' Program, noting its objectives extend beyond threatened mammal recovery, costs several million dollars a year (Possingham *et al*. 2004). In some cases, reintroductions to areas from which feral predators have been eradicated are at present the only proven means of securing populations of small to medium-sized mammals susceptible to feral cats and foxes (Woinarski *et al*. 2014; Doherty *et al*. 2017).

AWC is primarily funded by donations from supporters, supplemented by government investment in some key projects. AWC also receives considerable in-kind support from volunteers – typically, several hundred people volunteer a year across AWC sanctuaries, from a few hours to several weeks at a time. Volunteers help in a variety of ways with reintroduction projects, from contributing to general sanctuary management to assisting with monitoring surveys.

The future

AWC's reintroduction program has substantially improved the conservation of 11 threatened mammals in a cost-effective manner. Expansion of the program in the next few years will increase the scope of the program, the number of species protected, the number of secure populations and global population sizes of protected species.

Conclusions

Reintroductions of threatened mammals to areas from which introduced predators have been eradicated has proven to be one of the few effective conservation measures available for small to medium-sized Australian mammals. The success of AWC's reintroduction program is due to several factors, including:

- Our approach deals effectively with the primary threat to small to medium-sized mammals.
- Projects are delivered in a framework that requires accountability for outcomes (increased mammal populations) and expenditure.
- AWC invests in 'in-house' scientific expertise, with tight integration of science and management.
- Staff are based on-site to manage the reintroduction program reliably.
- AWC is supported by a well-established funding model and sound governance structure.

Acknowledgements

AWC's founder, Martin Copley, had the vision to establish a mammal reintroduction project at Karakamia, WA. The success of this project led to AWC developing its current network of reintroduction projects. AWC thanks supporters for meeting the financial costs of these projects, the large number of AWC staff (present and past) and volunteers whose hard work made the projects a reality, and the numerous other stakeholders including government agencies, zoos and other private nature conservation organisations for providing source animals, expertise, approvals and other support. Thanks also to John Woinarski, Stephen Garnett and David Lindenmayer for reviewing the manuscript.

References

Bradley K, Lees C, Lundie-Jenkins G, Copley P, Paltridge R, Dziminski M, *et al*. (Eds) (2015) '2015 greater bilby conservation summit and interim conservation plan: an initiative of the

Save the Bilby Fund'. IUCN SSC Conservation Breeding Specialist Group, Apple Valley CA, USA.

Claridge A, Seebeck J, Rose R (2007) *Bettongs, Potoroos and the Musky Rat-kangaroo.* CSIRO Publishing, Melbourne.

Doherty TS, Dickman CR, Johnson CN, Legge SM, Ritchie EG, Woinarski JCZ (2017) Impacts and management of feral cats *Felis catus* in Australia. *Mammal Review* **47**, 83–97. doi:10.1111/mam.12080

Fleming PA, Anderson H, Prendergast A, Bretz MR, Valentine LE, Hardy GES (2014) Is the loss of Australian digging mammals contributing to a deterioration in ecosystem function? *Mammal Review* **44**, 94–108. doi:10.1111/mam.12014

Hayward MW, L'Hotellier F, O'Connor T, Ward-Fear G, Cathcart J, Cathcart T, *et al.* (2012) Reintroduction of bridled nailtail wallabies beyond fences at Scotia Sanctuary – Phase 1. *Proceedings of the Linnean Society of New South Wales* **134**, A27–A37.

Hayward MW, Poh ASL, Cathcart J, Churcher C, Bentley J, Herman K, Kemp L, *et al.* (2015) Numbat nirvana: conservation ecology of the endangered numbat (*Myrmecobius fasciatus*) (Marsupialia: Myrmecobiidae) reintroduced to Scotia and Yookamurra Sanctuaries, Australia. *Australian Journal of Zoology* **63**, 258–269. doi:10.1071/ZO15028

Lundie-Jenkins G, Lowry J (2005) 'Recovery plan for the bridled nailtail wallaby (*Onychogalea fraenata*) 2005–2009'. Report to the Department of Environment and Heritage, Canberra. Environmental Protection Agency/Queensland Parks and Wildlife Service, Brisbane.

Marlow NJ, Thomas ND, Williams AAE, MacMahon B, Lawson J, Hitchen Y, *et al.* (2015) Cats (*Felis catus*) are more abundant and are the dominant predator of woylies (*Bettongia penicillata*) after sustained fox (*Vulpes vulpes*) control. *Australian Journal of Zoology* **63**, 18–27. doi:10.1071/ZO14024

McKenzie NL, Burbidge AA, Baynes A, Brereton RN, Dickman CR, Gordon G, *et al.* (2007) Analysis of factors implicated in the recent decline of Australia's mammal fauna. *Journal of Biogeography* **34**, 597–611. doi:10.1111/j.1365-2699.2006.01639.x

Pavey C (2006) 'National Recovery Plan for the Greater Bilby *Macrotis lagotis*'. Northern Territory Department of Natural Resources, Environment and the Arts, Darwin.

Possingham H, Jarman P, Kearns A (2004) Independent review of Western Shield – February 2003. *Conservation Science Western Australia* **5**, 2–18.

Short J (2009) 'The characteristics and success of vertebrate translocations within Australia: a progress report to Department of Agriculture, Fisheries and Forestry'. Wildlife Research and Management, Kalamunda, Western Australia, <www.wildliferesearchmanagement.com.au/Final%20Report_0609.pdf>.

Wayne AF, Maxwell MA, Ward CG, Vellios CV, Wilson I, Wayne JC, *et al.* (2015) Sudden and rapid decline of the abundant marsupial *Bettongia penicillata* in Australia. *Oryx* **49**, 175–185. doi:10.1017/S0030605313000677

Woinarski JCZ, Burbidge AA, Harrison PL (2014) *The Action Plan for Australian Mammals 2012*. CSIRO Publishing, Melbourne.

Yeatman GJ, Groom CJ (2012) 'National Recovery Plan for the woylie *Bettongia penicillata*'. Wildlife Management Program No. 51. Department of Environment and Conservation, Perth, WA.

29

The contribution of captive breeding in zoos to the conservation of Australia's threatened fauna

Dan Harley, Peter R. Mawson, Liberty Olds, Michael McFadden and Carolyn Hogg

Summary

The problem

1. Significant numbers of species in Australia are threatened with extinction.
2. Twenty-eight per cent of recovery plans for EPBC-listed vertebrates specify a captive breeding action.
3. Several species are likely to have become extinct or are at risk of extinction in the near future without viable insurance populations in captivity.

Actions taken to manage the problem

1. Captive breeding can fulfil three main conservation roles: (1) provision of insurance populations; (2) breeding animals for release to the wild; and (3) research to assist conservation efforts.
2. Insurance populations have been established for several species at high risk of extinction.
3. Captive breeding and release programs have been undertaken for many threatened species.

Markers of success

1. Extinctions of some species have been averted and some wild populations are increasing due to releases from captivity.
2. There is more engagement of the community in threatened species conservation.

Reasons for success

1. Good long-term planning, with clear timeframes and goals.
2. Ongoing funding, because captive breeding programs are expensive.
3. Strong collaborative partnerships.
4. Strong integration between captive breeding programs and *in situ* recovery actions.

Introduction

The increasing risk of species' extinction in Australia and globally is well documented, and a growing number of species require urgent management intervention if they are to be secured in the wild. The urgent need to tackle biodiversity loss has shaped the evolution of zoos, and today many are dedicated to achieving conservation outcomes (Zimmermann *et al.* 2007).

From the 1980s and 1990s, there was an increased focus on coordinated recovery programs for threatened Australian fauna. Zoos have played active roles in many of these programs from the outset, with contributions spanning more than two decades in several cases, such as the western swamp tortoise (*Pseudemydura umbrina*), numbat (*Myrmecobius fasciatus*), orange-bellied parrot (*Neophema chrysogaster*), helmeted honeyeater (*Lichenostomus melanops cassidix*) and the mainland eastern barred bandicoot (*Perameles gunnii*). The desire to achieve tangible conservation outcomes has seen some Australian zoos place increasing emphasis on local species with dedicated recovery programs. The potential contribution that zoos can make to conservation is reflected in 55 of 196 (28%) national recovery plans for threatened Australian vertebrates specifying at least one action relating to captive breeding, including 43% of mammal recovery plans.

Captive breeding undertaken by zoos for Australian threatened species has attempted to meet several objectives. A primary objective is to establish and maintain insurance populations for species that have become extinct in the wild or are in imminent danger of becoming extinct, such as the southern corroboree frog (*Pseudophryne corroboree*) (Fig. 29.1) and orange-bellied parrot. Insurance populations can 'buy time' while we await breakthroughs for the control of threats operating in the wild (Conway 2011). To offer recovery potential, captive insurance populations must have the capacity to re-populate the wild, and so key attributes include adequate genetic diversity and behavioural fitness of

Fig. 29.1. Corroboree frogs have been central to improving strategies to manage captive populations of threatened frogs in Australia (photo: Damian Goodall, Melbourne Zoo).

individuals. Other objectives for captive breeding programs include releases to the wild to augment existing populations or establish new populations, such as the regent honeyeater (*Anthochaera phrygia*), and supporting research into the development of improved or novel techniques to assist the conservation of species (e.g. Taggart *et al.* 2005; Brannelly *et al.* 2016). The ultimate measure of success for captive breeding programs for threatened species is prevention of extinction and/or recovery in the wild. Hence, the ultimate goal for both *in-situ* and *ex-situ* actions is the same – the restoration of a threatened species to a secure state in the wild. The most effective captive breeding programs are those strongly integrated with actions underway in the wild (Pritchard *et al.* 2012).

Another critical role that zoos play in conservation is in increasing public awareness about Australia's threatened species, particularly given that zoos attract millions of visitors each year and can provide unique opportunities to connect people with rarely seen species. Increasingly, zoos are drawing on social science and using behaviour change strategies to generate tangible conservation outcomes from their interactions with visitors (Lowry 2009).

There are several well-known and celebrated captive breeding success stories globally, measured either by prevention of extinction and/or recovery of wild populations: examples include the golden lion tamarin (*Leontopithecus rosalia*), black-footed ferret (*Mustela nigripes*), Przewalski's horse (*Equus ferus przewalskii*), Arabian oryx (*Oryx leucoryx*) and American bison (*Bison bison*) (WAZA 2012). Australia has its own captive breeding success stories, including the western swamp tortoise (see Chapter 23), southern corroboree frog, numbat and chuditch (*Dasyurus geoffroii*); however, these Australian examples typically have not received comparable national or international profile.

This chapter presents a summary of the directions for zoo-based captive breeding programs in Australia. Consideration is restricted to terrestrial vertebrates, although at least one invertebrate warrants mention – the Lord Howe Island phasmid (*Dryococelus australis*) (see Chapter 20). This chapter is predominantly focused on captive breeding programs led by Australia's larger zoos, but smaller institutions have had, and continue to make, an important contribution for several species.

Insurance – what would we have lost without zoos?

One of the major roles that zoos have played to support threatened species has been the establishment of insurance populations in captivity (Bowkett 2009). In practice, this can be challenging in terms of the numbers of individuals required to provide genuine insurance, the need for reliable outlets for young born (given limits to holding facilities), and the need to retain behavioural fitness and genetic diversity over time. Despite these challenges, and some failed attempts, such as the sharp-snouted day frog (*Taudactylus acutirostris*) and the Christmas Island forest skink (*Emoia nativitatis*) (Woinarski *et al.* 2017), Australian zoos play a critical insurance role for several species. The largest captive insurance populations for Australian threatened vertebrates include: Tasmanian devil (*Sarcophilus harrisii*) (Fig. 29.2), with >500 individuals managed across 37 captive facilities; orange-bellied parrot, with >300 birds held across 10 captive facilities; blue-tailed skink (*Cryptoblepharus egeriae*); and Lister's gecko (*Lepidodactylus listeri*), with >950 adults and >550 adults held across two captive facilities (Andrew *et al.* 2016); and southern corroboree frog and northern corroboree frog (*Pseudophryne pengilleyi*), with >400 adults and >460 adults, respectively, held across multiple captive facilities. Full-scale insurance, comprising hundreds of captive individuals, is generally more feasible for amphibians and reptiles where large numbers can often be housed at lower cost than is possible for birds and mammals.

Fig. 29.2. Captive populations of Tasmanian devils have been critical, not just as insurance populations, but as centres of research into disease management (photo: Amie Hindson, Healesville Sanctuary).

Insurance populations have been critical for preventing the extinction of the eastern barred bandicoot (mainland subspecies), helmeted honeyeater and western swamp tortoise. Other species that would be at imminent danger of extinction were it not for captive insurance populations include the orange-bellied parrot, brush-tailed rock-wallaby (Southern Evolutionarily Significant Unit) (*Petrogale penicillata*), Bellinger River turtle (*Myuchelys georgesi*), blue-tailed skink, Lister's gecko, southern corroboree frog, northern corroboree frog and yellow-spotted bell frog (*Litoria castanea*). New captive populations aiming to provide insurance against extinction are currently being established for the Baw Baw frog *Philoria frosti*) (Fig. 29.3), plains-wanderer (*Pedionomus torquatus*) and lowland Leadbeater's possum (*Gymnobelideus leadbeateri*). Future recovery of each of these species is likely to be reliant upon successful captive breeding and release programs.

Releases to the wild – what has been restored?

Zoos have been highly successful in breeding many of Australia's threatened species (Table 29.1). Key measures of success include the number of captive-bred animals available for release, the genetic and behavioural attributes of those individuals, and their post-release establishment success. In a review of 267 vertebrate translocations in Australia, Short (2009) found that 92 translocations (34%) used captive-bred animals.

Some reviews of translocation programs have identified that success rates tend to be higher with use of wild-born individuals rather than captive-bred individuals (Fischer and Lindenmayer 2000). However, a review of Western Australian cases concluded that the success rate for 39 conservation translocations involving exclusively captive-bred animals

Fig. 29.3. Owing to considerable husbandry work with frogs, saving a species such as the Baw Baw frog is much more feasible now than 20 years ago (photo: Damian Goodall, Melbourne Zoo).

(59%) was higher than that for 120 translocations using animals sourced from the wild (27%) (Morris *et al.* 2015). Short (2009) found little difference in the success rate for bird and mammal translocations sourced from captive versus wild stock.

The restoration of populations through wild-to-wild translocations may often be preferable to captive breeding because wild adaptations are more readily retained. However, in some instances, wild populations have declined to such an extent that the sustained removal of significant numbers of animals from a wild population for translocation is not desirable (Todd *et al.* 2002), and captive breeding may provide the only pathway to support translocations. In these situations, certain requirements must still be met for captive breeding to be a successful mechanism towards re-establishment of a viable wild population. These include sourcing an adequate number of founders from the wild that are representative of existing genetic diversity, rapid breeding and expansion of the captive population, a high percentage of founders producing young, production of sufficient young to meet the release objectives, and a high rate of survival for the captive-bred animals that are released to the wild. Each of these steps can present significant challenges.

Improving the fitness of captive-bred animals (e.g. modifying the behaviours that influence post-release survival and reproduction) is a major area for research in zoos (Reading *et al.* 2013). The helmeted honeyeater release program provides a good example. The survival rate for captive-bred birds released to the wild after 12 months had been ~40% before a pre-release predator awareness training regime, after which survival rates increased to ~80% (Helmeted Honeyeater Recovery Team *pers. comm.*). This improved result is likely to be attributable to several concurrent changes to the release strategy, but illustrates the benefits that pre-release training may provide.

Table 29.1. Threatened Australian terrestrial vertebrates with captive breeding conservation programs.

Data were not available for some species and some smaller breeding programs are not listed. The number of years refers to the duration in which zoos were involved in captive breeding. The number of animals released refers to the release of individuals specifically to meet conservation objectives, and excludes small numbers that may be released for other purposes (e.g. following rehabilitation). Multiple captive facilities refer to situations where more than one facility makes a significant captive breeding contribution that extends beyond holding a very small number of individuals. Table cells have been left blank where the information was not available.

Common name	Scientific name	EPBC status	Year commenced	No. years	Status	Multiple captive facilities	Purpose	Maximum no. adult individuals	Current no. adult individuals	No. animals bred	No. of release years	No. animals released	No. populations augmented	No. new populations or subpopulations*
Dibbler	*Parantechinus apicalis*	EN	1997	18	Ongoing	No	Release	24	13	838	18	885	0	1 (4)
Tasmanian devil	*Sarcophilus harrisii*	EN	2006	10	Ongoing	Yes	Insurance	550	500	699	4	130	2	2
Chuditch	*Dasyurus geoffroii*	VU	1992	8	Concluded	No	Release	30	0	315	9	315	0	3 (2)
Numbat	*Myrmecobius fasciatus*	VU	1984	23	Ongoing	No	Release	31	31	237	22	206	2	2 (4)
Leadbeater's possum (lowland ESU)	*Gymnobelideus leadbeateri*	CR	2012	5	Ongoing	No	Insurance and release	16	14	0	0	0	0	0
Mountain Pygmy possum (Mt Buller ESU)	*Burramys parvus*	EN	2006	8	Concluded	No	Release and research	99	35	88	1	13	1	0
Eastern barred bandicoot	*Perameles gunnii*	EN	1988	29	Ongoing	Yes	Insurance and release	68	58	896	21	644	2	3 (6)
Greater bilby	*Macrotis lagotis*	VU	1979	38	Ongoing	Yes	Release	130	112	1129	24	491	8	0
Quokka	*Setonix brachyurus*	VU	2009	7	Ongoing	Yes	Insurance	72	72	40	0	0	0	0
Tammar wallaby	*Macropus eugenii eugenii*	EX	2003	13	Concluded	Yes	Release and insurance	143	43	>294	5	138	0	1
Brush-tailed rock-wallaby (southern ESU)	*Petrogale penicillata*	VU	1996	20	Ongoing	Yes	Insurance and release	85	85	301	5	39	0	0 (1)
Brush-tailed rock-wallaby (central ESU)	*Petrogale penicillata*	VU	2004	13	Ongoing	Yes	Insurance and release	71	71	176	16	70	5	0
Yellow-footed rock-wallaby (SA)	*Petrogale xanthopus xanthopus*	VU	1980	37	Ongoing	Yes	Release	170	170	>1797	2	14	0	1

Common name	Scientific name	EPBC status	Year commenced	No. years	Status	Multiple captive facilities	Purpose	Maximum no. adult individuals	Current no. adult individuals	No. animals bred	No. of release years	No. animals released	No. populations augmented	No. new populations or subpopulations*
Warru (MacDonnell Ranges)	Petrogale lateralis	VU	2006	10	Ongoing	No	Release and insurance	28	6	16	4	22	1	1
Shark Bay mouse	Pseudomys fieldi	VU	1997	5	Concluded	No	Release	21	0	304	6	347	0	2 (1)
New Holland mouse (Vic)	Pseudomys novaehollandiae	VU	1996	9	Concluded	Yes	Release	86	0	166	3	33	1	0
New Holland mouse (ACT)	Pseudomys novaehollandiae	VU	2010	5	Concluded	No	Release	~70	0	100–150	2	~100	0	1
Greater stick-nest rat	Leporillus conditor	VU	1985	15	Concluded	Yes	Release	61	59		10	426		
Orange-bellied parrot	Neophema chrysogaster	CR	1986	31	Ongoing	Yes	Insurance and release	350	347	2060	18	543	1	0
Western ground parrot	Pezoporus flaviventris	CR	2014	1	Ongoing	No	Insurance and release	9	4	0	0	0	0	0
Helmeted honeyeater	Lichenostomus melanops cassidix	CR	1989	28	Ongoing	No	Insurance and release	34	34	371	16	253	0	2 (4)
Regent honeyeater	Anthochaera phrygia	CR	1995	22	Ongoing	No	Release	79	57	459	6	196	1	0
Malleefowl	Leipoa ocellata	VU	1988	1988	Concluded	Yes	Release	51	25	929	14	556	3	0
Black-eared miner	Manorina melanotis	EN	1995	9	Concluded	Yes	Insurance and research	50	0	≥36	1	45	1	0
Western Swamp tortoise	Pseudemydura umbrina	CR	1987	29	Ongoing	Yes	Insurance and release	50	50	985	22	723	2	2
Lancelin Island Skink	Ctenotus lancelini	VU	1995	3	Concluded	No	Release	9	1	140	3	152	0	1
Blue-tailed skink	Cryptoblepharus egeriae	CR	2011	5	Ongoing	Yes	Insurance	645	645	>900	0	0	0	0
Lister's gecko	Lepidodactylus listeri	CR	2011	5	Ongoing	Yes	Insurance	538	538	>500	0	0	0	0
Bellinger River turtle	Myuchelys georgesi	CR	2015	1	Ongoing	No	Insurance	16	16	0	0	0	0	0

Common name	Scientific name	EPBC status	Year commenced	No. years	Status	Multiple captive facilities	Purpose	Maximum no. adult individuals	Current no. adult individuals	No. animals bred	No. of release years	No. animals released	No. populations augmented	No. new populations or subpopulations*
Southern corroboree frog	Pseudophryne corroboree	CR	1997	20	Ongoing	Yes	Insurance and release	410	410	>7500	7	7021	0	5 (1)
Northern corroboree frog	Pseudophryne pengilleyi	CR	2003	13	Ongoing	Yes	Insurance and release	465	465	>3500	6	3734	4	0
Baw Baw frog	Philoria frosti	EN	2011	6	Ongoing	Yes	Insurance	1000	380	0	0	0	0	0
Spotted tree frog	Litoria spenceri	EN	2001	12	Concluded	Yes	Research, insurance and release	>1000	~220	≥1700	8	~1900	1	1 (2)
Booroolong frog	Litoria booroolongensis	EN	2007	4	Concluded	No	Insurance and release	176	26	>800	1	608	1	0
Green and golden bell frog	Litoria aurea	VU	1994	20	Concluded	No	Release	58	67	>27 570	11	27 332	2	0 (3)
Yellow-spotted bell frog	Litoria castanea	EN	2010	6	Ongoing	No	Insurance	14	7	0	0	0	0	0
Alpine tree frog	Litoria verreuaxii alpina	VU	2009	3	Concluded	Yes	Release	1340	19	0	1	1241	2	0 (2)
Sunset frog	Spicospina flammocaerulea	EN	2011	1	Concluded	No	Release	31	0	251	1	281	0	1
White-bellied frog	Geocrinia alba	EN	2009	7	Ongoing	No	Release	12	12	17	7	466	3	0
Orange-bellied frog	Geocrinia vitellina	VU	2010	5	Ongoing	No	Release	11	11	0	6	316	6	1

*Values in brackets represent failed attempts.

Releases to the wild of captive-bred animals have occurred for at least 25 species of threatened terrestrial vertebrate in Australia (Table 29.1). In 2015 alone, individuals of 13 threatened vertebrate species were released to the wild from zoo-based captive breeding programs in Australia. Notable examples have included the chuditch, mainland eastern barred bandicoot, tammar wallaby (*Macropus eugenii*), Shark Bay mouse (*Pseudomys fieldi*), helmeted honeyeater and western swamp tortoise. There have also been several failed attempts to stabilise declining wild populations using captive-bred animals. Ten years of releases of captive-bred orange-bellied parrots to Birchs Inlet (DELWP 2016), and 12 years of releases of captive-bred helmeted honeyeaters to Bunyip State Park (Helmeted Honeyeater Recovery Team *pers. comm.*) were both ultimately terminated without the establishment of self-sustaining populations. Some failures are to be expected, but if releases are well planned, supported by adequate monitoring and are conducted in an investigative manner, they can generate significant knowledge essential for improving subsequent releases. The process for successful recovery usually comprises a series of incremental steps, and robust programs have the ability to contend with setbacks, and to adapt.

Many successful translocations of captive-bred animals have been to areas in which primary threats are absent or effectively controlled, notably to islands and areas protected by predator or disease exclusion fences. Examples include the Tasmanian devil, eastern barred bandicoot, tammar wallaby, bridled nailtail wallaby (*Onychogalea fraenata*), mala (*Lagorchestes hirsutus*), numbat, burrowing bettong (*Bettongia lesueur*), brush-tailed bettong (*Bettongia penicillata*), eastern bettong (*Bettongia gaimardi*), greater stick-nest rat (*Leporillus conditor*), Lancelin Island skink (*Ctenotus lancelini*) and southern corroboree frog.

Planning and policy

National recovery plans guide strategic decision making on captive breeding for Australia's threatened species. Recovery planning aims to integrate *in-situ* and *ex-situ* actions, where appropriate, as a package of actions necessary to meet species recovery objectives.

Globally, the IUCN's Species Survival Commission (SSC) has developed guidelines that describe the key questions that should be considered before establishing a captive breeding program (IUCN/SSC 2014). The IUCN/SSC's Conservation Breeding Specialist Group has developed a holistic strategy, referred to as the 'One Plan Approach', to guide captive breeding for conservation (Byers *et al.* 2013). In the Australasian region, the Zoo and Aquarium Association (ZAA, Australasia) is the peak industry body, and provides coordination for the management of captive populations through the Australasian Species Management Program (Hogg 2013).

Before the establishment of a captive breeding program, it is important to clearly define the recovery model being applied and to ensure that the scale is matched to the objectives. For instance, the release of captive-bred helmeted honeyeaters from Healesville Sanctuary during the early 2000s played an important role in preventing the extinction of that taxon, but the size and survival rate of release groups was insufficient to promote population growth. Some early captive breeding programs lacked criteria for evaluating their success or failure, and greater emphasis is now placed on specifying clear objectives and targets.

Strong partnerships – the essential foundation

Strong partnerships underpin the success of captive breeding programs. Australian zoos work closely to coordinate the management of captive populations spread across many facilities. Translocation of animals between the wild and captivity requires coordination among ecologists, veterinarians, animal keepers, regulatory authorities, land managers (and often volunteers). Many facets of programs are reliant upon significant support from non-zoo staff. For instance, the release of captive-bred yellow-footed rock-wallabies to Aroona Sanctuary (Flinders Ranges) has seen complementary *in-situ* management efforts spanning 20 years (see Chapter 15).

Funding and resources

Captive-breeding is expensive, particularly when operating costs are considered over the multi-decadal scale necessary for some recovery programs. Many Australian captive-breeding programs require >A$200 000 annually for their operation, and some programs greatly exceed this figure (e.g. the annual cost of maintaining the captive insurance population for the Tasmanian Devil is ~A$1 200 000). Funding to support captive breeding programs has come from a range of sources, including the Australian and state governments, sponsorship by the corporate sector and philanthropy. Much of the captive breeding has also been self-funded by the zoos leading the work.

'Caring for country' – engaging and inspiring the public

The conservation community has yet to capture the hearts and minds of Australians at the scale required to halt the loss of biodiversity. No government agency has access to the public in the manner that zoos do, attracting millions of visitors each year. Hence, zoos have an opportunity to be the shopfront of the conservation sector.

The visibility and accessibility of zoo captive breeding programs provides a unique opportunity to engage and inspire the wider community. The focus of zoos is strongly tied to particular species, yet 'place' is an integral part of managing our environment sustainably (including preventing species loss). Contemporary Australia needs to learn from our traditional owners' 'connection with country', and find its connection with the natural environment. Zoos can play an important role in supporting this, particularly for increasingly urbanised human populations. Greater community support, and a major focus on species and place, is essential for achieving the vision of healthy wildlife populations secure in healthy landscapes.

Challenges and lessons learned

The many challenges facing captive breeding programs have long been well known (Snyder *et al.* 1996). These include: the number and quality of founders sourced from the wild; knowledge gaps surrounding a species' ecology and/or captive husbandry; genetic and demographic factors that affect long-term viability of the captive program; ensuring captive animals retain behaviours necessary for survival in the wild; constraints on enclosure capacity; small size of release groups for many threatened species (particularly birds and mammals); survival rates for captive-bred animals released to the wild; and the resources required to sustain captive breeding programs over numerous years.

Many lessons have been learned in facing these challenges through decades of successful (and unsuccessful) zoo-based captive breeding programs (Zimmermann *et al.* 2007). Early intervention with captive breeding – before wild populations have become critically small in size – greatly improves the likelihood of success. There is now wide recognition of the need for better planning and target setting. For any captive breeding program, key measures of success relate to the number of individuals bred, the number available for release and the retention of genetic diversity. Monitoring of release outcomes is also essential to evaluate the impact of a captive breeding program.

Captive breeding is not a pathway that produces rapid results. For example, the establishment of a captive breeding program for the plains-wanderer has taken more than 2 years of discussion and planning before sourcing founders from the wild. For species with complex life histories, or that inhabit specialised environments (e.g. alpine species), it may take several years to develop appropriate husbandry techniques (e.g. it took 13 years to achieve reliable breeding success for the southern corroboree frog in captivity). It is important that a high proportion of captive founders breed successfully, and that representation in subsequent generations is not skewed towards a few individuals (Ballou et al. 2010). Minimising the number of generations in captivity is also an important goal (Frankham 2008). Furthermore, it is important to recognise that the creation of a captive breeding population does not reduce the urgency required to alleviate threats in the wild.

It may take several years to increase the size of a captive population so that it is capable of supporting releases to the wild. Knowledge gaps surrounding effective release methods may reduce the impact of initial releases. Hence, it can easily take 6 years or more before a captive breeding program is in a position to deliver tangible benefits to wild populations. Factoring in time required for threat alleviation in the wild, it is reasonable to expect that most captive breeding programs (for conservation) would run for something in the order of 20–30 years.

When to initiate captive breeding?

Captive-breeding programs for threatened species have typically been triggered by an increase in extinction risk to a species in the wild and the presence of an active recovery program. However, deciding **when to intervene** with captive breeding is often not straightforward (IUCN/SSC 2014).

In practice, early intervention rarely occurs, because recovery programs often retain some optimism that it may be possible to reduce threats in the wild, and focus resources towards this outcome. The high costs associated with captive breeding also restrict opportunities for early intervention. Yet a significant problem emerges with delaying, because the opportunity to collect genetically diverse founders with high levels of fitness may be lost. There are some notable Australian examples where action was left too late, such as the Christmas Island pipistrelle (*Pipistrellus murrayi*), where the delays in implementing an emergency rescue plan involving captive breeding may have contributed to the extinction (Martin et al. 2012; Woinarski et al. 2017). In contrast, although the threats in the wild remain for the Tasmanian devil, orange-bellied parrot and both corroboree frogs, because strong captive insurance populations are in place, there is little likelihood that these species will be lost.

Future directions

Improving the 'fitness' of captive-bred individuals is a major theme in conservation breeding programs. This can be supported in a variety of ways, ranging from husbandry methods through to the manipulation of the captive environment (Reading et al. 2013). Mate choice, enrichment techniques to promote behaviours important for survival, research into individual variation and personality traits within populations are all major research themes.

There is also increasing focus on the genetic management of captive populations (e.g. Hogg et al. 2015). For example, the helmeted honeyeater is managed as part of a 'captive-wild metapopulation', whereby the populations in captivity and the wild are linked through annual translocations in both directions (see Chapter 24). The operational resources to effectively implement such programs are significant. For several small populations, **genetic rescue** is being investigated via managed levels of gene-pool mixing among populations to broaden their genetic base without swamping unique local diversity. Although this may involve wild-to-wild translocations in some instances, such as the mountain pygmy possum (*Burramys parvus*), for some species it may be desirable to undertake gene-pool mixing in captivity where matings and rates of introgression can be closely managed. Cryopreservation of reproductive material (e.g. sperm, eggs) offers a way of storing genetic material from threatened species for the future, and in so doing may offer a means to increase the size of founder groups.

Zoos provide a unique opportunity to draw on veterinary expertise that can be applied to the management (and risk analysis) of wildlife disease. Disease considerations form a key part of the captive management strategies for the Tasmanian devil and orange-bellied parrot. Disease risk assessments associated with translocations have recently been completed for the regent honeyeater, plains-wanderer, eastern barred bandicoot and Bellinger River turtle. There is also increasing research focus on stress levels and other heath parameters for animals held in captivity (e.g. Hogan et al. 2012).

Conclusions

The need to increase conservation impact has emerged as one of the strongest drivers of change across the zoo sector. Given the considerable resources required, major captive-breeding programs will continue to be focused on species at greatest risk of extinction, but early intervention remains a key ingredient underpinning likelihood of success. The most successful captive breeding programs are strongly integrated with *in-situ* recovery actions, framed within long-term recovery plans, work to clear timeframes, have clear goals, attempt to apply effective adaptive management and embrace new technologies. There are many factors that can setback or delay the progress of captive breeding programs, and it is important to have realistic expectations. Typically, they do not yield results for several years, sometimes longer.

In Australia, the number of wild populations that have been established or restored using captive-bred individuals is relatively modest (see Table 29.1). This is not surprising given the difficulty in effectively mitigating threatening processes impacting wild populations. The conservation impact of captive-bred individuals has been greatest for releases to islands and sites protected by predator exclusion fences. Although population recovery within historic ranges remains elusive, some recovery programs have been highly successful in preventing extinction and retaining the potential for future recovery.

Captive breeding is not an easy solution to conservation problems. It is neither fast, nor cheap. Yet for some species, such as the southern corroboree frog and orange-bellied parrot, captive breeding is all that stands between survival and extinction. Despite the many challenges, there are notable success stories in Australia that demonstrate the conservation value that captive breeding can provide to support the ultimate goal – **recovery of secure, self-sustaining wildlife populations in the wild**.

Acknowledgements

The success of captive breeding programs is attributable to the passion, dedication and commitment of numerous animal keepers, veterinarians and operational/support staff based at Australian zoos. Several people contributed information for Table 29.1, including Lauren Florisson, Camille Goldstone-Henry, Deon Gilbert, Marissa Parrot, Chris Hartnett, Deborah Ashworth, Rohan Clarke, Dave Hunter, Matt West, Kym Birgan, Peter Myroniuk, Barbara Wilson, Daniel Gowland, Diana Fisher and ZAA studbook keepers. Thanks are extended to funding bodies, both government and philanthropic, who have generously supported captive breeding programs over many years.

References

Andrew P, Cogger H, Driscoll D, Flakus S, Harlow P, Maple D, Misso M, *et al.* (2016) Somewhat saved: a captive breeding program for two endemic Christmas Island lizard species, now extinct in the wild. *Oryx*, (online early) 1–4. doi:10.1017/S0030605316001071

Ballou JD, Lees CM, Faust LJ, Long S, Lynch C, Bingaman-Lackey L, *et al.* (2010) Demographic and genetic management of captive populations. In *Wild Mammals in Captivity*. (Eds DG Kleiman, KV Thompson and CK Baer) pp. 219–252. University of Chicago, Chicago IL, USA.

Bowkett AE (2009) Recent captive-breeding proposals and the return of the Ark concept to global species conservation. *Conservation Biology* **23**, 773–776. doi:10.1111/j.1523-1739.2008.01157.x

Brannelly LA, Hunter DA, Skerratt LF, Scheele BC, Lenger D, McFadden MS, *et al.* (2016) Chytrid infection and post-release fitness in the reintroduction of an endangered alpine tree frog. *Animal Conservation* **19**, 153–162. doi:10.1111/acv.12230

Byers O, Lees C, Wilcken J, Schwitzer C (2013) The One Plan Approach: the philosophy and implementation of CBSG's approach to integrated species conservation planning. *WAZA Magazine* **14**, 2–5.

Conway WG (2011) Buying time for wild animals with zoos. *Zoo Biology* **30**, 1–8.

DELWP (2016) 'National Recovery Plan for the Orange-bellied Parrot *Neophema chrysogaster*'. Department of Environment, Land, Water and Planning, Canberra.

Fischer J, Lindenmayer DB (2000) An assessment of the published results of animal relocations. *Biological Conservation* **96**, 1–11. doi:10.1016/S0006-3207(00)00048-3

Frankham R (2008) Genetic adaptation to captivity in species conservation programs. *Molecular Ecology* **17**, 325–333. doi:10.1111/j.1365-294X.2007.03399.x

Hogan LA, Lisle AT, Johnston SD, Robertson H (2012) Non-invasive assessment of stress in captive numbats, *Myrmecobius fasciatus* (Mammalia: Marsupialia), using faecal cortisol measurement. *General and Comparative Endocrinology* **179**, 376–383. doi:10.1016/j.ygcen.2012.09.020

Hogg CJ (2013) Preserving Australian native fauna: zoo-based breeding programs as part of a more unified strategic approach. *Australian Journal of Zoology* **61**, 101–108. doi:10.1071/ZO13014

Hogg CJ, Ivy JA, Srb C, Hockley J, Lees C, Hibbard C, Jones M (2015) Influence of genetic provenance and birth origin on productivity of the Tasmanian devil insurance population. *Conservation Genetics* **16**, 1465–1473. doi:10.1007/s10592-015-0754-9

IUCN/SSC (2014) 'Guidelines on the use of *ex situ* management for species conservation'. Version 2.0. IUCN Species Survival Commission, Gland, Switzerland.

Lowry R (2009) Visitor based conservation campaigns. *International Zoo Educators Journal* **45**, 11–14.

Martin TG, Nally S, Burbidge AA, Arnall S, Garnett ST, Haywood MW, et al. (2012) Acting fast helps avoid extinction. *Conservation Letters* **5**, 274–280. doi:10.1111/j.1755-263X.2012.00239.x

Morris K, Page M, Kay R, Renwick J, Desmond A, Comer S, et al. (2015) Forty years of fauna translocations in Western Australia: lessons learned. In *Advances in Reintroduction Biology of Australian and New Zealand Fauna*. (Eds DP Armstrong, MW Hayward, D Moro and PJ Seddon) pp. 217–235. CSIRO Publishing, Melbourne.

Pritchard DJ, Fa JE, Oldfield S, Harrop SR (2012) Bring the captive closer to the wild: redefining the role of ex situ conservation. *Oryx* **46**, 18–23. doi:10.1017/S0030605310001766

Reading RP, Miller B, Shepherdson D (2013) The value of enrichment to reintroduction success. *Zoo Biology* **32**, 332–341. doi:10.1002/zoo.21054

Short J (2009) 'The characteristics and success of vertebrate translocations with Australia'. Department of Agriculture, Fisheries and Forestry, Canberra.

Snyder NFR, Derrickson SR, Beissinger SR, Wiley JW, Smith TB, Toone WD, et al. (1996) Limitations of captive breeding in endangered species recovery. *Conservation Biology* **10**, 338–348. doi:10.1046/j.1523-1739.1996.10020338.x

Taggart DA, Schultz D, White C, Whitehead P, Underwood G, Phillips K (2005) Cross-fostering, growth and reproductive studies in the brush-tailed rock-wallaby, *Petrogale penicillata* (Marsupialia: Macropodidae): efforts to accelerate breeding in a threatened marsupial species. *Australian Journal of Zoology* **53**, 313–323. doi:10.1071/ZO05002

Todd CR, Jenkins S, Bearlin AR (2002) Lessons about extinction and translocation: models for eastern barred bandicoots (*Perameles gunnii*) at Woodlands Historic Park, Victoria, Australia. *Biological Conservation* **106**, 211–223. doi:10.1016/S0006-3207(01)00247-6

WAZA (2012) Fighting extinction. *WAZA Magazine* Vol. 13.

Woinarski JCZ, Garnett ST, Legge SM, Lindenmayer DB (2017) The contribution of policy, law, management, research, and advocacy failings to the recent extinctions of three Australian vertebrate species. *Conservation Biology* **31**, 13–23. doi:10.1111/cobi.12852

Zimmermann A, Hatchwell M, Dickie L, West C (2007) *Zoos in the 21st Century: Catalysts for Conservation?* Cambridge University Press, Cambridge, UK.

30

Mobilising resources for the recovery of threatened species

Samantha Vine, Linda Bell and Allan Williams

Summary

The problem
1. Recovery of threatened species may often require considerable resources, but resources may often – at least initially – be limited.
2. Some potential funding sources may show little appreciation of conservation values relative to other values and consequently show little disposition to contribute resources.

Actions taken to manage the problem
1. Three case studies show that it is critical to recognise that there may be a wide range of resources ('raw ingredients') that can (and may need to) be contributed to threatened species recovery, and to think strategically and creatively about resource provision, including what may motivate investment by different sectors.

Markers of success
1. One case study demonstrates that state-based funding for recovery actions increased 10-fold
2. Two other case studies demonstrate that appropriate mix of committed individuals and problem-solving capability catalysed access to hitherto unavailable funding.

Reasons for success
1. Government was persuaded to invest long-term and substantial funding because a sound business case was made that highlighted return on investment, and strategic prioritisation.
2. A core team of committed individuals with practical problem-solving ability can catalyse change and investment in conservation.
3. Especially when working with landholders, threatened species management is most likely to succeed when it is flexible and adaptive.

Introduction

There are many threatened species in Australia. We have, in many instances, estimated how many individuals of each threatened species are left, where they occur, and what factors have led to the decline of these taxa. We also have some great ideas on what we can do to bring these species back from the brink. Generally, we also should be able to at least estimate how much it will cost to recover threatened species (e.g. McCarthy et al. 2008). Now we just need the wherewithal to do it. How difficult can it be to secure support for such a good cause?

The 'good cause' is the first hurdle. **We** know it is, but do the people holding the potential purse strings know it? Do they even have the same values in deciding it is a worthy enough cause to support financially? Complications arise in assigning species values, which may be influenced by the ecological niche it occupies, cultural value or scientific significance. However, it may also be that the species is 'cute' enough to capture public support, or that benevolence towards it could tap a community wellspring of goodwill or serve to influence other matters unknown to its proponents. The connection between resourcing the recovery of a species and the support to provide those resources may often be more emotional than logical.

Is it a good enough cause? Arguably, humanity's actions are mass-producing threatened species and extinctions. There is now a buyer's market of good causes for which a rational approach may assist in prioritisation, but also for which a willingness to support may not necessarily be logical.

Furthermore, although there are some basic principles to recovering threatened species, the range of circumstances, needs and actions is breathtakingly broad, even at the individual species level. These factors directly affect costs and may determine the level to which it is necessary to 'pitch' a case for resources – and what that 'pitch' will be.

Mobilising resources

Despite the significant time that has been invested in listing and drafting the 445 or so national recovery plans (DoE 2017), the implementation of recovery plans has been rated as poor nationally (DEWHA 2009). Resources currently allocated to protect most of Australia's threatened species are grossly inadequate for the task of preventing extinction and improving the conservation status of species most in need (Nimmo et al. 2016).

Resource mobilisation is a term most often described as a sociological theory of understanding social movements. Here we use the term in a broader sense, meaning to make available resources. For example, The World Bank defines 'resource mobilisation' as the process by which resources (i.e. inputs that are used in the activities of a program such as the physical, financial, human and social resources) are solicited by a program (World Bank 2007). By using the term 'mobilising', we suggest that resources are available, but that they are either not being used, or being directed elsewhere. The resources available to conserve species, the raw ingredients, include but are not limited to:

- funding to implement projects or programs
- funding to implement projects or programs
- instruments, such as legal mandates, legal instruments, policy, and institutional strategies
- actors; such as people (including, for example, volunteers and professionals working in threatened species conservation, the public, and people in governments or institutions)
- technology and knowledge.

These resources interrelate to varying degrees and can be optimised to increase their conservation impact, for example by:

- leadership
- political will
- social licence/social interest
- good planning
- good governance, particularly around decision making and accountability
- access to expertise, knowledge sharing and capacity building
- stability and corporate knowledge that ensures program continuity and efficiency
- good engagement
- capitalising on opportunities
- creating momentum
- productive relationships
- connections and networks
- recognising diversity of approaches, expertise and knowledge base.

In this chapter, we present three case studies that demonstrate different ways in which these 'ingredients' have been mixed together to mobilise resources for the conservation of threatened species.

Case study 1. Making a business case: the New South Wales Government's Saving our Species Program

At a minimum, the community expects that governments will prevent further extinctions of plants and animals and is hopeful that recovery of species under threat is possible. Recovering threatened species is a substantial challenge. With the average risk of extinction for birds, mammals and amphibians still increasing globally (SCBD 2014), more and more species are being listed as threatened with extinction across Australia and internationally. With funds usually limited and recovery standards very high, in most cases, only the most threatened, politically sensitive, charismatic or highly valued species receive funding. In New South Wales, over 1000 entities are listed on the schedules of the *Biodiversity Conservation Act 2016 (NSW)*.

Saving our Species (SoS), is the overarching state government framework for the strategic management of all threatened species in New South Wales. The program adopts the recommendations of an independent review of the actions needed to conserve threatened species in NSW: the Priorities Action Statement Review (2011–2012). The review found that there were problems with New South Wales recovery plans: they took years to develop and there were insufficient funds for complete implementation; the amount of government investment in preparing the plans was disproportionately high compared with investment in recovery actions; in many cases, actions were general and not site-specific; and objectives and performance indicators were not clearly defined, resulting in inadequate monitoring and reporting.

The Saving our Species program considers the balance between planning and implementation. It is an innovative approach for dealing with the growing numbers of plants and animals in NSW that are faced with extinction. The program provides more opportunities for community involvement and to align everyone's efforts under a single banner, rather than having isolated programs with competing priorities and funding. The SoS objective is to maximise the number of threatened species that are secure. SoS recognises that, given current political priorities, we will not have adequate resources to recover many species and that it is more practical to establish the goal of there being a 95% probability of having a viable population of the species in the wild in 100 years from now. The first step was to allocate species into management streams according to their characteristics, our knowledge and imminent threats – for example, species characterised by small populations that require actions at local sites, and species with declining populations that may require actions across broader landscapes. The SoS program categorises threatened species into six management streams: (1) those that require site management; (2) iconic species; (3) data deficient species; (4) species that require management across broader landscapes; (5) partnership species; and (6) species for which the management response is to 'keep watch'. We are currently resolving our approach to threatened ecological communities and key threatening processes.

SoS species projects include, where possible, the minimum number of sites required to secure the species at least cost. Prioritisation extends to prioritisation between streams and between site-managed species. The prioritisation is based on cost effectiveness (benefit × likelihood of success/cost).

The SoS program received funding of A$7.9 million from the NSW Government for 2012–2013 to 2015–2016. A total of A$4.8 million was allocated to the implementation of species conservation projects. This funding was delivered to implement the findings of the review. Experts within the agency and Professor Hugh Possingham, his team, and representatives from the New Zealand Government were able to build a convincing case for how the recommendations for the review could be implemented and that the approach would be cost effective and deliver conservation outcomes. Being clear about what actions are required, what will be delivered and a commitment to monitoring, evaluation and reporting provides greater transparency and will help us to be able to show what can be achieved with targeted investment.

Since then, the Office of Environment and Heritage has publicly exhibited over 500 projects for individual species, as well as the scientific and program justification for the changes that we are making. This approach has been supported by Environment Ministers who could see the value of a cost-effective approach and have successfully supported the importance of threatened species conservation within government.

The NSW Government is investing A$100 million over 5 years in saving more threatened animals and plants from extinction. The funds were allocated to SoS from 2016–2021. In 2016–17, 240 species projects were being delivered; this is complimented by A$41 million over 10 years to reintroduction and A$10 million over 10 years for the Environmental Trust SoS Partnership Program.

A large amount of investment has occurred in program development. The next steps are to find out what people are doing for threatened species conservation across New South Wales, and to consider the long-term sustainability of the program funding, given that the work for conservation is ongoing.

Case study 2. Creating momentum: bringing people together for threatened mallee birds

Why are more resources available for some species than for others? Certainly, charismatic and iconic species tug at the heart (and purse) strings, but it is more than that. It is about the people behind the recovery effort.

It can be difficult to pinpoint exactly where and when decisions are made to dedicate resources, particularly funding, to a recovery action or set of actions. However, in many instances, the first resources to be dedicated to the recovery effort are people's time and energy. Once a group of people come together and agree on what needs to happen, this clarity and commitment to purpose tends to generate a momentum that pulls in other

Fig. 30.1. Mallee emu-wrens became a poster child for mallee ecosystems threatened by fire after they were lost from South Australia in 2014 (photo: Dean Ingwersen).

resources. The development of the Threatened Mallee Bird Recovery Program provides a good demonstration:

In the summer of 2014, fires raged through the Murray Mallee of south-eastern Australia, burning more than 90% of two conservation reserves and extirpating important populations of endangered bird species. Bronzewing Flora and Fauna Reserve (in Victoria) was home to the 'insurance' population of the Endangered black-eared miner (*Manorina melanotis*). Extensive fire in that reserve meant the global population of black-eared miner became almost exclusively confined to the Bookmark Biosphere area of South Australia. At about the same time, an important population of Endangered mallee emu-wren (*Stipiturus mallee*) (Fig. 30.1) was also lost from Billiatt Conservation Park (in South Australia), resulting in the extirpation of the species in that state. The only remaining population of mallee emu-wren known at the time was confined to one contiguous area that covered Murray-Sunset and Hattah-Kulkyne National Parks (Victoria), meaning that the emu-wren and miner were vulnerable to a reserve-wide fire extinction event. Furthermore, these and other recent fires destroyed large tracts of habitat for other nationally threatened mallee bird species including the red-lored whistler (*Pachycephala rufogularis*) and malleefowl (*Leipoa ocellata*).

These fires left staff at BirdLife Australia shocked. However, rather than becoming despondent, a small team got together and asked: 'What can we do?' The idea of an emergency summit was born. Although it was met by resistance initially – as unbudgeted and untested initiatives often are – such was the passion of individuals for the cause that a small amount (~A$6000) of funding was donated from the local BirdLife Branch at Horsham.

BirdLife Australia hosted a summit in early May 2014 to identify the urgent actions needed to prevent extinction of the threatened mallee bird species. The summit was well attended by fire and mallee species experts from universities, as well as representatives from all three state and the Australian Government. It provided an important forum for sharing information on latest developments and research, including recent modelling indicating that the majority of Victoria's threatened mallee bird habitat would be lost should the fire management policies at that time continue.

As well as developing a budget for the urgent actions needed to prevent species at imminent risk of extinction, summit participants agreed on the need to work together to plan and coordinate implementation of threatened mallee bird recovery efforts. Between early August and February 2015, we convened a series of workshops for threatened mallee birds. The planning process for the Threatened Mallee Birds Project used the Open Standards for the Practice of Conservation, developed by the Conservation Measures Partnership (CMP 2013), to help conservation teams systematically plan, implement and monitor their conservation initiatives in an adaptive management framework. Specifically, we used the Conservation Action Planning (CAP) framework, the Nature Conservancy's version of Open Standards (TNC 2007). The current iteration of the Threatened Mallee Bird CAP (Boulton and Lau 2015) focuses on emergency, short-term actions (1–5 years). Work Plans are used to drive implementation of the CAP, outlining the expected duration and timeframe for each action, the likely agencies or individuals to undertake the work, and an approximate cost and source of possible funding. Partners have committed substantial funding and internal resources to implement priority actions, and the model continues to nurture collaborative working relationships, continuously building trust and reinforcing our shared commitment to achieving agreed goals.

Getting people together built a response to the crisis and together we created enough momentum to trigger widespread action: The National Recovery Plan for the mallee emu-wren, red-lored whistler and western whipbird (*Psophodes nigrogularis*) was adopted by the Australian Government in May 2016. The Australian Government also funded pre-translocation studies for the mallee emu-wren, which included: a translocation plan; surveys of bird numbers and potential habitat and support for La Trobe University to undertake fine-scale mapping of emu-wren habitat in potential source and release sites; and enabled Zoos South Australia to develop the capacity to house emu-wrens. The project also led to the discovery of a second wild population of emu-wrens in Wyperfeld National Park, bringing us closer to our objective of securing five wild populations.

With funding from the Mallee Catchment Management Authority (CMA), BirdLife produced two educational resources focused on threatened mallee birds – a 'Mallee Wing Thing' aimed at Primary School children and a 'Mallee Bird ID and Conservation Guide', made available to visitors to mallee parks, volunteer fire fighters, local farmers, CMAs and schools. Further funding was then secured from the Mallee CMA to develop an education toolkit to accompany the 'Mallee Wing Thing' and BirdLife staff have successfully run professional development workshops for teachers in the Murray Mallee to encourage the use of the Mallee Wing Thing in school programs.

The Partnership was also instrumental in securing funding for a 3-year mallee emu-wren fire ecology project to be undertaken by La Trobe University: research that will directly contribute to improvements in the implementation and evaluation of bushfire management by delivering more detailed information on how fire affects the abundance of emu-wrens. Dr Jenny Lau of BirdLife Australia prevented the scheduled burning of critical emu-wren habitat and was invited onto the newly established Mallee Fire Advisory Committee in Victoria where she successfully negotiated an agreement that all burning of mallee emu-wren habitat in Hattah-Kulkyne National Park be postponed until comprehensive surveys of the species distribution and density in the Park had been completed.

BirdLife's advocacy for evidence-based fire management of threatened mallee bird habitat contributed to a change in fire management policy in Victoria from an area-based target for its planned burning program to a more strategic, risk-based approach. This means there is a greater focus on burns to reduce risk to life and property, and a reduction in burning of important habitat distant from communities.

As this example demonstrates, conservation is about people. Bringing people together can create enough momentum to overcome inertia and mobilise resources. Momentum needs to be maintained, which requires leadership, trust and working together towards a common goal. Conservation Action Planning provides a good framework for adaptive program management. With this approach, the people involved in the recovery effort have agreed on where we are going, how we are going to get there and how we will know when we have arrived. Actions are prioritised and there is a clear line of sight, which maintains clarity of purpose. The approach has leveraged well over half a million dollars towards the recovery of mallee birds (and much more in-kind resources) in just a few short years. But, more importantly, it has brought people together: bold, aspirational people; leaders in their field. Being part of such a team is truly inspirational.

Case study 3. Partnerships for success: the northern hairy-nosed wombat

The northern hairy-nosed wombat (*Lasiorhinus krefftii*) is one of the world's rarest large mammals, with the most recent count (2013) estimating 196 individuals. It was once widespread from central Queensland to Victoria, but never abundant in historic times because it requires specific soil conditions for its deep and extensive burrows (Horsup 2004). Drought, competition from introduced grazers, habitat change, persecution and feral dogs reduced the total population to an estimated 35 individuals in 1983, all at a single location: Epping Forest in central Queensland.

Epping Forest National Park was declared in 1971, with dog-proof fencing installed in 2002. Dams have been established to reduce the effects of drought. However, although these measures facilitated a steady growth in the population at Epping Forest, the risks of localised disease, a natural disaster or potential adverse habitat change prompted establishment of other colonies (Horsup 2004). In 2006, a trial translocation to an unpopulated area of Epping was successful. Extensive habitat and soil suitability research and surveys were conducted to identify new sites, with nearly 150 sites (mainly across the Brigalow Belt) surveyed. The most suitable site identified was on a grazing property near St George, 620 km from Epping Forest.

The translocation was enabled by two partnerships in 2009: the landholders generously agreed to allow the use of 130 ha of land as the Richard Underwood Nature Refuge, and Glencore mining company (then Xstrata) provided A$3.5 million for the translocation, including 'start-up' burrows for the colonising wombats, supplementary water supply, predator-proof fencing and a ranger base.

There has been a steady increase of the Epping Forest wombat population. The Richard Underwood Nature Refuge population initially bred successfully, but lost two adult males to disease believed to be induced by injuries from territorial fights. The remaining male does not appear interested in breeding, so a new male was recently introduced from Epping Forest. A new, third translocation site is being sought. The long-term aim is to establish three sites in Queensland and, desirably a further colony in another state. The wombat's recovery has been carefully planned, with an emphasis on adaptation and flexibility. These were key attributes essential to forming the partnerships with landholders, Glencore and volunteers upon whom the success of the program to date has been founded.

Nonetheless, the key threats to the northern hairy-nosed wombat will not be overcome in the wider landscape for the foreseeable future. However, although more needs to be known about the species' social behaviour and physiology, and buffel grass (*Cenchrus ciliaris*) (a potential new threat to the species), the prognosis for the species in suitably protected enclaves is good, providing resources can be secured to maintain existing sites and establish new colonies. However, fenced, well-managed enclaves with effective monitoring and research are costly. Fortunately, the risk of losing Australia's largest remaining wombat species is clearly seen as an even greater and unacceptable cost by a wide spectrum of interests across our society. The establishment of a critically needed second colony of wombats was enabled by a diverse range of community interests – Glencore, the Underwood family and numerous volunteer organisations and individuals – who related to the vision of securing this species.

Conclusion

The amount of government funding available for species protection may often be one of the best predictors of successful recovery (Gerber 2016). The first case study demonstrates that, by championing a coherent business case within government, the Saving our Species Program has successfully mobilised a 10-fold increase in funding available for threatened species management in that state.

In case study 2 we see that, even with a very modest financial outlay, bringing the right people together can create enough momentum to make change, and in turn mobilise even more resources over time (in excess of A$500 000 to date). This case study demonstrates the importance of leadership, trust and working together towards a common goal.

Finally, the northern hairy-nosed wombat case demonstrates that, although careful planning is critical, flexibility and the ability to adapt is essential in any productive partnership, especially when working with landholders and local communities. In particular, the 2009 translocation was a success because landholders generously provided the use of 130 ha of land and a mining company provided the A$3.5 million for the translocation.

In these case studies, the 'raw ingredients' – that is, resources such as funds, instruments, actors, technology and knowledge – were mixed in different ways. Overwhelmingly though, it is the people involved that influence how well these ingredients interrelate. That, and a good plan that everyone can get behind.

Acknowledgements

Threatened Mallee Birds: the leadership and energy provided in particular by Vicki Jo Russell, Chair of the TMB CAP implementation team, Jen Lau (co- chair), Rebecca Boulton and Dean Ingwersen, BirdLife Australia; Jill Flemming, Victorian Department of Environment, Land, Water and Planning; Wendy Stubbs, Chris Hedger and Peter Copley, South Australian Department of the Environment, Water and Natural Resources; Simon Nally and the Threatened Species Commissioner's office, Australian Government; Simon Taylor LaTrobe University; Liberty Olds, Zoos South Australia.

References

Boulton RL, Lau J (2015) 'Threatened Mallee Birds Conservation Action Plan, Report June 2015'. Report to the Threatened Mallee Birds Implementation Team, BirdLife Australia, <http://birdlife.org.au/documents/TMB-CAP-first-iteration-June_2015.pdf>.

CMP (2013) *Open Standards for the Practice of Conservation.* Version 3.0. Conservation Measures Partnership, <http://cmp-openstandards.org/wp-content/uploads/2014/03/CMP-OS-V3-0-Final.pdf>.

DEWHA (2009) 'Assessment of Australia's Terrestrial Biodiversity 2008. Report prepared by the Biodiversity Assessment Working Group of the National Land and Water Resources Audit for the Australian Government'. Department of the Environment, Water, Heritage and the Arts, Canberra.

DoE (2017) 'Recovery Plans adopted under the EPBC Act in Species Profile and Threats Database'. Department of the Environment, Canberra, <http://www.environment.gov.au/sprat>.

Gerber LR (2016) Conservation triage or injurious neglect in endangered species recovery. *Proceedings of the National Academy of Sciences of the United States of America* **113**, 3563–3566. doi:10.1073/pnas.1525085113

Horsup A (2004) 'Recovery Plan for the Northern Hairy-nosed Wombat *Lasiorhinus krefftii* 2004–2008'. Environmental Protection Agency/Queensland Parks and Wildlife Service, Brisbane.

McCarthy MA, Thompson CJ, Garnett ST (2008) Optimal investment in conservation of species. *Journal of Applied Ecology* **45**, 1428–1435. doi:10.1111/j.1365-2664.2008.01521.x

Nimmo D, Cunningham S, Lindenmayer DB, Mac Nally R, Woinarski J (2016) Mass coral bleaching is a symptom of ecosystem collapse across Australia. *The Conversation*, <http://theconversation.com/great-barrier-reef-bleaching-is-just-one-symptom-of-ecosystem-collapse-across-australia-58579>.

SCBD (2014) 'Global Biodiversity Outlook 4'. Secretariat of the Convention on Biological Diversity, Montréal, Canada.

TNC (2007) 'Conservation Action Planning handbook: developing strategies, taking action and measuring success at any scale'. The Nature Conservancy, Arlington, VA, USA, <https://www.conservationgateway.org/Files/Pages/action-planning-cap-handb.aspx>.

World Bank (2007) Resource mobilization and financial management. In *Sourcebook for Evaluating Global and Regional Partnership Programs – Indicative Principles and Standards.* (Ed. Independent Evaluation Group–World Bank) pp. 83–86. Washington DC, USA, <http://siteresources.worldbank.org/EXTGLOREGPARPROG/Resources/grpp_sourcebook_chap13.pdf>.

31

Reporting on success in threatened species conservation: the national policy context

Peter Latch

Summary

The problem
1. Although threatened species recovery program successes are reported, there is no national system to coordinate reporting and build a story of conservation effort and success and report to the Australian and international communities.

Actions taken to manage the problem
1. A new policy approach under the Threatened Species Strategy provides the foundation to enhance recovery planning practices and facilitate national reporting by recovery teams on outcomes for threatened species.

Markers of success
1. Robust and collaborative planning processes drive recovery effort.
2. Improvements in conservation outlook for many species.
3. Annual reporting against targets under the Threatened Species Strategy has commenced.

Reasons for success
1. Community awareness, support and involvement in recovery efforts.
2. Collaborative and well-governed recovery programs.
3. Recovery programs informed by legislation, policy and planning frameworks protecting threatened species.

Introduction

Under Australia's federal system of government, responsibility for environmental protection and regulation is collectively undertaken by the Australian, state and territory governments. The respective roles and responsibilities of each level of government in Australia are established to ensure a coordinated approach is undertaken on biodiversity conservation issues, acknowledging that the different levels of government have different, but complementary, roles to play. Certain matters, particularly those related to Australia's international obligations, fall under the jurisdiction of the Australian Government, while the state and territory governments have primary constitutional and legal responsibility for the management of land, water and biodiversity within their jurisdictions.

At local, regional, state, territory, national and international levels, various levels of government in collaboration with partners, implement a broad range of policies and programs designed to tackle major threats to biodiversity, arrest the decline in threatened species and ecological communities and promote their longer term recovery. These policies and programs are established in strong governance and legislative frameworks. National biodiversity policies and strategies are developed through cooperative arrangements among the jurisdictions. Many international agreements are closely aligned with, and are informed by, this conservation effort. Implicit in this framework is the need to report on efficacy of conservation effort, at multiple temporal and spatial scales.

National law and policy protecting threatened species

Noting the legal responsibilities of states and territories in managing biodiversity within their jurisdictions, three key pieces of legislation and policy for which the Australian Government has responsibility, collectively provide a strategic framework to identify, protect and recover threatened species and to report on conservation progress to the Australian and global communities.

International biodiversity agreements

Conservation of biodiversity is a global concern, and national action is needed to support international cooperation. Australia participates in the development and implementation of many international agreements and conventions responding to biodiversity conservation and sustainable use issues (see NRMMC 2010). These agreements impose obligations and require a range of actions to be undertaken to deal with matters of concern to the international community.

One of the most significant multilateral agreements is the Convention on Biological Diversity, to which Australia has been a Contracting Party since 1993. The Convention on Biological Diversity is a comprehensive, binding agreement covering the use and conservation of biodiversity and obliges all parties to develop and implement national biodiversity strategies and action plans and for parties to report on national implementation of the Convention.

Australia agreed with other countries in 2010 to implement the Strategic Plan for Biodiversity 2011–2020, including its 20 measurable, time-bound Aichi Biodiversity Targets (CBD 2010). Aichi Target 12 states that by 2020 the extinction of known threatened species has been prevented and their conservation status, particularly of those most in decline, has been improved and sustained. The achievement of Target 12 is linked to progress towards many of the other Aichi Targets such as the identification of sites for Protected Areas (Target 11).

National biodiversity policy

Australia's Biodiversity Conservation Strategy 2010–2030 (NRMMC 2010) is the guiding framework for conserving the country's biodiversity. It was developed to fulfil Australia's international obligations under the Convention on Biological Diversity, and supports the alignment of national priorities and outcomes with ongoing international efforts to conserve biodiversity.

The Strategy outlines national priorities for action to help stop the decline in biodiversity and identifies 10 national targets, one of which is to establish a long-term national biodiversity monitoring and reporting system. The Strategy functions as a policy umbrella over other more specific national frameworks relevant to the conservation of threatened species, such as Australia's Strategy for the National Reserve System 2009–2030.

The collective actions of all governments in implementing the Strategy provides the basis for the Australian Government to report on biodiversity conservation efforts every 5 years through the national State of the Environment Report (Cresswell and Murphy 2017) and on obligations under the Convention on Biological Diversity (DoE 2014).

National environmental law

The *Environment Protection and Biodiversity Conservation Act 1999* (EPBC Act) is the Australian Government's primary environmental legislation and provides a national scheme for the conservation of biodiversity and environmental and heritage protection. The EPBC Act is significantly guided by, and is the legislative vehicle for, delivering Australia's international obligations for protecting and conserving the country's biodiversity.

The EPBC Act identifies and protects matters of national environmental significance. 'Threatened species and ecological communities' is one of nine matters of national environmental significance and are therefore subject to the Act's regulatory and approval processes. They can also receive protection through the protection of other matters of national environmental significance: for example, where they occur in areas such as world heritage properties, Ramsar wetlands, Commonwealth marine areas, and the Great Barrier Reef Marine Park.

The EPBC Act also provides a framework to plan for the long-term recovery of listed species and ecological communities through the development of recovery plans and conservation advices. They provide a planned and logical framework to guide collaborative investment in, and participation by, all levels of government, non-government organisations, research organisations and community groups.

At the time of listing, conservation advice is prepared that provides guidance on priority research and conservation actions necessary to protect the newly listed species or ecological community. A more comprehensive recovery plan may also be developed and these are generally prepared where the listed species or ecological community has complex management needs due to its ecology, the nature of threats affecting it, or the number of stakeholders affected by or involved in implementing the necessary recovery actions.

At 13 April 2017, 1710 species and 77 ecological communities were listed as either critically endangered, endangered or vulnerable under the EPBC Act (Fig. 31.1). Recovery plans were in place for 42% of species and 32% of ecological communities, while conservation advices were in place for all others.

The EPBC Act also lists key threatening processes and, where required, Threat Abatement Plans are prepared that provide for the research, management and any other actions necessary to reduce the impact of a listed key threatening process on native species and ecological communities.

Measuring and reporting success: the challenges

Australia's legal and policy regime affords a level of protection to those species and ecological communities that are formally listed as threatened under legislation. Noting that the list of threatened species is skewed towards a relatively data-rich subset of all species that may be threatened (Walsh *et al*. 2013), it provides a significant foundation upon which longer term conservation planning and action is directed.

Assessing the overall effectiveness of the conservation response to the listing of threatened species and ecological communities through the planning and subsequent

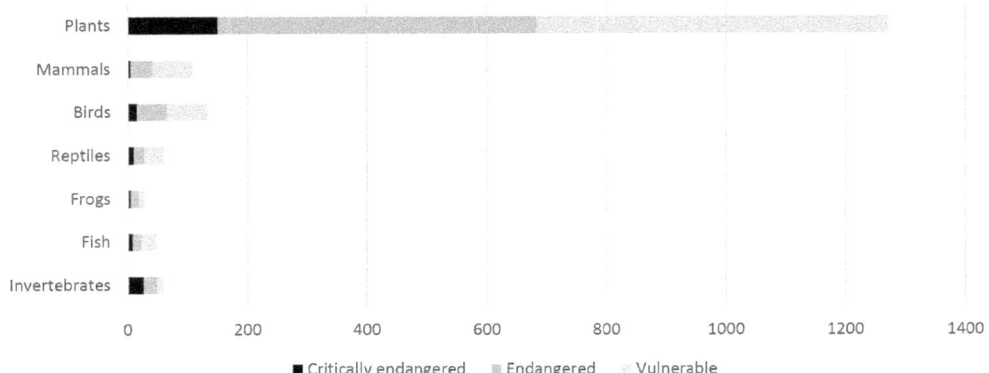

Fig. 31.1. The number of species listed nationally as threatened (either Critically Endangered, Endangered or Vulnerable), under the *Environment Protection and Biodiversity Conservation Act 1999* at 13 April 2017. Of the 1710 species listed, 13% were Critically Endangered, 40% Endangered and 47% Vulnerable. Of all species listed, 74% were plants. In addition to species listed, 77 threatened ecological communities – 34 Critically Endangered, 41 Endangered and 2 Vulnerable – were also listed nationally.

implementation of action is, however, challenging. Biodiversity declines continue to be reported globally (Tittensor *et al.* 2014) and nationally (Woinarski *et al.* 2015). Although the Australian community invests substantial resources in biodiversity conservation, national reviews have reported on the difficulty of measuring the efficacy of that overall effort, largely because there are few appropriately designed national datasets that measure and report on trends (DEWHA 2009; Morton and Tinney 2012; Commonwealth of Australia 2016; Cresswell and Murphy 2017). Australia's Biodiversity Conservation Strategy target to establish a national biodiversity monitoring and reporting system by 2015 was not met; the Strategy's first 5-year review acknowledging that in the absence of such a system it 'made sourcing reliable national level information challenging' (Commonwealth of Australia 2016).

Considerable research has also assessed the effectiveness of such current management and policy responses, recommending a range of approaches and actions to improve threatened species conservation planning, governance, monitoring, evaluation and reporting processes (Watson *et al.* 2010; Bottrill *et al.* 2011; Ortega-Argueta *et al.* 2011; Martin *et al.* 2012; Szabo *et al.* 2012; Lindenmayer *et al.* 2013; Walsh *et al.* 2013; Woinarski *et al.* 2014; McDonald *et al.* 2015; Evans *et al.* 2016).

Notwithstanding these assessments of shortcomings in reporting of conservation outcomes, there have been many reported conservation success stories in Australia where appropriately resourced and implemented recovery programs, supported by dedicated people, have led to improved conservation outcomes (DEWHA 2009; Garnett *et al.* 2011; Szabo *et al.* 2012; SECRC 2013; Woinarski *et al.* 2014). Further examples are illustrated in this book and in Fig 31.2.

Nationally, Birdlife Australia, through its substantial volunteer network, collects and manages a wealth of data that contribute to national and global assessments on the status of birds (Szabo *et al.* 2012; Birdlife Australia 2015). The work of Birdlife Australia represents a significant contribution to the building of a national information set that could be interpreted and used to inform decision making and policy and report to the community.

Although individual successes are reported, there is no coordinated system to bring these stories together to establish a national narrative on threatened species recovery suc-

Fig. 31.2. One of many successful recovery programs underway across Australia, the Save the Tasmanian Devil Program, coordinates recovery effort for the Endangered Tasmanian devil (*Sarcophilus harrisii*). The Program was established in 2003 by the Australian and Tasmanian Governments in response to declines in the devil population threatened by the devil facial tumour disease. With the long-term aim of securing an enduring and ecologically functional population of Tasmanian devils in the wild, major successes to date include an insurance population, research to advance understanding of the disease, monitoring, island translocations, major partnerships and community engagement. The current focus is to establish significant disease-free populations in the wild, determine the status and trends of the wild population and develop techniques for managing diseased populations (see: http://www.tassiedevil.com.au; photo: Save the Tasmanian Devil Program).

cess. Apart from the challenges in reporting against statutory and other obligations, the absence of such a system may constrain broader community engagement in ongoing conservation investment because evidence of success or other signals of positive results that arise from conservation interventions are important motivators for engendering 'hope' and more effort by the community (Garnett and Lindenmayer 2011). Because most recovery programs are long-term endeavours with long-term recovery goals, identifying and celebrating measures of success as steps in the pathway towards longer term outcomes could be important motivators to keep the community engaged and allow progressive reporting on outcomes.

In the absence of national reporting on threatened species recovery success, reports on biodiversity trends are often limited to the use of indirect measures, such as the numbers of listed threatened species (DEWHA 2009; DoE 2014; Cresswell and Murphy 2017). Maintaining and assessing changes in such lists is important for statutory and other reporting and serves as an important communication tool, such as communicating the significance of biodiversity loss. These lists, however, have limited value as indicators in tracking overall trends and the extent to which conservation management is adequate (Possingham *et al.* 2002; Walsh *et al.* 2013).

Measuring and reporting success: opportunities

In response to ongoing concerns about the state of threatened species and acknowledging that a more strategic response was needed, the Australian Government established a new policy approach in 2015 – the Threatened Species Strategy (Australian Government 2015). The appointment of a Threatened Species Commissioner is also central to the Australian Government's new policy approach and the Commissioner facilitates national action including through implementation of the Strategy.

Building upon the existing legal and policy framework, the Strategy establishes for the first time a mechanism under which priority action areas and targets are established against which the Government is to report. For example, the identified 20 bird, 20 mammal and 30 plant priority species are to have improved population trajectories by 2020, with these species selected in part because of the urgency of their conservation need. Accounts of progress towards these targets are now provided annually; of the Strategy's 26 first-year targets, 21 were met, four were partially completed and one was not achieved (Australian Government 2016).

The Strategy reflects the need for collaboration and coordination, building on and celebrating success to date, while also recognising new and reinvigorated approaches are needed to better deliver effective outcomes. The importance of monitoring, evaluation and reporting as an important part of any recovery effort is also identified in the Strategy, and an associated key target is to improve recovery planning practices.

Recovery plans and conservation advices are recognised as critical national planning processes to facilitate national action on threatened species and ecological communities, to engage communities, to monitor progress and to report on outcomes and conservation success. Securing the long-term recovery of threatened species and ecological communities is a challenging task, involving many individuals, organisations and agencies. The development and implementation of recovery and other conservation plans are therefore necessarily collaborative processes that involve multiple stakeholders. A key mechanism to assist this process is to ensure effective governance (Martin *et al.* 2012).

The Strategy recognises the importance of the recovery team – a collaboration of partners brought together to plan and/or coordinate the implementation of a recovery program – as a governance model to coordinate implementation of recovery programs and to report more frequently, consistently and rigorously on progress and effectiveness. National recovery team governance guidelines are being established in 2017 and will provide a framework for establishing and operating effective recovery teams.

A national reporting framework is also being established in 2017, and will allow recovery teams to report on progress in achieving the objectives of a recovery program. Consistent reporting across national recovery teams will build a national snapshot of species' conservation efforts over time and will contribute to other national and international reporting requirements.

Although the initial focus will be on the 70 Strategy priority species, the implementation of the Strategy, and key learnings that might arise, represent an opportunity to build a more comprehensive national reporting framework. The Strategy's governance and reporting systems could serve as a model on which to expand a representative monitoring and reporting program covering a larger set of Australia's threatened species and ecological communities and to capture reporting through other governance systems apart from recovery teams.

National reporting should link to international reporting requirements against the Aichi Targets so choice of appropriate indicators would be important (Nicholson *et al.* 2012). Models of international reporting are in place, such as the Living Planet Index (Collen *et al.* 2009).

Informing these initiatives is research being undertaken in the Threatened Species Recovery Hub of the National Environmental Science Program. A key aspect of this Australian Government-funded research Program is that the research meets key policy and planning needs of the Australian Government and other stakeholders. 'Better understanding, measuring and reporting on the condition and trend of threatened species' is a key research priority for the Program.

Importantly, any such program should also be integrated under a broader biodiversity monitoring and reporting program and align with other relevant programs in place across jurisdictions (SECRC 2013). Australia's Biodiversity Conservation Strategy is to be revised in response to its 5-year review (Commonwealth of Australia 2016). A key finding includes the need to better align the national strategy against the Aichi Targets; opportunities therefore exist to integrate a threatened species reporting system into an updated national strategy.

These opportunities could also link to broader changes in the management of biodiversity information across jurisdictions. National policy reform in the management of biodiversity information is underway to improve the Australian Government's capacity to monitor, detect and predict change in the environment. The National Environmental Information Infrastructure managed by the Bureau of Meteorology, for example, is an information platform designed to improve discovery, access and re-use of nationally significant environmental data. The Atlas of Living Australia and its integration of biological records and related environmental layers has potential use in establishing Australia wide biodiversity monitoring (Zerger *et al.* 2013).

Conclusion

Government and community effort and investment in threatened species protection and management is significant, yet national reviews have consistently found it difficult to measure efficacy of that effort, largely because there are few appropriately designed national datasets that measure trends in biodiversity and report on implementation success.

A national legal and policy framework, linked to international obligations, is in place that identifies and aims to protect threatened species and ecological communities, with various requirements to report on efficacy of conservation effort. Although there are many examples of individual recovery program successes, there is no national system to coordinate reporting of national effort and progressively build a national story of conservation effort and success. It is therefore difficult to report to the community, to governments or to international bodies on how our threatened species and ecological communities are faring, and how effective our management interventions are at conserving them.

The Threatened Species Strategy represents a new national approach to the conservation of threatened species and ecological communities and identifies priorities, action areas and measurable and reportable targets. Improving recovery practices will build upon learnings to date, focus on priority species, build strong governance arrangements with recovery teams and develop a national model of reporting on recovery outcomes. Opportunities exist to build the Strategy model and align with other programs to capture a larger, national representative set of species that can better, and more effectively, meet national and international reporting requirements. Other national policy initiatives provide opportunities to further integrate threatened species success measures within national reporting frameworks.

References

Australian Government (2015) *Threatened Species Strategy*. Australian Government, Canberra.
Australian Government (2016) 'Threatened Species Strategy – Year One Report'. Australian Government, Canberra.
Birdlife Australia (2015) 'The state of Australia's birds 2015: Headline trends for terrestrial birds'. Birdlife Australia, Melbourne.

Bottrill MC, Walsh JC, Watson JE, Joseph LN, Ortega-Argueta A, Possingham HP (2011) Does recovery planning improve the status of threatened species? *Biological Conservation* **144**, 1595–1601. doi:10.1016/j.biocon.2011.02.008

CBD (2010) 'COP 10 Decision X/2. Strategic Plan for Biodiversity 2011–2020'. Tenth meeting of the Conference of the Parties to the Convention on Biological Diversity, Aichi Prefecture, Japan.

Collen B, Loh J, Whitmee S, McRae L, Amin R, Baillie JEM (2009) Monitoring change in vertebrate abundance: the Living Planet Index. *Conservation Biology* **23**, 317–327. doi:10.1111/j.1523-1739.2008.01117.x

Commonwealth of Australia (2016) 'Report on the review of the first five years of Australia's Biodiversity Conservation Strategy 2010–2030'. Commonwealth of Australia, Canberra.

Cresswell ID, Murphy HT (2017) 'Australia state of the environment 2016: biodiversity'. Independent report to the Australian Government Minister for the Environment and Energy. Australian Government Department of the Environment and Energy, Canberra.

DEWHA (2009) 'Assessment of Australia's Terrestrial Biodiversity 2008'. Report prepared by the Biodiversity Assessment Working Group of the National Land and Water Resources Audit for the Australian Government, Canberra.

DoE (2014) 'Australia's Fifth National Report to the Convention on Biological Diversity'. Department of the Environment, Canberra.

Evans DM, Che-Castaldo JP, Crouse D, Davis FW, Epanchin-Niell R, Flather CH, *et al.* (2016) 'Species recovery in the United States: increasing the effectiveness of the Endangered Species Act'. Issues in Ecology Report Number 20. Ecological Society of America, Washington DC, USA.

Garnett ST, Lindenmayer DB (2011) Conservation science must engender hope to succeed. *Trends in Ecology & Evolution* **26**, 59–60. doi:10.1016/j.tree.2010.11.009

Garnett S, Szabo J, Dutson G (2011) *The Action Plan for Australian Birds 2010*. CSIRO Publishing, Melbourne.

Lindenmayer DB, Piggott MP, Wintle BA (2013) Counting the books while the library burns: why conservation monitoring programs need a plan for action. *Frontiers in Ecology and the Environment* **11**, 549–555. doi:10.1890/120220

Martin TG, Nally S, Burbidge AA, Arnall S, Garnett ST, Hayward MW, *et al.* (2012) Acting fast helps avoid extinction. *Conservation Letters* **5**, 274–280. doi:10.1111/j.1755-263X.2012.00239.x

McDonald JA, Carwardine J, Joseph LN, Klein CJ, Rout TM, Watson JEM, *et al.* (2015) Improving policy efficiency and effectiveness to save more species: a case study of the megadiverse country Australia. *Biological Conservation* **182**, 102–108. doi:10.1016/j.biocon.2014.11.030

Morton S, Tinney A (2012) 'Independent review of Australian Government environmental information activity: Final report'. Department of Sustainability, Environment, Water, Population and Communities, Canberra.

NRMMC (2010) *Australia's Biodiversity Conservation Strategy 2010–2030*. Natural Resource Management Ministerial Council, Department of Sustainability, Environment, Water, Population and Communities, Canberra.

Nicholson E, Collen B, Barausse A, Blanchard JL, Costelloe BT, Sullivan KM, *et al.* (2012) Making robust policy decisions using global biodiversity indicators. *PLoS One* **7**, e41128. doi:10.1371/journal.pone.0041128

Ortega-Argueta A, Baxter G, Hockings M (2011) Compliance of Australian threatened species recovery plans with legislative requirements. *Journal of Environmental Management* **92**, 2054–2060. doi:10.1016/j.jenvman.2011.03.032

Possingham HP, Andelman SJ, Burgman MA, Medellín RA, Master LL, Keith DA (2002) Limits to the use of threatened species lists. *Trends in Ecology & Evolution* **17**, 503–507. doi:10.1016/S0169-5347(02)02614-9

SECRC (2013) *Senate Inquiry – Effectiveness of Threatened Species and Ecological Communities' Protection in Australia*. Senate Environment and Communications References Committee, Senate Printing Unit, Parliament House, Canberra.

Szabo JK, Butchart SHM, Possingham HP, Garnett ST (2012) Adapting global biodiversity indicators to the national scale: a Red List Index for Australian birds. *Biological Conservation* **148**, 61–68. doi:10.1016/j.biocon.2012.01.062

Tittensor DP, Walpole M, Hill SL, Boyce DG, Britten GL, Burgess ND, et al. (2014) A mid-term analysis of progress toward international biodiversity targets. *Science* **346**, 241–244. doi:10.1126/science.1257484

Walsh JC, Watson JEM, Bottrill MC, Joseph LN, Possingham HP (2013) Trends and biases in the listing and recovery planning for threatened species: an Australian case study. *Oryx* **47**, 134–143. doi:10.1017/S003060531100161X

Watson JE, Bottrill MC, Walsh JC, Joseph LN, Possingham HP (2010) 'Evaluating threatened species recovery planning in Australia'. Prepared on behalf of the Department of the Environment, Water, Heritage and the Arts by the Spatial Ecology Laboratory, University of Queensland, Brisbane.

Woinarski JCZ, Burbidge AA, Harrison PL (2014) *The Action Plan for Australian Mammals 2012*. CSIRO Publishing, Melbourne.

Woinarski JCZ, Burbidge AA, Harrison PL (2015) Ongoing unraveling of a continental fauna: decline and extinction of Australian mammals since European settlement. *Proceedings of the National Academy of Sciences of the United States of America* **112**, 4531–4540. doi:10.1073/pnas.1417301112

Zerger A, Williams KJ, Nicholls M, Belbin L, Harwood T, Bordas V, et al. (2013) 'Biodiversity profiling: components of a continental biodiversity information capability'. Environmental Information Program Publication Series, No. 2. Bureau of Meteorology, Canberra.

32

More than hope alone: factors influencing the successful recovery of threatened species in Australia

Stephen T. Garnett, Peter Latch, David B. Lindenmayer, David J. Pannell and John C.Z. Woinarski

Summary

The case studies in this book show that recovery of Australian threatened species is an achievable objective, and that many in the community are devoted to that objective. Successful recovery may take many decades and require many steps. The successful cases share some characteristics: (1) the recovery effort is strategically planned; (2) the recovery effort is implemented by an effective and representative group of stakeholders; (3) in many cases, there is an enduring recovery 'champion' who brings continuity and long-term commitment to the cause; (4) the legislative and policy settings appropriately support, and provide a commitment to, recovery effort; (5) resources (often from many sources) are adequate and continuous; (6) the recovery effort is well-grounded in evidence (particularly about the species' biology and its threats), monitors progress, and is receptive to change where needed; (7) the story of the species, its conservation need, and its recovery progress is well told.

Introduction

The history of Australian conservation is one of mixed fortunes. Much has been lost; many environments and species now persist only as insecure vestiges of their former extent, abundance or ecological importance. Over most of the period of post-European settlement, Australian nature has been accorded little value, and the colonisers have worked diligently to replace or 'improve' it. But there have also been inspiring counter examples of individuals and organisations recognising the significance, wonder and uniqueness of Australian species and environments, communicating that worth effectively, and working assiduously to conserve nature. Over the last few decades, these efforts have increasingly been rooted in national and state/territory policy and law that now recognises a responsibility to attempt to avoid further extinctions (Commonwealth of Australia 2015). Such policy is also consistent with international commitments to maintain biodiversity, and explicitly to prevent further extinctions (SCBD 2010). Although the number of formally recognised threatened species in Australia continues to grow, there are also more conservation successes and an expanding and robust body of knowledge and practice about species' recovery.

Of course, success is a protean concept. From a narrow perspective, few threatened species have been de-listed or downlisted nationally as a result of successful implementation of management actions. However, more broadly, success may involve: preventing or forestalling what otherwise would have been an inevitable extinction; the first tentative trends for population increase after decades of decline; the gradual removal of pressure arising from one or more seemingly intractable threats; or the coalescing of an otherwise disparate group of individuals and organisations to work together to protect threatened species. Ultimate success builds on such steps – the recovery of most threatened species in Australia is likely to be protracted and incremental.

There is no reason why Australia should lose more species – as a country, we are blessed with both sound governance and sufficient wealth to retain all that has survived two centuries of western colonisation. The most important message from the chapters in this book is hope – that threatened species recovery is possible, and is happening. The science and practice of threatened species management are becoming increasingly sophisticated, targeted and successful. There is now a fertile legacy of experience that can be applied from individual cases to threatened species more generally. With innovative ideas and support from managers and the general public, there are now many examples of the persistence or increase of species that would otherwise have been lost. In many cases, this effort is spearheaded by Indigenous groups or based on the management approaches and skills of Indigenous Australians. The application of two knowledge streams (western scientific and Indigenous cultural) presents exciting and fertile new options for the management of Australian environments. These are causes for celebration: our community has taken the opportunity to respect and enhance our natural legacy, rather than to leave to our descendants a less diverse, less healthy and less beautiful world.

So how has this been achieved? Are there common patterns among the eclectic examples we have provided here? Can the lessons be applied universally? In Chapter 1, we set out seven questions about threatened species recovery in Australia. We seek to answer these questions below. Of course, it may have been easier to identify characteristics had we counterpointed our examples with others exemplifying failure, but such cases and characteristics of failures have (appropriately) received much attention elsewhere (Black *et al.* 2011; Woinarski *et al.* 2017).

What characterises the threatened species that have been recovered?

We cast a wide net when seeking examples for this book, including plants, invertebrates and every class of vertebrate. The first lesson from the case studies is that recovery is possible for threatened species in any taxonomic group, large and small species, long-lived and ephemeral species, and species from any environment.

Although charismatic species are often the subject of campaigns, charisma can come in many forms. For example, two chapters (6, 19) describe plants characterised by spininess, which is hardly endearing; pygmy blue-tongue lizards rarely leave the cover of spider holes (Chapter 7) and most threatened mammals are shy and largely nocturnal. Yet all species described have fervent supporters. The lesson is that all species have a story and possess fascination, suggesting that priority for investment should closely consider the risk of extinction rather than simply preconceptions of charisma.

Many chapters describe long-lived species. Longevity may indeed have helped the species considered to endure neglect and many threats. But recovery of species with fleeting life histories can also succeed – for example, individual Lord Howe Island phasmids (Chapter 20) live less than 2 years. Certainly long life would not have helped any of the Australian vertebrates rendered extinct in the last 5 years (Woinarski *et al.* 2017). For these reasons, we propose that longevity is not a suitable criterion for allocating resources for species recovery.

To what extent have community contributions been instrumental to success?

Community involvement was the characteristic most commonly cited by chapter authors as a reason for recovery or as a marker of success (for 25 of 28 chapters). Such support has been delivered in diverse ways. For humpback whales, it was in the form of advocacy once the public had come to appreciate the likelihood of extinction (Chapter 2). For malleefowl, conservation engages diverse rural communities and is supported by volunteers spread across the nation (Chapter 8). Many community organisations and keen individuals have contributed to recovery of threatened species by propagating plants for restoration of habitat. Other programs involve the public in supplementary feeding, nest monitoring and pest eradication.

The monetary value of volunteer contributions is estimated in Fig. 32.1, though community contributions transcend budgetary considerations. Certainly more Australian species would be extinct but for the voluntary efforts of thousands of people. A spin-off benefit is that their involvement expands the group advocating for conservation of threatened species, leading to further support through policy and legislation.

Species protection and recovery are not the only benefits that can result from public participation in threatened species recovery. Community members who contribute may be more healthy and content, reducing the cost of health services. In socially disadvantaged remote and rural communities, contributions to threatened species recovery may bring much-needed employment, education benefits, support for local businesses and give communities a sense of purpose and a fulfilling connection to their country (Wilson *et al.* 2017).

What sort of people feel passionately about threatened species and what motivates them? We suspect many are retired and enjoy the opportunity to be involved with like-minded people in outdoor activities for a good cause. Although traditional volunteering may be declining, episodic, corporate and digital volunteering is expanding along with closer partnerships between government and volunteer organisations and greater professionalism in volunteer coordination (McLennan *et al.* 2015). There is also a cadre of younger people involved in threatened species conservation, some introduced to it by government programs or through schools.

To what extent has recovery depended on champions?

After community involvement, species' champions come through as the next most important component of successful species recovery (22 chapters). Good champions tend to be

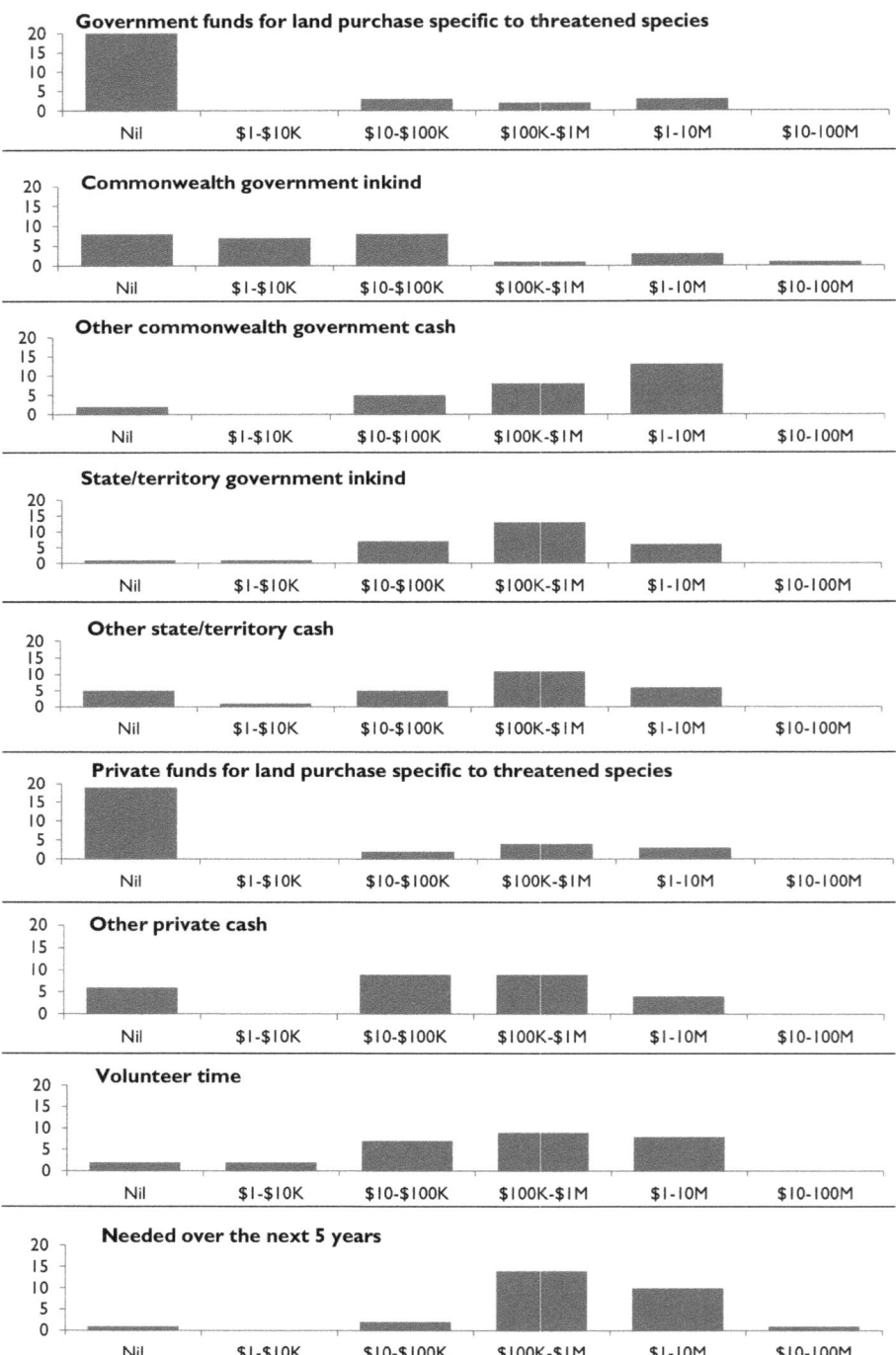

Fig. 32.1. Funds (in Australian dollars) provided to successful threatened species recovery programs described in Chapters 2–28, and total needs anticipated for the next 5 years. The vertical axis is the number of case studies, K is thousands, M is millions (in 2016). Note that the life of the programs varied markedly across the cases.

personally invested in conservation of a species, while at the same time empowering others to contribute. Successful recovery programs create the conditions under which species' champions might emerge, support them when they do, guide them when championship tends towards ownership and exclusion, and try to ensure that there is a succession plan.

The contributions made by species' champions can be extraordinary. Many have dedicated their career to ensuring a species survives, often eschewing other rewards so their species can prosper. The role that champions play in conservation programs is poorly understood (Sutton 2015) and warrants more detailed analysis. What it was in their background that led them into their life's work has not been studied, at least not in a way that can produce lessons for nurturing the emergence of future champions.

We also know little about degrees of success among species' champions, largely because there is an understandable reluctance to look closely at projects that have failed. Analyses of extinction (Woinarski *et al*. 2017) have identified institutional failings, but have been shy about articulating individual failings.

There is an interesting divergence among the champions in this book. Some champions have been located in the government agencies with explicit fiduciary responsibilities for species' conservation. Other champions have been outside government agencies. There are weaknesses and strengths in both positions. Champions within government may bring to bear the authority and resources of their agencies, but may also be hampered by broader institutional processes and obligations. Champions outside agencies may be able to advocate without political restraint, but may be ineffective at galvanising agency support.

How has policy, law and project governance helped deliver success?

Although government policy was critical to what could be the most successful of Australia's threatened species recovery efforts, such as the humpback whale, fewer than half of the other chapters explicitly acknowledged government policy as underpinning their success. However, it has done so in nearly every case by embedding a requirement to protect threatened species in legislation and then funding programs to ensure policy compliance.

On the other hand, the governance of recovery programs was considered critical to their success. Almost all recovery stories described in this book were of species with recovery plans. Most were implemented by recovery teams. Usually these teams included the champions described above, but many of the champions would not have been as successful without being part of a team. Some teams have been running effectively for decades, using the corporate memory of their long-serving members to build on success and avoid repeating failed approaches. Important features of teams include skill diversity, redundancy (so that absence of key individuals is not debilitating), longevity, expertise, well-established linkages to bodies responsible for management and a capacity to draw on skills and resources beyond team membership (Holmes *et al*. 2016). Such teams welcome periodic external evaluation and are receptive to suggested improvements in their governance and programs. They also have a vision of the future for the species and a realistic plan of how to get there. Guidelines currently being drafted by the Australian Department of Energy and the Environment capture many of these points.

National legislative change now means that recovery plans do not need to be established for all threatened species, and most threatened species have neither recovery plans nor active recovery teams. Hence, across Australia's threatened species, there is an increasing absence of the glue and strategic direction for recovery.

How much funding is required for successful recovery of a species?

Authors of every case study in this book estimated the amount and source of funds spent to date, and that is needed over the next 5 years (Fig. 32.1). There is great variation – some species have cost little to recover; others have been more complex, expensive and time-consuming, and have needed and attracted many millions of dollars from the Australian and state governments, private business and in volunteer time. Rarely has support taken the form of land purchase, either by governments or with private cash, the main exception being the land purchased by the Australian Wildlife Conservancy, mostly for reintroduction of threatened mammals (Chapter 28).

Private cash support is the most variable funding category. Some relatively expensive programs, such as Arid Recovery and the sanctuaries run by the Australian Wildlife Conservancy, are almost entirely privately funded. Some species have benefited from smaller amounts provided through donations to non-government organisations. A more important role of non-government organisations, including local community groups and national organisations dedicated to conservation, has been through the provision of volunteer time. We costed this uniformly at A$25 per hour (2016), but this is almost certainly an underestimate of the costs borne by volunteers, including the opportunity cost of their time, transport costs and other material contributions.

Requirements to sustain conservation gains are often substantial. The case studies here report some notable conservation successes, but even these advances may be fragile, and most cases require an ongoing commitment. For example, the eradication of pests from Macquarie Island needs ongoing support to ensure quarantine protocols are followed by all those visiting the island and to monitor ecological responses to the pest eradication to ensure there are no unintended consequences (Chapter 3). Similarly the ban on harvest of humpback whales requires ongoing diligence by Australian governments to avoid backsliding on agreements. Other programs need ongoing funds to cement gains and to expand programs so that species can be downlisted and ongoing costs institutionalised into routine land management activities. As noted above, some will need ongoing investment for the foreseeable future. Failure to provide this investment increases the risk of extinction and the squandering of past investments.

Compared with other countries, Australian governments have generally chosen to allocate only very limited funding to conservation relative to the conservation needs and biodiversity responsibilities (Waldron *et al.* 2013). It may be that this allocation can be increased if the value and potential benefits of investment are well demonstrated, and a strategic, long-term and marketable package of investment is constructed, as has happened in New South Wales (Chapter 30). This is an important lesson, for much of the past investment in conservation in Australia has been reactive, *ad hoc* and short term. Governments, and other investors, may also be more likely to invest if the case is well made that such investment is likely to deliver success and multiple benefits (e.g. for employment opportunities, tourism or carbon capture and storage). This book shows that such investment can achieve many and lasting benefits, and also that such investment is a necessary component for recovery success.

Is the progress described now entrenched?

The conditions that have caused species to become threatened have mostly developed over a period of two centuries. For some species, the new environment is now so different, and so adverse, that recovery to a pre-European state is unlikely. Others needed only the major threats to be lifted for their populations to return to what are believed to be historic numbers. The question is whether we can now relax – whether any of species have been 'saved' and can persist without ongoing intervention. The answer is no. As noted in the section on funding, even populations of humpback whales and Macquarie Island require some ongoing investment to maintain the gains achieved, although, in most cases, such needs for ongoing funding may be minor compared with the funding that brought the initial major gains.

Recovery for both these projects is arguably more entrenched than for most others in that the legislation for protecting cetaceans is particularly strong and Macquarie Island quarantine is embedded in protocols that are part of recurrent funding. Similarly, private funding for the Australian Wildlife Conservancy reserves is at least partly derived from investment dividends rather than drawing down a fixed reserve of capital. For all the remaining case studies, however, the funding is outside mainstream budgets and requires active endeavour to be maintained from year to year. For example, conservation effort for the forty-spotted pardalote has required volunteers to maintain the momentum during long periods of institutional disinterest and financial drought (Chapter 14).

Although at least 20 of the species discussed in this book occur in protected areas, the chapters highlight that protection of land alone, while often necessary, is very rarely enough for threatened species' conservation – many threatened species need active management. Protection of land removes the threat of acute habitat loss, which may be particularly critical for species with highly localised distributions in diminishing habitats that are coveted by other potential land users, but predators, weeds, fire and many other threats are usually tenure blind.

How do we now prevent more species becoming threatened?

Apart from recovering threatened species, could we have reduced the number of recruits to the threatened species lists had we acted earlier? There is much we can learn about early intervention from the examples in this book. In terms of the features that drove success, an early alert to the community could slow declines before a species crosses threatened species thresholds. In this sense, an official Near Threatened category, as adopted by the IUCN Red List, would act as an early warning and initiate community attachment before action is needed, and possibly before more substantial investments are needed to achieve recovery. Research on species before they are actually threatened is also more likely to demonstrate how a healthy population behaves and uses the landscape – threatened species often live in the forgotten corners of the landscape, not because they thrive there but because those are the places from which they have not been displaced. Early research can also highlight threatening processes and what can be done to ameliorate them, with modelling providing scenarios of what might happen and the consequences of prophylactic action.

Over recent decades, governments have oscillated in their focus on threatened species relative to other approaches and priorities for conservation (NRMMC 2010). The cases presented here show that management actions taken for the recovery of threatened species bring collateral benefit to many other species and many environments. In many cases, research undertaken for threatened species recovery helps to identify threats that affect many other components of biodiversity, which are also helped by subsequent management of those threats.

The need for research, monitoring and effective communication

The situation and requirements of every threatened species are idiosyncratic, but there are some common threads connecting all cases. It may seem self-evident, but threatened species recovery rests heavily on evidence about their ecology, distribution and abundance, particularly the 'weak spots' in their life history, about the workings and impacts of different threats, and about the options to control those threats. Furthermore, recovery practitioners need to know how they are progressing, through monitoring of population size, and reproductive success or threat mitigation, which allows refinement of management practices and priorities. Hence, research and monitoring are critical components for recovery. Communication is also essential to connect those involved in recovery, attract required investment, interest the broader community and disseminate lessons learnt from a single case to the many other cases that may benefit.

Conclusion

Although our cases are diverse, it is difficult to evaluate the representativeness of the species included. Many had high profiles, or were suggested to us by those who were willing to share their successes and experiences. We suspect that our examples represent only a minority of threatened species. Unfortunately, many others have no active recovery plans, recovery teams, recovery efforts or investment. In many cases, there is little or no information on their current trends, and limited information on the management they most require. In part, this neglect is because there is no national commitment to an accounting system that tracks trends for threatened species, or national cataloguing of efforts devoted to the recovery of individual threatened species. Such a systematic overview is much needed, and would help to contextualise the case studies presented here. Although the efforts described here for the case studies presented show that much has been and is being done, with tangible benefits, there is much more that needs to be done.

Pessimism pervades conservation biology (Morton 2017), but this absence of hope can lead to despair and inaction. This book does not seek to sugar coat the state of Australia's natural environments (Cresswell and Murphy 2017) or gloss over existing deficiencies and biases in threatened species listing, funding and recovery effort (Walsh *et al.* 2012); the advances we describe still need to be applied far more widely. We show instead that threatened species recovery can be successful. We draw important lessons from the cases we consider, and identify lessons that could kick-start efforts for all of Australia's threatened species. Even more important is the message from these examples of the necessity and triumph of hope. Many Australians are actively involved in the conservation of threatened species, and their remarkable successes are cause for celebration and a foundation for the future.

References

Black SA, Groombridge JJ, Jones CG (2011) Leadership and conservation effectiveness: finding a better way to lead. *Conservation Letters* **4**, 329–339. doi:10.1111/j.1755-263X.2011.00184.x

Commonwealth of Australia (2015) 'Threatened species strategy.' Australian Government, Canberra.

Cresswell ID, Murphy HT (2017) 'Australia State of the Environment 2016: biodiversity'. Independent report to the Australian Government Minister for the Environment and Energy, Australian Government Department of the Environment and Energy, Canberra.

Holmes TQ, Head BW, Possingham HP, Garnett ST (2016) Strengths and vulnerabilities of Australian networks for conservation of threatened birds. *Oryx*. (Online early) doi.org/10.1017/S003060531600045

McLennan B, Whittaker J, Handmer J (2015) *Emergency Volunteering in Australia: Transforming, not Declining*. Centre for Risk and Community Safety and Bushfire and Natural Hazards CRC, Melbourne.

Morton S (2017) On pessimism in Australian ecology. *Austral Ecology* **42**, 122–131. doi:10.1111/aec.12410

NRMMC (2010) 'Australia's Biodiversity Conservation Strategy 2010–2030.' Natural Resource Management Ministerial Council, Department of Sustainability Environment Water Population and Communities, Canberra.

SCBD (2010) 'Strategic Plan for Biodiversity 2011–2020 and the Aichi Biodiversity targets: living in harmony with nature.' Secretariat of the Convention on Biological Diversity, Montréal, Canada.

Sutton AE (2015) Leadership, management and outcomes of wildlife reintroduction programs: findings from the Sea Eagle Recovery Project. *PeerJ* **3**, e1012. doi 10.7717/peerj.1012

Waldron A, Mooers AO, Miller DC, Nibbelink N, Redding D, Kuhn TS, et al. (2013) Targeting global conservation funding to limit immediate biodiversity declines. *Proceedings of the National Academy of Sciences of the United States of America* **110**, 12144–12148. doi:10.1073/pnas.1221370110

Walsh JC, Watson JEM, Bottrill MC, Joseph LN, Possingham HP (2012) Trends and biases in the listing and recovery planning for threatened species: an Australian case study. *Oryx* **47**, 131–143.

Wilson M, Mirchandani D, Shenouda R (2017) Older-person volunteering in rural and regional Australia: recruitment, retention, and health benefits. *Educational Gerontology* **43**, 139–146. doi:10.1080/03601277.2016.1269546

Woinarski JC, Garnett ST, Legge SM, Lindenmayer DB (2017) The contribution of policy, law, management, research, and advocacy failings to the recent extinctions of three Australian vertebrate species. *Conservation Biology* **31**, 13–23. doi:10.1111/cobi.12852

Index

Acanthocladium dockeri 179–87
Action Plan for Australian Mammals 2012 275
Adelotus brevis 36
aerial baiting
    foxes 138, 239
    Macquarie Island 15, 16, 17–18, 20, 21
Agreement for the Conservation of Albatrosses and Petrels (ACAP) 25, 28
Agrotis infusa 152
Aichi Biodiversity Targets 306
albatrosses 14, 24–5
    conservation status 25
    killed as bycatch 26, 29
    tracking and foraging distributions 27–8
    wandering 14, 24
    white-capped 14, 24
Alectryon oelifolius 142
Alinytjara Wilurara (AW) Natural Resource Management Board 239, 243
Allocasuarina
    leuhmannii 86
    verticillata 76
American bison 283
Anangu, concerns over removal of warru for captive breeding 242
Anangu Pitjantjatjara Yangkunytjatjara (APY) Lands *see* APY Lands
Anas superciliosa superciliosa 17
Antarctic krill 9
Antarctic prions 14, 15
Antarctic terns 16, 20
Antarctic Treaty Consultative Meetings 8
Apis melifera 78
Aptenodytes patagonicus 16
APY Land Management
    concerns over removal of warru for captive breeding 242
    managed warru funding 244
    as part of Warru Recovery Team 239
APY Lands
    buffel grass and fire 240–1
    warru recovery 237–46
Arabian oryx 283
Arid Recovery, SA 259–67
    community involvement 263
    conservation management 261–6
    funding 266
    future 266
    infrastructure 265–6
    initiation of the project 260–1, 263
    local mammal extinctions near Roxby Downs 260
    monitoring sites 264
    native predators reintroduced to improve anti-predator behaviour 264, 266
    people, agencies, governance and accountability 263
    planning and policy 263
    research and biology 262
    threats from overgrazing, rabbits and predators 260, 263–4
    volunteers 263, 266, 267
Arid Recovery Reserve, SA
    bird trends 264
    enclosures 261
    natural increase of plains mouse 262, 264
    population trends 262, 265
    predator-exclusion fencing 261, 262, 263, 264, 265
    rabbits, feral cats and foxes removal 261, 263–4
    reintroductions
        burrowing bettong 261, 262, 264, 265
        greater bilby 262, 264, 274
        greater stick-nest rat 261
        numbats 262
        quolls and dingoes 264, 266
        strategy 264
        western barred bandicoots 261, 264
        woma pythons 262
    reptile trends 264
Asparagus aethiopicus 194
Atlas of Living Australia 311
Atrichornis clamosus 95–104
Atriplex 142
Australasian Species Management Program 289
Australian Antarctic Division 14, 18
Australian Botanic Garden, Mount Annan 162, 163
Australian Fisheries Management Authority 27, 28
Australian lungfish 33, 35, 36

Australian National Botanic Gardens   170, 173, 175
Australian National Parks and Wildlife Service   224
Australian National University   118, 119, 130
Australian Research Council Linkage Projects   230, 233, 234
Australian Trout Foundation   204
Australian Whale Sanctuary   9
Australian Wildlife Conservancy
  conservation of critical weight range mammals   269–78
  conservation management   272–8
  feral cats and foxes removal sites   272
  funding   277–8, 321
  future   278
  infrastructure   277
  management   276
  monitoring   276–7
  people, agencies, governance and accountability   276
  planning and policy   275
  predator-proof fencing   261, 262, 263, 264, 265, 272–3, 277
  reasons for success   278
  reintroduction program   271–2
    bridled nailtail wallaby   274–5
    greater bilby   274–5
    woylie or brush-tailed bettong   273, 275
  research, biology, identification of key threats   274
  volunteer involvement   278
  wildlife sanctuaries in which predators are excluded   270

Balls Pyramid
  banning of recreational climbing on   191
  Lord Howe Island phasmid found on   190–1, 193
  nocturnal fauna surveys   191
  targeted weed control of five-leaved morning glory   193–4
bandicoots
  eastern barred   249–57, 282, 289
  golden   260
  pig-footed   260
  western barred   261, 264
Banrock Station, spiny daisy translocation   182–3, 184, 185, 186
Banrock Station Environmental Trust   185
Baw Baw frog   284, 285
Beecroft Weapons Range, eastern bristlebird   116, 117, 118, 120, 121
bell miner   229, 231–2
Bellinger River turtle   284
bettong
  brush-tailed   138, 273, 275
  burrowing   260, 261, 262, 264, 265, 277, 289
  eastern   289
Bettongia
  gaimardi   289
  lesueur   260, 277, 289
  penicillata   138, 289
BHP Billiton   260, 263, 266
bilbies
  greater   260, 262, 264, 274–5, 277
  lesser   274
Billiatt Conservation Park, SA, fire effect on mallee emu-wrens   299
Biodiversity Conservation Act 2016 (NSW)   160, 297
Biodiversity Conservation Strategy 2010–2030 (Australia)   306–7, 308, 311
Birchs Inlet   289
BirdLife Australia   20, 88, 90, 91, 130, 308
  educational resources focused on threatened mallee birds   300
  evidence-based fire management of threatened mallee bird habitat   300–1
  summit to prevent extinction of mallee bird species   299–300
Bison bison   283
bitou bush   116, 117–18
black duck   17
black-eared miner   299
black-footed ferret   283
black-footed rock-wallaby, APY Lands   136, 237–46
  buffel grass management   240–1, 244, 245
  captive-bred warru returned to APY Lands   241–2
  conservation management   239–45
  distribution   238, 242, 243
  early conservation focus   239
  as Endangered   238
  fox control effects   239
  funding   239, 243–4
  future   244–5
  insurance population and captive management   241–2, 245
  monitoring   242–3
  population genetics   242
  radio-tracking   243
  recovery partners   243–4
  South Australian Warru Recovery Team/Program   239–40, 241, 242, 243, 244, 245
  staff   239, 240, 241, 243–4
  supplementary feeding and water   241
  surveys   238
  threats from foxes and feral cats   239–40, 243, 244, 245
  Warru Rangers role   239, 240, 241
blackberry   193
Blackwood River National Park, WA, white-bellied frog habitat   210

blue petrel   14, 16, 20
blue-tailed skink   283, 284
blue tongue lizard   139
bluebush   142
Bogong High Plains, mountain pygmy-
    possums   149, 155
bogong moths   152, 155
Booderee National Park, NSW
    bitou bush management   116, 117–18
    eastern bristlebird in   116, 117, 118, 120, 121
    feral predator control program   118, 119
Bookmark Biosphere, SA, black-eared miner
    population   299
Bounceback program, South Australia   136,
    137–42
    common fauna reintroductions   138–9, 143
    fauna monitoring   140–2
    fox, goat, rabbit and weed control   139, 140–2
    funding   140, 142–3
    Indigenous support for   139
    infrastructure   142
    landholder and volunteer involvement   142
    locations in South Australia   138
    people, agencies, governance and
        accountability   139–40
    planning and policy   137–9
    recovery indicators   138
    success rate   138
    vegetation monitoring   142
bridled nailtail wallaby   277
    reintroduction programs   274–5, 289
bristlebird
    eastern   115–22
    rufous   96
    western   96
Bronzewing Flora and Fauna Reserve, Vic, fire
    effects on 'insurance' population of black-eared
    miner   299
brown skua   16
brown stringybark   86, 87, 88, 89
Bruny Island, forty-spotted pardalote on   128, 130
brush-tailed bettong   138
    reintroduction program   273, 275, 289
brush-tailed rock-wallaby   284
buffel grass
    as threat to northern hairy-noses wombat   302
    as threat to warru   240–1, 244, 245
bullock bush   142
buloke   86, 87, 88, 91
Bunyip State Park, unsuccessful helmeted
    honeyeater introductions   231, 233, 289
Burramys parvus   147–56, 292
burrowing bettong   260, 261, 289
    over-abundance   262, 264
    over-grazing by   264, 265
    population estimation   277
bycatch of seabirds in longline fisheries   23–30

Cacatua sanguinea   78
Caladenia
    arenaria   160–6
    concolor   160–6
Callitris glaucophylla   162
Calyptorhynchus
    banksii graptogyne   85–93
    lathami halmaturinus   75–83
Canis familiaris   14, 136, 239
Cape Arid National Park, WA   100
cape petrel   14, 16, 20
Capparis mitchellii   142
Capra hircus   66, 136
captive breeding programs
    Australian terrestrial vertebrates   286–9
    Baw Baw frog   284
    Bellinger River turtle   284
    black-footed rock wallaby   241–2
    blue-tailed skink   284
    brush-tailed rock-wallaby   284
    challenges and lessons learned   290–1
    chuditch   283, 289
    eastern barred bandicoot   251, 252, 253, 255,
        282
    failures of release to the wild   289
    funding and resources   290
    greater bilby   262
    helmeted honeyeater   229, 231, 282, 285, 289
    improving 'fitness' of captive-bred
        individuals   282–3, 285, 291
    karri frog   210
    Leadbeater's possum   284
    Lister's gecko   284
    Lord Howe Island phasmid   192–3, 195, 196,
        283
    mountain pygmy-possum   155
    Norfolk Island green parrot   107
    northern corroboree frog   284, 291
    numbat   282, 283
    orange-bellied parrot   282, 284, 289, 291
    overseas successes   283
    plains-wanderer   284, 291
    planning and policy   289
    and pre-release predator awareness
        training   285
    primary objectives   282–3
    regent honeyeater   283
    Shark Bay mouse   289
    southern corroboree frog   282, 283, 284, 291
    strong partnerships underpin success of   289–
        90
    tammar wallaby   289
    Tasmanian devil   284, 289, 290, 291
    western ground parrot   99, 101
    western swamp tortoise   218, 219, 220, 225,
        282, 283, 284
    when to intervene?   291

white-bellied frog   210–11
yellow-spotted bell frog   284
zoos contribution to conservation of Australia's threatened fauna   281–93
see also translocations and reintroductions
Caring for Country Woodlands Birds for Biodiversity Program   128
Caring for our Country Program (Australian Government)   112, 142, 165, 185, 195, 256
carpet python   139
cascade tree frog   36
Catharacta lonnbergi   16, 17
cats *see* feral cats
cattle   66
Cenchrus ciliaris   240
Chaeropus ecaudatus   260
Chamaecytisus prolifer   193
Charlotte Pass, mountain pygmy-possums   151, 155
Christinus guentheri   191
Christmas Island forest skink   283
Christmas Island pipistrelle   291
Chrysanthemoides monilfera ssp. *rotundata*   116
chuditch *see* western quoll
Churchill Island, eastern barred bandicoots   252, 253, 255
climate change
    impact on glossy black-cockatoos   81, 88
    impact on mountain pygmy-possums   154–5
    impact on pygmy bluetongue lizards   61
    impact on spiny rice-flower   51
    impact on western swamp tortoise   224
    impact on white-bellied frog   213
cockatoos
    glossy black   75–83
    red-tailed black   85–93
'Cockies helping Cockies helping Cockies' project   89
cod
    Mary River   33, 35–6, 37, 38, 40
    Murray   201
    trout cod   199–205
Commission for the Conservation of Antarctic Marine Living Resources   27
common brushtail possum
    as competitors to glossy black-cockatoos   78, 79, 80, 82
    reintroductions   138, 143
common starling   107
communication, need for   322
Community Conservation   224
community contributions
    in success   317
    see also community involvement under specific topics, e.g. spiny daisy
Conservation Action Planning (CAP) framework   300, 301
Conservation Measures Partnership   300

Conservation Volunteers Australia   254
Convention of Antarctic Marine Living Resources   8
Convention on Biological Diversity   2, 306
Convention on the Conservation of Migratory Species of Wild Animals   8, 25
Convention on International Trade in Endangered Species (CITES)   8
crest-tailed mulgara   260
crimson rosella   107
crimson spider orchid   160, 161
    conservation management   162–6
    description   162
    as Endangered   160
    population change   165
    population distribution   160, 162
    propagation   163
    rabbit threat   163
    Recovery Plan   162, 163
Cryptoblepharus egeriae   283
Ctenotus lancelini   289
Currawinya National Park, Qld, bilby population collapse   277
Cyanoramphus
    cookii   105–13
    erythrotis   14
    spp., New Zealand   111, 112
Cyprinus carpio   200

Dacelo novaeguineae   127, 230
daisy, spiny   179–87
Daption capense   14
Dasycercus cristicauda   260
Dasyornis
    brachyterus   115–22
    broadbenti litoralis   96
    longirostris   96
Dasyurus geoffroii   138, 266, 283
Department of Biodiversity, Conservation and Attractions (WA)   223
Department of Conservation and Land Management (WA)   220
Department of Defence   116, 118, 119, 120
Department of Environment, Water and Natural Resources (SA)   60, 143, 182, 183, 184, 185, 239, 243, 260, 263
Department of Parks and Wildlife (WA)   210, 211, 212, 224, 276
Department of the Environment and Energy (Cth)   47, 48
desert stringybark   86, 87, 88, 89
dingoes   136, 239, 264
Diomedea exulans   14, 24
Diuris callitrophila   160–6
diving petrel   20
dogs, used for rabbit and rodent detection, Macquarie Island   15, 16, 18, 19
drooping she-oak   76, 77, 81

Dryococelus australis 189–97

East coast Australian humpback whales subpopulation, monitoring 7, 9
eastern barred bandicoot, mainland subspecies 249–57
   captive breeding and release programs 251, 252, 253, 255, 282, 289
   community involvement 252
   conservation management
      to 1989 252
      to 2005 252
      to present 252–3
   distribution 250, 251, 252
   as Endangered 250
   funding 256
   future 256
   genetic diversity 253, 254, 255
   infrastructure 255–6
   insurance population 284
   metapopulation management 255
   monitoring 255
   people, agencies, governance and accountability 254
   population size 251-2, 253
   predator-proof fencing 251, 255
   Recovery Program/Plan/Team 252–3, 254, 255, 256, 257
   release sites and reintroductions 252–3, 254
   research and biology 253–4
   threats from foxes and feral cats 251, 252, 253–4, 255
   volunteer involvement 254, 256
eastern bettong 289
eastern bristlebird, central coastal NSW and Jervis Bay 115–22
   Beecroft Weapons Range 116, 117, 118, 120, 121
   Booderee National Park 116, 117, 118, 119, 120, 121
   conservation management 116–21
   as Endangered 116
   feral predator-control program 118, 119, 121, 122
   fire effects 117, 120, 121
   funding 118–19, 121
   future 121
   geographical populations 116, 117
   higher level education programs 120–1
   impact of bitou bush removal 117–18
   key features in success 118–21
   partner organisations 120
   site-level infrastructure 120
   staff 118, 119, 120
   statistical support and analysis 119–20
   study area and long-term monitoring programs 116–18
   surveys 117
   translocation 118
   volunteer support 120
eastern brown snake 56
eastern grey kangaroo 255
Eastern Tuna and Billfish Fishery (ETBF) 26
   bycatch reduction strategies 27
echidnas 139, 142
Ellen Brook Nature Reserve, WA, western swamp tortoise 218, 219, 222, 223, 224
Elusor macrurus 33
Emoia nativitatis 283
Emydura inspectata see *Pseudemydura umbrina*
Endangered Species Program (Australian Government) 175
Endangered Species Protection Act 1992 128
Environment Protection and Biodiversity Conservation Act 1999 2, 307
   framework 307
   listed species
      eastern barred bandicoot, mainland subspecies 250
      Endangered orchids, NSW 160
      forty-spotted pardalote 129
      glossy black cockatoo, South Australian subspecies 77
      helmeted honeyeater 228
      humpback whale 9
      Lord Howe Island phasmid 191
      malleefowl 66
      mountain pygmy-possum 149, 154
      Norfolk Island green parrot 106
      pygmy bluetongue lizard 56, 60
      red-tailed black cockatoo 86, 88
      spiny daisy 180
      trout cod 200
      Tumut grevillea 170
      western swamp tortoise 218–19
      white-bellied frog 208
      Wimmera rice-flower 44
      yellow-footed rock wallaby 136
   and non-target mortality of a threatened species 18
   number of species listed under 307, 308
   threatened species, Mary River catchment 34
Environmental Protection Act 1986 (WA) 212
Environmental Trust SoS Partnership Program (NSW) 298
   for the Wild Orchids Project 164–5
Eolophus roseicapillus 78, 130
Epping Forest National Park, Qld, northern hairy-nosed wombats 301
Equus ferus przewalskii 283
Eucalyptus
   arenacea 86
   baxteri 86
   camphora 230, 232
   cladocalyx 78
   viminalis 126

Euphausia superba 9
euro 139
European carp 200
European honey bees 78
European rabbit fleas 14
extinct Australian species 1, 14, 36
extinctions, campaigns to prevent 2
Eyre Peninsula, SA 77

Falls Creek, mountain pygmy-possums 149, 150, 154
feral cats
    control and eradication
        Macquarie Island 14, 15
        SA 239–40, 243, 244, 245, 262
    exclusion from AWC sanctuaries 270, 272
    predation
        on black-footed rock-wallabies 239–40, 243, 244
        on eastern barred bandicoots 251, 255
        on eastern bristlebirds 118
        on malleefowl 66, 71
        on mountain pygmy-possums 151, 152–3, 155
        on Norfolk Island green parrots 107, 111
        on western ground parrots 99, 102, 103
        on yellow-footed rock-wallabies 137
Ficus
    macrophylla 193
        subsp. *columnaris* 193
    platypoda 241
Fitzgerald River National Park, WA 98, 99
five-leaved morning glory 193–4
fixed transects 255
flesh-footed shearwater 26, 27
Fleurieu Peninsula, SA 77, 81
Flinders Island, forty-spotted pardalote 128, 131
Flinders Ranges, yellow-footed rock-wallabies 137, 140, 290
Flinders University 57, 60
Flora and Fauna Guarantee Act 1988 (Vic) 149, 154
Forest Grove National Park, WA, white-bellied frog habitat 210
forty-spotted pardalote, Tasmania 125–32
    Bruny Island 128, 130
    community involvement 128
    competition 127, 130, 131
    conservation management 127–30
    current and former range 126
    Flinders Island 128, 131
    funding 130
    future 130–1
    habitat protection 128, 129
    Maria Island National Park 128
    monitoring 129–30
    nesting box installation 128, 129, 130, 131

    recovery planning and policy 129
    Recovery Program 127–8, 130, 131
    reliance on white or manna gum 126–7, 128, 130
    research 128
    as Threatened 129
    threats 128
Foundation for Australia's Most Endangered Species (FAME) 143, 185
foxes
    control measures, SA 138, 139, 140, 239, 252, 262
    exclusion from AWC sanctuaries 270, 272
    predation
        on arid area mammals 260, 262
        on black-footed rock-wallabies 239–40, 244
        on eastern barred bandicoots 251, 252, 253–4, 255
        on eastern bristlebirds 118
        on malleefowl 66, 71
        on western ground parrots 99
        on western swamp tortoises 222
        on yellow-footed rock-wallabies 136
French Island, eastern barred bandicoots 252, 253, 256
freshwater mullet 33, 35, 36–7
Friends of Arid Recovery 263
Friends of the Helmeted Honeyeater 228, 230, 233, 234, 235
Friends of the Western Ground Parrot 101, 103
frogs
    Baw Baw 284, 285
    cascade tree 26
    giant barred 33, 35, 36, 37, 38–9
    karri 210
    northern corroboree 283, 284, 291
    orange-bellied 210
    sharp-snouted day 283
    southern corroboree 282, 283, 284, 289, 291
    southern day 36
    southern gastric brooding 36
    tusked 26
    white-bellied 207–14
    yellow-spotted bell 284
funding
    how much is needed? 320
    of successful threatened species recovery programs 318, 320
    see also funding under specific topics, e.g. noisy scrub-bird

galahs 78, 130
galaxiids 204
Gallirallus australis 14, 15
geckoes 264
    Lister's 283, 284

genetic rescue   151, 232, 233, 234, 292
Geocrinia
  alba   207–14
    captive conservation project, Perth Zoo   211
  rosea   210
  vitellina   210
giant barred frog   33, 35, 36, 37, 38
  community connection to   38–9
giant petrel
  conservation status   25
  Macquarie Island   18
  tracking and foraging distributions   28
gillnet fisheries, seabird deaths and injuries   24
Glencore mining company   301
glossy black cockatoo, Kangaroo Island   75–83
  artificial nest boxes   78, 79, 80
  brushtail possum predation prevention   78, 79, 80, 82
  comparison of South Australians subspecies with eastern Australian subspecies   76
  conservation management   77–82
  current and presumed historical range   76–7
  dealing with an unbalanced ecosystem   77–8
  as Endangered   77
  feeding on drooping she-oak seeds   76
  funding and staff   80–1, 82
  future   82
  key elements to success   80–2
  landholder involvement   82
  monitoring   78, 79, 80
  Recovery Program/Team   78, 79, 80, 81–3
  reproductive rate/success   77–8, 79–80
  research   81
  results of 20 years of continuous management   79–80
  threats from predation, competition and land clearance   78, 80
  volunteer programs and activities   79, 81, 82
Glyciphila melanops   99
goannas   264
  sand   139, 142
goats   66, 136, 137
golden bandicoot   260
golden lion tamarin   283
Gould's mouse   260
Gould's petrel   2
governments
  funding of conservation needs and biodiversity responsibilities   320
  responsibility for environmental protection and regulation   2, 305–7
greater bilby   260, 277
  population trend, Scotia Wildlife Sanctuary   276
  reintroductions   262, 274–5
greater stick-nest rat   260, 261, 289
Greening Australia   60, 184

Greenpeace anti-whaling campaigns   8
Grevillea wilkinsonii   169–76
grey petrel   14, 15, 16, 20
ground asparagus   194
Gymnobelideus leadbeateri   232, 284

Halobaena caerulea   14
Hamilton, Vic, eastern barred bandicoots   252, 253, 254
Hattah–Kulkyne National Park, Vic, mallee emu-wrens   299, 300
Healesville Sanctuary, helmeted honeyeater captive breeding program   231, 289
helmeted honeyeater   227–35
  biodiversity benefits from Recovery Program   232
  Bunyip State Park, introduction failure   231, 233
  captive breeding and release   229, 231, 282, 289
  competition with bell miners   229, 231–2
  conservation management   229–34
  as Critically Endangered   228
  funding   233–4
  future   234
  genetic management   229, 232, 234
  habitat protection and restoration   230, 232, 234, 235
  insurance population   284
  key threats   229–30
  keys to success   233–4
  low rate of juvenile recruitment   230–1
  political support   233
  population size and trend   228, 232
  protection of nests from predators   230
  Recovery Team/Plan/Program   229, 231, 232, 233, 234, 235
  research and monitoring   233
  restricted distribution   228–9
  supplementary feeding   229, 231, 234
  survival rate of captive-bred following pre-release predator training   285
  translocation program   231, 233, 234
  volunteer involvement and community engagement   229, 230, 231, 233, 234, 235
  Yellingbo Nature Conservation Reserve, Vic   228–34
honeyeaters
  helmeted   227–35, 282, 289
  regent   283
  tawny-crowned   99
  yellow-tufted   228, 231
hopping-mouse
  long-tailed   260
  spinifex   262
house mice, Macquarie Island   14, 15
  eradication project   15–16
  impact on ecosystem   17

humpback whale Australian subpopulations
    conservation management   8–10
    East and West coast subpopulations   6, 9–10
    funding   320, 321
    future   10
    monitoring   9–10
    planning and policy   8–9
    recovery   5–11
    research, biology and threats   9
    as Vulnerable species   9
    whaling impact   6–8, 10
humpback whales
    global protection   8
    IWC ban on killing   8
Huperzia australiana   16

iconic species   38–9, 155
Idalia National Park, bridled nailtail wallaby introduction   274
idnya   138, 139
iga   142
Ikara-Flinders Ranges National Park, SA   138, 139, 140, 141
Insektus   192
insurance populations   111, 112, 241–2, 256, 266, 282, 283–4, 299, 309
international biodiversity agreements   306
International Whaling Commission (IWC)
    ban on killing humpback whales in Southern Hemisphere   8, 10
    establishment   8
invasive species eradication, Macquarie Island   13–21, 320
Ipomoea cairica   193–4
Isoodon auratus   260
IUCN Red List   20, 143, 190, 200, 218
    Near Threatened category   321
IUCN/SSC's Conservation Breeding Specialist Group, guide to captive breeding for conservation   289

Kalka – Pipalyatjara Indigenous Protected Area   244
Kangaroo Island, glossy black cockatoo   75–83
kangaroos
    eastern grey   255
    red   139
    western grey   139
Karakamia Wildlife Sanctuary, WA, mammal reintroductions   272, 273
karri frog   210
kelp gull   17
king penguin   16
Kosciuszko National Park, NSW, mountain pygmy-possums   148, 149, 151–2, 154, 155
Kowree Tree Farm Group   89, 90

La Trobe University   300

Lagorchestes hirsutus   289
Lancelin Island skink   289
Larus dominicanus   17
Lasiorhinus krefftii   301–2
Lathamus discolor   128
laughing kookaburra   127, 230
Leadbeater's possum   232, 284
Leipoa ocellata   65–73, 299
Leontopithecus rosalia   283
Lepidodactylus listeri   283
Leporillus conditor   260, 289
Leptospermum lanigerum   230
Lerista spp.   264
lesser bilby   274
Lichenostomus melanops
    cassidix   227–35, 282
    gippslandicus   231, 232
linear transects   212
Lister's gecko   283, 284
Litoria
    castanea   284
    pearsoniana   36
little corella   78, 80
Living with Threatened Species Program   36, 37, 40
lizards
    blue tongue   139
    pygmy bluetongue   55–62
    sleepy   139
long-tailed hopping-mouse   260
Long-term Ecological Network (LTERN)   119, 121
longline fisheries, seabird bycatch management   23–30
Longline Fishing Threat Abatement Plan (TAP)   25–6, 27, 28
Lord Howe Island
    planned reintroduction of phasmids to   193, 194, 196
    providence petrel   2
    rodent eradication program   193, 195–6
    weed eradication and habitat recovery   194
Lord Howe Island banyan   193
Lord Howe Island Board plant nursery, phasmid captive breeding program   192, 193
Lord Howe Island boobook owl   196
Lord Howe Island gecko   191
Lord Howe Island melaleuca   190, 193
    smothered by five-leaved morning glory   193–4
Lord Howe Island phasmid   189–97
    captive breeding program   192–3, 195, 196, 283
    conservation management   192–5
    as Critically Endangered   190, 191
    description   190
    extirpated by black rats by the 1930s   190
    funding   195
    future   195–6
    limited genetic diversity   192, 196

media involvement and community
    engagement   194–5
palatability trials of Lord Howe Island
    plants   193
plans to reintroduce to Lord Howe Island   193,
    194, 196
population size   191
Recovery Plan/Team   191, 196
rediscovered on Balls Pyramid   190–1, 193
risks of reintroduction to Lord Howe
    Island   196
targeted weed control of known habitat on Balls
    Pyramid   193–4
Lord Howe Island skink   191

Maccullochella
    macquariensis   199–205
    mariensis   33
    peeli   201
Macquarie Island, invasive species
    eradication   13–21
    bird breeding response   16, 20
    conservation management strategies   14–21
    environmental impact of pest eradication
        16–17
    factors in success   18
    funding and budget   19, 21, 320, 321
    future   21
    introduced mammals   14, 15–16
    management and planning   17–18, 19, 20
    measures of success   20–1
    monitoring   18
    pest eradication project   15–16, 17–18, 20–1
    project wins awards   21
    regulatory and logistical challenges   17–18
    support from New Zealand staff   19–20
    vegetation response   16, 20
Macquarie Island cabbage   16
Macquarie Island parakeet   14, 21
Macquarie Island rail   14, 21
Macronectes halli   17
Macropus
    fulginosus   139
    giganteus   255
    robustus   139
    rufus   139
Macrotis
    lagotis   260
    leucura   274
Maireana   142
mala   289
Mallee Catchment Management Authority   300
Mallee Cliffs, NSW, planned wildlife
    introductions   274
mallee emu-wren   299, 300
malleefowl   65–73, 299
    Adaptive Management Project   71, 72
    biology   66

conservation management   66–72
distribution   66
fox and cat control   71
funding   71–2
future   72
infrastructure   71
key threats   66–7
monitoring and data management   68–70, 71,
    72
nest mound monitoring   68, 71
on-ground management   70–1
people, governance and accountability   67–8
planning and policy   67
reasons for success   72–3
Recovery Plan/Team   67–8, 70, 71, 72
volunteer support   69, 70, 71
as Vulnerable   66
Malleefowl Recovery Program   68
manna gum   126–7
Manorina
    melanocephala   127, 130
    melanophrys   229, 231–2
    melanotis   299
Maria Island National Park, Tas, forty-spotted
    pardalote   128
mark–recapture methodology   140, 141, 243, 255,
    277
Mary River Catchment Coordinating
    Committee   36, 40
Mary River catchment, recovery of threatened
    aquatic species   33–41
    community engagement   36, 37, 38–9, 40
    conservation management   35–40
    effectiveness of actions   37
    future   40
    landholder role   35, 36, 37, 38, 39, 40
    location and threats   34, 36
    recovery actions   36–7
    riparian zone improvement   39, 40
    stream frogs   36
    success factors   37–8
        commitment: role of key individuals and
            organisational longevity   39–40
        community: involvement has created better
            plans and better implementation   40
        connections: use of iconic species and iconic
            habitat   38–9
Mary River cod   33, 35, 36, 38, 40
    community connection to   38
    fingerlings release   37, 38
    increased range   37
    Recovery Plan   35–6, 38, 40
Mary River Cod Recovery Program/Team   35, 37,
    40
    organisational support and funding   35–6
    reasons for program's success   37–40
Mary River Threatened Species Recovery Plan   36,
    37, 38, 40

community engagement 37
funding and investment 37
recovery actions 37
Mary River turtle 33, 35, 36, 37, 38
community connection to 38
conservation project 36
scholarship program 36
masked owl 195, 196
measuring and reporting success
challenges 307–9
opportunities 309–11
megapodes 65–6
Megaptera novaeangliae, Australian
population 5–11
Melaleuca
howeana 190–1
squarrosa 230
Melbourne Strategic Assessment 45, 46
Melbourne Zoo
Lord Howe Island phasmid captive breeding program 192, 193, 195
palatability trials of Lord Howe Island plants (for plasmids) 193
miners
bell 229, 231–2
black-eared 299
noisy 127, 130
Mixophyes iteratus 33
mobilising resources for threatened species recovery 295–302
bringing people together for threatened mallee birds 298–301, 302
the 'good cause' 296
making a business case: NSW Government's Saving our Species Program 297–8, 302
partnerships for success: northern hairy-nosed wombat 301–2
resources available to conserve species 296–7
see also funding under specific topics, e.g. spiny daisy
Mogumber Nature Reserve, WA, western swamp tortoise 220, 224
Monarto Zoo, SA
greater bilby captive breeding 262
pygmy bluetongue lizard captive breeding 61
warru captive breeding 241, 242, 243
Moore River Nature Reserve, WA, western swamp tortoise 220
Morelia spilota 139
Moreton Bay fig 193
morning glory, five-leaved 193–4
mountain plum-pine 148, 151, 152
mountain pygmy-possum 147–56
captive breeding program 155
climate change effects 154–5, 156
conservation management
NSW 151–2, 153
Victoria 149–51, 153–4

conservation status 148–9
distribution 148, 149–51, 152
drought effects 152
as Endangered 149
environmental requirement 154–5
feeding on bogong moths 152, 155
feeding on mountain plum-pine 148, 151
fossil and sub-fossil remains 148
funding 154
future 154–5
gene-pool mixing 151, 292
as icon of the Australian Alps 155
key threats from habitat loss, predators and ski slope development 150, 151, 152–3, 155
planning and policy 153–4
population numbers 149, 152, 153
possum crossings 150, 151, 152
predator-control program 151, 152–3, 155
Recovery Plans/Priority Action Statements 153, 154
translocations 151
volunteer involvement 154
mountain swamp gum 230
mouse
Gould's 260
plains 262, 264
Shark Bay 289
see also house mice
Mt Blue Cow, mountain pygmy-possums 149, 150, 151, 152–3, 155
feral cat control 152–3
funding 154
monitoring and recovery program 154
Recovery Team and Plan 153
Mt Bogong to Mt Higginbotham region, mountain pygmy-possums 130
Mt Buller, mountain pygmy-possums 148, 149, 150–1, 152, 154, 155
monitoring and recovery program 154
Recovery Plan 151
Mt Gibson Wildlife Sanctuary, WA
cost of introduction projects 277
mammal reintroductions 272, 273
Mt Hotham, mountain pygmy-possums 148, 149, 150, 154, 155
Mt Rothwell Biodiversity Interpretation Centre, eastern barred bandicoots 252, 254, 255, 256
mulga snake 262
mullet, freshwater 33, 35, 36–7
Multi-Regional Malleefowl Project 70
Murray cod 201
Murray–Darling Basin, trout cod recovery 199–205
Murray Mallee, fire effects on threatened mallee birds 299
Murray–Sunset National Park, Vic, mallee emu-wrens 299
Mus musculus see house mice

Mustela nigripes 283
Myrmecobius fasciatus 262, 282
Myuchelys georgesi 284

national biodiversity policy 306–7
National Environmental Information Infrastructure 311
national environmental law 307
National Environmental Science Program 310–11
National Landcare Program 184, 185
national law and policy protecting threatened species 306
National Malleefowl Monitoring Database 70, 72
National Malleefowl Monitoring Manual 68
National Malleefowl Recovery Team 67–8, 70, 71
National Parks and Wildlife Act 1972 (SA) 56, 136, 180, 238
National Parks and Wildlife Act 1974 (NSW) 148
National Parks and Wildlife Service (NSW) 153, 154, 171, 173, 191
National Recovery Plan
    for crimson spider orchid 163
    for mallee emu-wren 300
    for mountain pygmy-possum 154
    for red-lored whistler 300
    for the spiny rice-flower 46
    for trout cod 202, 204
    for western whipbird 300
national reporting framework 310
native fig 241
Native Fish Australia 203
native orange 142
Natural Heritage Trust 71, 128, 184, 185, 224, 244
Nematoceras
    dienemum 16
    sulcatum 16
Neoceratodus fosteri 33
Neophema chrysogaster 282
nesting box installation 78, 79, 128, 129, 130
New Zealand Department of Conservation, Island Eradication Advisory Group 19, 20
Ninox novaeseelandiae
    albaria 196
    undulata 196
noisy miner 127, 130
noisy scrub-bird 95–104
    biology 99
    conservation management 96–103
    as Endangered 96
    fire risk 98
    funding 102–3
    future 103
    habitat management and translocation 97, 98, 99, 101, 102
    historical and current distribution 96, 97, 111
    infrastructure 99–100
    people 100–1, 103
    recovery actions 97–8
    Recovery Team 98, 102, 103
    social support and governance 102
    volunteer involvement 101, 103
Norfolk Island, providence petrel 2
Norfolk Island boobook owl 196
Norfolk Island green parrot 105–13
    biology 109
    captive breeding program 107
    conservation management 106–12
    distribution 110
    as Endangered 106
    funding 112
    future 112–13
    infrastructure and technology 111–12
    management and monitoring 108–9, 111–12
    nesting success 108, 109, 111–12
    people, agencies, governance and accountability 110–11
    planning and policy 110
    population surveys 107, 108, 111
    predation threats and competition 107, 109, 111, 112
    recovery team 112–13
    translocation 111, 112
    volunteers 111, 112
Norfolk Island National Park 108, 110, 111, 112, 113
Norfolk Island Region Threatened Species Recovery Plan 106, 110
northern corroboree frog 283, 284, 291
northern giant petrel 17–18
northern hairy-nosed wombat 301–2
Norway rats, eradication 15
Notechis scutatus 230
Notomys
    alexis 262
    longicaudatus 260
numbats 262, 277, 282, 283, 289

Oaklands diuris 160
    conservation management 162–6
    description 161–2
    as Endangered 160
    'Orchid in a Teacup' sculpture, Oaklands 166
    population change 165
    population distribution 160, 162
    propagation 163
    Recovery Plan 162, 163–4
    and strategic grazing regimes 164
Oceanites spp. 14
Office of Environment and Heritage (NSW) 162, 164, 165, 171, 176
    Saving our Species Program 153, 154, 165, 173, 174, 175, 297–8
Olary Ranges, SA, yellow-footed rock-wallabies 136, 139, 140
Oligosoma lichenigera 191
Onychogalea fraenata 289

Open Standards for the Practice of
    Conservation   300
orange-bellied frogs   210, 211
orange-bellied parrot   282, 283, 284, 289, 292
orchids
    crimson spider   160, 161, 162, 163, 165
    Oaklands diuris   160, 161–2, 164, 165
    sandhill spider   160, 161, 163, 164, 165
orchids, Endangered species, southern
    NSW   159–67
    conservation management   162–6
    funding   164–6
    future   166
    management   165
    monitoring   165
    people, agencies, governance and
        accountability   164
    planning and policy   163–4
    pollinators and mycorrhizal partners   163, 164, 166
    propagation   162–3, 166
    reasons for success   166
    reintroductions   166
    research, biology, identification of key
        threats   163
Oryctolagus cuniculus see rabbits
Oryx leucoryx   283
owls
    Lord Howe Island boobook   196
    masked   195, 196
    Norfolk Island boobook   196

Pachycephala rufogularis   299
Pandorea doratoxylon   241
pardalotes
    forty-spotted   125–32
    spotted   127
    striated   127
Pardalotus
    punctatus   127
    quadragintus   125–32
    striatus   127
Parks Australia   116, 118, 119, 120, 121
parrots
    Norfolk Island green   105–13
    orange-bellied   282, 283, 284, 289, 292
    swift   128
partnerships for success
    Arid Recovery   259–67
    Australian Wildlife Conservancy   270, 276
    Bounceback program   137, 143, 144
    captive breeding programs in zoos   289–90
    Conservation Measures Partnership   300
    eastern bristlebird   120, 121, 122
    Endangered orchids, NSW   162–3, 164, 165–6
    forty-spotted pardalote   127, 131
    helmeted honeyeater   227–35

Mary River catchment   38–9
mountain pygmy-possum   147, 153
Norfolk Island green parrot   106, 111
northern hairy-nosed wombat   301–2
red-tailed black cockatoos   90
spiny daisy   182–3, 184, 185, 186, 187
spiny rice-flower   47, 51
trout cod   203, 204
warru   243–4
western swamp tortoise   224
white-bellied frog   211
Passeromyia longicornis   128
Pedionomus torquatus   284
Perameles
    bougainville   261
    gunnii   249–56
Perca fluviatilis   200
Perth Zoo   210, 224, 276
    Geocrinia species 'head-start' captive
        conservation project   211
    western swamp tortoise captive breeding
        program   219, 220, 223, 224
pest eradication project, Macquarie Island
    13–21
Petaurus breviceps   130
petrels
    blue   14, 16, 20
    cape   14, 16, 20
    diving   20
    giant   17–18, 25, 28
    Gould's   2
    grey   14, 15, 16, 20
    providence   2
    soft-plumed   14, 20
    storm   14
    white-headed   14, 15
Petrogale
    lateralis   136, 237–45
    penicillata   284
    xanthopus ssp. *xanthopus*   135–44, 241
Pezoporus flaviventris   95–104
Phasmid: Saving the Lord Howe Island Stick Insect
    (children's book)   194, 195
phasmid, Lord Howe Island   189–97, 283
Phillip Island (near Norfolk Island), translocation
    of Norfolk Island green parrot to   111, 112
Phillip Island, Vic, eastern barred
    bandicoots   252, 253, 256
Philoria frosti   284
pig-footed bandicoot   260
Pilliga, NSW, planned wildlife introductions   273, 274
Pimelea and Her Grassland Friends (children's
    book)   49
Pimelea Conservation Trust   47–8, 50, 51
Pimelea spinescens   44, 46
    Action Statement for   46

Recovery Team 46–7, 48, 49, 50, 51
subsp. *pubiflora* 44, 45, 48
    distribution 44, 45
subsp. *spinescens* 44, 45, 48
    distribution 45
    genetic variation 49
    Policy Statement on Significant Impact
        Guidelines for 46
    surveys 50
    translocation 49
Pinkie and Pete (children's book) 61
Pipistrellus murrayi 291
plains mouse 262, 264
plains-wanderer 284, 291
Platycercus elegans 107
Pleurophyllum hookeri 16
Poa
    annua 16
    foliosa 16
Podocarpus lawrencei 148, 151
policy, law and project governance 319
    see also planning and policy under specific
        topics, e.g. Tumut grevillea
possum crossings 150, 151, 153
possums
    common brushtail 78, 79, 80, 82, 138, 143
    Leadbeater's 232, 284
    see also mountain pygmy-possum
predator-proof fencing
    Arid Recovery Reserve, SA 261, 262, 263, 264, 265
    Australian Wildlife Conservancy 261, 262, 263, 264, 265, 272–3
    eastern barred bandicoot, mainland
        subspecies 251, 255
prevention of species becoming threatened 321
Priority Action Statements (NSW)
    Endangered orchids 164
    mountain pygmy-possum 153
    review 297
    Tumut grevillea 173, 174
Private Forest Reserves Program 128, 130
Procellaria 28
    cinerea 14
Protected Areas on Private Land Program 128, 130
Protecting Victoria's Environment – Biodiversity 2037 51
providence petrel 2
Przewalski's horse 283
Pseudechis australis 262
Pseudemydura umbrina 217–26, 282
Pseudomys
    australis 262
    fieldi 289
    gouldii 260
Pseudonaja textilis 56

Pseudophryne
    corroboree 282
    pengilleyi 283
Psophodes nigrogularis 300
    nigroagularis 96
    oberon 96
Pterodroma mollis 14
Puffinus carneipes 26
purse-seine fisheries, seabird deaths and
    injuries 24
Pygmy Bluetongue Conservation Association 61
pygmy bluetongue lizard 55–62
    biology 58–9
    burrows 57, 58, 59, 60
    conservation management 58–61
    considered possibly extinct 56
    distribution 56–7
    as Endangered 56
    future 61
    local community involvement 61
    monitoring 58–9
    population management 59, 60
    Recovery Plan/Team 57–8, 60, 61
    rediscovery near Burra, SA 56
    supporting groups and funding 60, 61
    translocation 61
pygmy-possum, mountain 147–56, 292
pythons
    carpet 139
    woma 262

Queensland Department of Primary Industries
    (Fisheries) 35–6
quolls, western 138, 142, 143, 264, 266, 283, 289

rabbit calicivirus disease 260, 267
rabbit haemorrhagic disease 20, 137, 139, 140
rabbits 66, 260
    Arid Recovery Reserve, SA 260, 261
    control measures, SA 137, 138, 139, 140, 141, 255
    eradication campaigns, Antarctic islands 15
    impact on Endangered orchids 163
    Macquarie Island 14–16, 17, 20
rainbow lorikeets 130
Rallus philippensis macquariensis 14
rats
    Lord Howe Island 190
    Macquarie Island 14, 15–16, 17
    Norfolk Island 107, 111
    see also stick-nest rat
Recovery Plan for Stream Frogs of South-East
    Queensland 36
recovery plans/programs
    and government policy/legislation 319
    see also recovery plans/programs under specific
        topics, e.g. Lord Howe Island phasmid

recovery teams
  importance of   310
  national guidelines   310
  see also recovery teams under specific topics, e.g. malleefowl
red fox see foxes
red kangaroo   139
red-lored whistler   299
red-tailed black-cockatoo   85–93
  annual count by volunteers   88–9
  breeding success rates   91
  broader public awareness   89, 92
  conservation management   88–9
  as Endangered   86
  fire effects   87, 88, 89–90
  funding   91
  future   91–2
  governance   90–1
  land clearance effects   87–8
  landholder involvement   89, 90
  landscape restoration projects   89, 90, 91
  nesting habitat   87, 88
  partner organisations   91
  Recovery Team   88, 89, 90–2
  research enhancing success   89–90
  selective feeding in buloke and stringybark   86–7, 88, 89, 91
  south-eastern subspecies, distribution   86, 87
redfin   200
regent honeyeater   283
reintroductions see translocations and reintroductions
resource mobilisation   296
Restoring Riparian Resilience Project   40
Rheobatrachus silus   36
rice-flower, spiny   43–52
Richard Underwood Nature Refuge, northern hairy-nosed wombat   301
riparian zone improvement   203
  Mary River catchment   39, 40
  to support galaxiids   204
rock-wallabies
  black-footed   136, 237–46
  brush-tailed   284
  yellow-footed   135–44
Royal Botanical Gardens Victoria   48, 162–3
Rubus fruticosa   193
rufous bristlebird, western subspecies   96

saltbush   142
sand goannas   139, 142
sandhill spider orchid   160, 161, 164
  conservation management   162–6
  description   161
  as Endangered   160
  population change   165
  population distribution   160, 161, 163

propagation   163
Recovery Plan   162, 163
and strategic grazing regimes   164
sandsliders   264
Sarcophilus harrisii   283, 309
Save the Tasmanian Devil Program   309
Saving our Species Program (NSW Government)   153, 154, 165, 173, 174, 175, 297–8
  categorisation of threatened species   298
  prioritisation and funding   298
scented paperbark   230
'Schools helping Cockies helping Cockies' project   89
Scotia Wildlife Sanctuary, NSW, mammal reintroductions   273, 274, 275, 276
scrub-bird, noisy   95–104
seabird bycatch management in longline fisheries   23–30
  bird species affected   24–5, 26–7
  conservation management   25–8
  data gathering   26–7
  funding   28
  future   28–9
  Longline Fishing TAP   25–6, 27, 28
  mitigation solutions   27, 28, 29
  monitoring   27, 28
  planning and policy   25–6
  success achieved   29–30
  tracking of seabirds   27–8
seabird conservation, Australian government approach to   25
Seven Creeks, total fishing closure   201
Shark Bay mouse   289
sharp-snouted day frog   283
she-oaks, drooping   76, 77, 81
shearwaters, flesh-footed   26, 27
sheep   66
ship rats, Macquarie Island   14, 15–16, 17
skinks
  blue-tailed   283, 284
  Christmas Island forest   283
  Lancelin Island   289
  Lord Howe Island   191
skuas   17, 18
  brown   16
sleepy lizard   139
snakes
  eastern brown   45
  mulga   262
  tiger   230
Society for Growing Australian Native Plants   170, 175
soft-plumaged petrel   14, 20
South Australian Museum   57, 60
South Australian Warru Recovery Team   239–40, 241, 242, 243, 244, 245

national recognition   244
South Pacific Regional Environment
  Programme   8
southern bluefin tuna   27
southern corroboree frog   282, 283, 284, 289, 291
southern day frog   36
southern gastric brooding frog   36
Southern Hemisphere, humpback whale kills   7
Soviet illegal whaling   7, 8
spear bush   241
species' champions, contributions made by
  317–19
Spilopsyllus cuniculi   14
spinifex hopping-mouse   262
spiny daisy   179–87
  conservation management   181–6
  as Critically Endangered   180
  current and historical distribution   181, 183, 185
  funding   184, 185–6
  future   186–7
  lack of genetic diversity   183
  landholder and council involvement   181–2
  link to famous Burke and Wills story   180, 183, 187
  monitoring   184–5
  partnership with Banrock Station   182–3, 184, 185, 186, 187
  people, agencies, governance and accountability   184
  propagation   182, 183
  Recovery Team/Program/Plan   181–2, 183, 184–5, 186, 187
  research, biology and identification of key threats   183
  roadside sites   181, 182, 183
  seed production and viability   181
  translocation sites   182–3, 184, 186–7
  volunteer and community involvement   186, 187
spiny rice-flower   43–52
  conservation management   45–51
  as Critically Endangered   44
  distribution, Victoria   44, 45
  funding   47–8
  future   51
  genetic variation   48, 51
  learnings   49–50
  monitoring   50–1
  planning and policy   46
  project management   48, 50
  Recovery Team   46–7, 48, 49, 50, 51
  research, biology and ecology   45–6, 48
  stakeholder support and involvement with the program   50
  threats   46
  translocation   49

spotlight transects   140, 277
spotted pardalote   127
State Recovery Plan (sandhill spider orchid)   163
Sterna vittata   20
stick-nest rat, greater   260, 261, 289
Stilbocarpa polaris   16
Stipiturus mallee   299
storm petrel   14
Strategic Plan for Biodiversity 2011–2020   306
Strategy for the National Reserve System 2009–2030   307
striated pardalote   127
stringybarks   86, 87, 88, 89, 91
strip-transect population monitoring   277
Sturnus vulgaris   107
'success' in threatened species recovery   2–3
  challenges of measuring and reporting   307–9
  community contributions role   317
  factors influencing   315–22
  funding requirements for   318, 320
  and need for ongoing investment in entrenched projects   320–1
  opportunities for measuring and reporting success   309–11
  policy, law and project governance role   319
  and prevention in more species becoming threatened   321
  reporting on, the national policy context   305–11
  species' champions role   317–19
sugar glider   130
sugar gums   78
swift parrot   128

Tachyglossus aculeatusi   139
tammar wallaby   289
Tasman parakeet *see* Norfolk Island green parrot
Tasmania Parks and Wildlife Service   15, 16, 19, 20
Tasmanian devil   283, 289, 290
  disease considerations in captive management   284, 292
  Save the Tasmanian Devil Program   309
Taudactylus
  acutirostris   283
  diurnus   36
tawny-crowned honeyeater   99
Terrestrial Ecosystem Research Network (TERN)   119
threatened Australian species   1
Threatened Mallee Bird Recovery Program
  development   299–301
  use of Conservation Action Planning (CAP) framework   300, 301
threatened mallee birds, bringing people together for   298–301

threatened species
    how to prevent species becoming
        threatened? 321
    as Matters of National Environmental
        Significance 2
Threatened Species Commissioner 112, 154, 310
threatened species conservation, governance
    for 2, 305–7
Threatened Species Conservation Act 1995
    (NSW) 149, 153, 170, 191
Threatened Species Network 130, 185
Threatened Species Protection Act 1995 (Tas) 129
threatened species recovery
    characteristics of threatened species that have
        been recovered 316–17
    factors influencing 315–22
    mobilising resources for 295–302
    need for research, monitoring and effective
        communication 322
    questions 3
    reporting on success, national policy
        context 305–11
    'success' in 2–3
Threatened Species Recovery Hub (of the National
    Environmental Science Program) 275, 310
Threatened Species Strategy (Australian
    Government) 45, 46, 106, 154, 244, 311
    function 310
    importance of recovery team 310
Thunnus maccoyii 27
Tiaro and District Landcare 36, 38
tiger snake 230
Tiliqua
    adelaidensis 55–62
    scincoides 139
Tiverton (property near Mortlake), Vic., eastern
    barred bandicoot site 252, 253, 255, 256
tortoises, western swamp 217–26, 282, 283, 284,
    289
Trachydosurus rugosus 139
Trachystoma petardi 33
traditional owner involvement
    warru 237–46
    yellow-footed rock-wallabies 137, 139
translocations and reintroductions
    bridled nailtail wallaby 273–4, 289
    brush-tailed bettong 273, 275, 289
    burrowing bettong 261, 262, 264, 265, 289
    captive-bred animals from zoos 284–5, 289
    common brushtail possum 138, 143
    common fauna, Bounceback program 138–9,
        143
    eastern bettong 289
    eastern bristlebird 118
    Endangered orchids, NSW 166
    greater bilby 262, 264, 274–5
    greater stick-nest rat 261, 289

helmeted honeyeater 231, 233, 234
Lord Howe Island plasmid 193, 194, 196
mala 289
mallee emu-wren 300
mountain pygmy-possum 151
noisy scrub-bird 97, 98, 99, 101, 102
Norfolk Island green parrot 111, 112
northern hairy-nosed wombat 301
numbats 262, 289
pygmy bluetongue lizard 61
southern corroboree frog 289
spiny daisy 182–3, 184, 186–7
spiny-rice flower 49
tammar wallaby 289
Tasmanian devil 289
western barred bandicoots 261, 264
western swamp tortoise 219, 220–1, 223,
    224–5
white-bellied frog 211
wild populations vs captive-bred 285–6
woma pythons 262
woylie 273
trawl fisheries
    bycatch mitigation strategies 28
    seabird deaths and injuries 24
tree lucerne 193
Trees for Life 184, 186
Trichoglossus moluccanus 130
Trichosurus vulpecula 78, 138
trout cod, Murray–Darling Basin 199–205
    angler involvement 203, 204
    changes in distribution over time 201, 203
    conservation management 200–3
    education of anglers 201, 204
    as Endangered 200
    fishery closures 201, 202
    funding 203–4
    future 204
    monitoring 203
    Murray River population 203
    planning, people, governance 202–3
    population decline factors 200–1
    population recovery 200, 203
    recovery actions 201–2, 203, 204
    Recovery Team/Program/Plan 202–3, 204
    recreational fishery for 204
    stocking program 203, 204
    threats from habitat destruction and introduced
        species 200
    threats from over-fishing 200, 201, 203
Trust for Nature 47
Tumut grevillea 169–76
    conservation management 170–5
    distribution and enhanced plantings 170,
        171–2, 176
    as Endangered 170
    fire and flood threat 171, 172, 173

funding   175–6
future   175–6
landholder involvement   172, 174, 175, 176
monitoring   174–5
people, agencies, governance, and accountability   173–4
planning and policy   173
plantings in Goobarragandra River valley   171, 176
population numbers   170, 171, 174–5
propagation   173
Recovery Plan/Program/Priority Action Statement   171, 172, 173, 174, 176
recruitment and re-establishment   173
research, biology, identification of key threats   172–3
Tumut Shire Council   171
turtles
   Bellinger River   284
   Mary River   33, 35, 36, 37, 38
tusked frog   36
Twin Swamps Nature Reserve, WA, western swamp tortoise   218, 219, 220, 221, 222, 223, 224
Two Peoples Bay Fauna Reserve/Wildlife Sanctuary, WA   96–7, 100
Tyto novaehollandiae castanops × *novaehollandiae*   195

University of Adelaide   260, 263
University of Melbourne   71, 72
University of Western Australia   210, 218, 220, 224

Varanus gouldii   139
virlda   138, 139
volunteers
   role in success   317
   see also volunteers under specific topics, e.g. Norfolk Island green parrot
Vulkathunha–Gammon Ranges National Park, SA   138
Vulpes vulpes see foxes

WA Wildlife Research Centre   220
wallabies
   bridled nailtail   274–5
   tammar   289
   see also rock-wallabies
wandering albatross   14, 24
warru   136, 237–46
Warru Rangers   239, 240, 241
weka, Macquarie Island   14, 15
West coast humpback whales Australian subpopulation, monitoring   9–10
Western Australian South Coast Threatened Birds Recovery Team   96, 99, 101, 102, 103
western barred bandicoot   261, 264

western bristlebird   96
western grey kangaroo   139
western ground parrot   95–104
   biology and monitoring   99–100
   captive breeding program   99, 101
   conservation management   96–103
   as Critically Endangered   96
   fire and feral cat predation effects   99, 102, 103
   funding   101, 102–3
   future   103
   historical and current distribution   97, 98
   infrastructure   99–100
   people   101–2, 103
   Recovery Team   99, 101, 102, 103
   social support and governance   102
   volunteer involvement   102, 103
Western Mining Corporation (WMC)   260, 266
western quoll   138, 142, 143, 264, 266, 283
'Western Shield' Program (WA)   277–8
western swamp tortoise   217–26
   artificial reproductive techniques   220
   Captive Breeding Management Committee   220, 223
   captive breeding program   218, 219, 220, 225, 282, 283, 284, 289
   climate change effects   224, 225
   community involvement   224
   conservation management   219–24
      mid-1950s to mid-1960s   219
      mid-1960s to mid-1980s   219–20
      to present   220–1
   conservation reserves   218, 219, 220, 221–2, 223–4
   as Critically Endangered   218–19, 225
   funding   224, 225
   future   224–5
   insurance population   284
   Management Program   223
   monitoring   220, 223–4
   people, agencies, governance and accountability   223
   planning and policy   223
   population number   218, 219, 220, 222, 223
   Recovery Program/Team/Plan   218, 223, 224, 225
   reintroduction/translocation of captive populations   219, 220–1, 223, 224–5
   research and biology   221–2
   restricted distribution   218
   threat from habitat loss and foxes   218, 219, 222–3
western whipbird   96, 300
Whale Protection Act 1980   8
whaling, impact on humpback whale Australian subpopulations   6–8, 10
whaling stations, Australia   7, 8
whistler, red-lored   299

white-bellied frog   207–14
   biology and habitat requirements   208–9
   climate change effects   213
   conservation management   208–13
   as Critically Endangered   208
   description   208
   distribution, south-western WA   208, 209–10
   evolution of a captive program   210–11
   funding   213
   future   213
   governance and planning   210
   in-situ management   211–12
   landholder involvement   212
   monitoring   212–13
   population surveys   210
   Recovery Plan/Program   208, 210, 213
   susceptibility to threats   211–12, 213
   translocation of egg masses from 'head-start' program   211, 213
   wetland habitat and land tenure   209–10, 213
white-capped albatross   29
white cypress pine   162, 164
white gum   126–7, 130
white-headed petrel   14, 15
Wilabalangaloo Reserve, SA, spiny daisy translocation   183, 184, 185, 186
Wild Orchids Project   166
Wildlife Conservation Act 1950 (WA)   208
Wildlife Conservation Fund   185
Wimmera rice-flower   44, 49, 50
woma python   262
wombats, northern hairy-nosed   301–2
Woodlands Historic Park, Vic, eastern barred bandicoots   253, 254, 255
woolly tea-tree   230
Working on Country program   239, 244
World Wildlife Fund   36
woylie, reintroduction program   273, 275
Wyperfeld National Park, Vic, mallee emu-wrens   300

Yellingbo Nature Conservation Reserve, Vic, helmeted honeyeater   228–34
yellow-footed rock-wallabies   135–44
   Bounceback program   136, 137–43
   competition from goats and rabbits   136, 137
   conservation management   136–43
   cross-foster pouch-young warru   241
   feral animal control effects   136, 137, 138, 139, 140
   future   143
   improved conservation status   143
   landowner and volunteer contribution   137
   monitoring   140–2
   planning an policy   137–8
   population surveys   136, 140
   predation by foxes, dingoes and cats   136, 137
   release of captive-bred animals   290
   research, biology, identification of key threats   136
   traditional owner involvement   137, 139
   as Vulnerable   136
yellow-spotted bell frog   284
yellow-tufted honeyeater   228, 231
Yookamurra Wildlife Sanctuary, SA, mammal reintroductions   272, 273

Zoo and Aquarium Association, management of captive populations   289
zoos, overseas captive breeding successes   283
zoos, captive breeding contribution to conservation of Australia's threatened fauna   281–92
   challenges and lessons learned   290–1
   funding and resources   290
   future   291–2
   insurance populations held   283–4
   planning and policy   289
   primary objectives   282–3
   releases to the wild of captive-bred animals   284–9
   role in increasing public awareness of Australia's threatened species   283, 290
   strong partnerships for success   289–90
   translocations of captive-bred animals   284–5, 289
   when to initiate captive breeding?   291
   see also specific zoos, e.g. Monarto Zoo, SA
Zoos SA   60, 61, 239, 241, 243, 300
Zoos Victoria   154
   eastern barred bandicoot captive breeding   253, 256
   Lord Howe Island phasmid captive breeding   193, 194

www.ingramcontent.com/pod-product-compliance
Lightning Source LLC
Chambersburg PA
CBHW042356030426
42337CB00029B/5124